Thermodynamics in Geology

PALLAS PAPERBACKS

The purpose of the PALLAS program is to make available at reasonable prices, to students and other general readers, selected broad-interest Reidel books on the humanities and the natural and social sciences.

PALLAS titles in print:

1. Wolff, *Surrender and Catch*
2. Fraser, *Thermodynamics in Geology*
3. Goodman, *The Structure of Appearance*
4. Schlesinger, *Religion and Scientific Method*
5. Aune, *Reason and Action*

A PALLAS PAPERBACK / 2

Thermodynamics in Geology

Proceedings of the NATO Advanced Study Institute held in Oxford, England, September 17-27, 1976

edited by

DONALD G. FRASER

Dept. of Geology and Mineralogy, University of Oxford, England

PALLAS paperbacks

D. Reidel Publishing Company

Dordrecht-Holland / Boston-U.S.A.

Published in cooperation with NATO Scientific Affairs Division

Library of Congress Cataloging in Publication Data

Nato Advanced Study Institute, Oxford, 1976.
Thermodynamics in geology.

(NATO advanced study institutes series : Series C, Mathematical and physical sciences ; v. 30)
Bibliography: p.
Includes index.
1. Geology—Congresses. 2. Thermodynamics—Congresses. I. Fraser, Donald G., 1949- II. Title. III. Series.
QE509.N37 1976 551 77-3242
ISBN 90-277-0794-4
ISBN 90-277-0834-7 pbk. (Pallas edition)

2-0794/0834-1279-700NC

Published by D. Reidel Publishing Company
P.O. Box 17, Dordrecht, Holland

Sold and distributed in the U.S.A., Canada, and Mexico
by D. Reidel Publishing Company, Inc.
Lincoln Building, 160 Old Derby Street, Hingham, Mass. 02043, U.S.A.

First published in 1977 in hardbound edition by Reidel in NATO Advanced Study Institutes Series C, Volume 30

Printed in The Netherlands

TABLE OF CONTENTS

PREFACE

It has long been realized that the mineral assemblages of igneous and metamorphic rocks may reflect the approach of a rock to chemical equilibrium during its formation. However progress in the application of chemical thermodynamics to geological systems has been hindered since the time of Bowen and the other early physical-chemical petrologists by the recurring quandary of the experimental geologist. His systems are complex and are experimentally intractable, but if they were not so refractory they would not be there to study at all. It is only recently that accurate measurements of the thermodynamic properties of pure, or at least well-defined minerals, melts and volatile fluid phases, combined with experimental and theoretical studies of their mixing properties, have made it possible to calculate the equilibrium conditions for particular rock systems.

Much work is now in progress to extend the ranges of composition and conditions for which sufficient data exist to enable such calculations to be made. Moreover the routine availability of the electron microprobe will ensure that the demand for such information will continue to increase.

The thermodynamic techniques required to apply these data to geological problems are intrinsically simple and merely involve the combination of appropriate standard state data together with corrections for the effects of solution in natural minerals, melts or volatile fluids. However the vocabulary of the subject often appears foreign to geologists and there has been a lack of up-to-date text books available for use in courses on thermodynamics in geology. The N.A.T.O. Advanced Study Institute, of which this volume is the Proceedings, set out to review some of the progress which has been made in the field, and to survey some topics of current research. In particular, however, the Institute functioned as a teaching medium in which the various review lecturers supplemented their talks with study problems worked out during afternoon teaching sessions.

The volume "Thermodynamics in Geology" is therefore more than just the proceedings of a conference, since each review lecturer was asked to write a chapter suitable for use as a teaching text. The book begins with a chapter on calorimetry (Navrotsky) followed by two chapters on the experimental determination of activity-composition relationships of mineral solid solutions by phase equilibrium experiments (Wood) and by high temperature solution calorimetry (Newton). After chapters on the nature of activity-composition relationships (Powell) and the expression of non-ideal behaviour using Margules' equations (Grover), a review of experimental techniques available for determining site occupancy is given (Whittaker).

The extraction of thermodynamic data from phase diagrams together with an analysis of accuracy and precision is considered by Anderson and Chatterjee and this section is followed by a discussion of the properties of volatile phases. The properties of supercritical solutions are reviewed by Holloway, and this is followed by a chapter on metamorphic solutions by Eugster and a review of studies of fluid inclusions is given by Touret.

Two specific examples of the application of thermodynamic methods to geological problems are considered by Wones and Dodge (biotites as monitors of granitic melts) and by El Goresy and Woermann (opaque minerals as oxygen barometers and geothermometers). These are followed by a section on the thermodynamic properties of melts. Following a review of the properties of "simple" molten salts (Kleppa) the nature and properties of silicate melts are considered by Fraser. This is followed by a chapter by Nicholls on the treatment of activities in natural silicate melts and a discussion of trace element behaviour by O'Nions and Powell.

Finally the kinetics of carbonate dissolution in sea-water are considered (Broecker and Takahashi) and a review of non-equilibrium thermodynamics in petrology is given by Fisher.

That so many authors should co-operate in this way to produce the collection of chapters in this volume, many provided with study problems and worked solutions, is a tribute to the enthusiasm and good-will of all those who participated in the Institute. We all hope that those reading the book will find it both useful and enjoyable.

OXFORD D.G. Fraser
December 1976

ACKNOWLEDGEMENTS

The chapters in this volume were written by the review lecturers at the NATO Advanced Study Institute held in the Department of Geology and Mineralogy, University of Oxford from 17th - 27th September 1976. I should like to thank all the lecturers for their care in preparing both their lectures and the review chapters and study problems presented here.

The Institute itself grew out of a number of one week vacation courses for British graduate students held in the Department of Geology, University of Manchester. These courses were organized and run by Drs. R.K. O'Nions, R. Powell, B.J. Wood and myself and I gratefully acknowledge their contributions which made the Institute possible. I particularly wish to thank Bernie Wood for his good advice on numerous occasions. I also thank the Natural Environment Research Council for financial support of the courses.

The Institute could not have been held without the good-will and co-operation of the technical and academic staff of the Department of Geology, University of Oxford and I thank them all for their assistance before, during and after the Institute. I particularly thank Valerie Miles for handling all the correspondence, Peter Deussen for help with draughting and design, and our conference secretary, Dominique, for her charm, good humour and cheerful efficiency.

Finally, I should like to thank my wife, Anna, both for her help and for her forbearance, and Pat Jackson for her skill in producing camera-ready typescript to a tight schedule.

All the members of the Institute thank NATO for its generous financial support.

D.G.F.

GEOLOGICAL APPLICATIONS OF HIGH TEMPERATURE REACTION CALORIMETRY

Alexandra Navrotsky

Department of Chemistry and
Center for Solid State Science, Arizona State
University, Tempe, Arizona 85281, U.S.A.

1. INTRODUCTION

Accurate thermodynamic data for minerals can be obtained by several methods; from analysis of phase relations, by measurement of oxidation-reduction equilibria using either gas mixtures or electrochemical cells with solid-electrolytes, and by calorimetry. The last technique can be subdivided into two major categories: the determination of heat capacities, both at cryogenic temperatures and at above room temperature, and the measurement of enthalpies of chemical reactions. Because silicate minerals are generally quite unreactive at room temperature, their equilibria must be studied at temperatures above 300°C, and, quite often, in the range 1000-1600°C. Solution calorimetry of silicates using aqueous hydrofluoric acid as a solvent has been carried out at temperatures of 50 - 80°C (Torgeson and Sahama, 1948; Neuvonen, 1952; King, 1952; Hovis, 1971). In this paper I will describe high temperature solution and reaction calorimetry of minerals and related substances. This technique has been applied to minerals only since the mid 1960's (Yokokawa and Kleppa, 1964b; Navrotsky and Kleppa, 1968; Holm and Kleppa, 1966) but has proved itself to be both versatile and widely applicable to problems of geological interest. A more detailed paper covering the same general subject is in press (Navrotsky, 1977).

We use a twin microcalorimeter of the Tian-Calvet type, (Calvet and Prat, 1956) patterned with certain modifications after the high temperature versions in use by Kleppa (1972). It consists of two sample chambers each one inch in diameter, each surrounded by a thermopile of 56 Pt-Pt13Rh thermocouples in series The two thermopiles are then connected in opposition and embedded

D. G. Fraser (ed.), Thermodynamics in Geology, 1-10.

in a massive (∿300 lb.) block of Hastelloy X, which is maintained at constant and uniform temperature by a cylindrical main heater and top and bottom plate heaters.

The low level d.c. voltage from the thermopiles is sent to an amplifier and then to a recorder and electronic integrator to obtain both a graphical record of the experiment and a numerical value of the area under the calorimetric peak. This area is proportional to the heat effect generated by the reaction.

The calorimeter will perform with high precision to about 900°C. It is possible to measure heat effects as small as ∿0.5 calorie in magnitude with a precision of 2 to 5%. Heat effects of 5-200 calories can be obtained to within a standard deviation of ∿1%. This gives the necessary precision and sensitivity for solution calorimetry in oxide melts, where heats of solution of 0-30 kcal/mole are measured on samples of 50-100 m.y. (0.1-1.0 mmoles). Calorimetric solvents used, molten oxide mixtures suitable for dissolving solid silicate samples, will be described below.

Advantages of oxide calorimetry are (1) ease of dissolution of compounds rich in Al_2O_3, MgO, and other refractories, (2) relatively small enthalpies of solution which are independent of variations in the amount of solute, the presence of other solutes, or small changes in melt composition, and (3) the ability to work with small amounts of sample (200-400 mg total for several duplicate runs) which enables one to study, for example, phases synthesized at high pressure in a piston-cylinder apparatus.

2. CALORIMETRIC SOLVENTS AND OXIDE MELTS

Enthalpies of mixing in "simple" fused salts, such as alkali halide mixtures, have been studied extensively by high temperature calorimetry (Kleppa, 1966; and this volume). However, in oxide systems where one component is MoO_3, V_2O_5, P_2O_5, B_2O_3, SiO_2, or GeO_2, the tendency toward complex formation is so strong that essentially no free Mo^{6+}, V^{5+}, B^{3+}, P^{5+}, Si^{4+} or Ge^{4+} ions are present. Rather, the melt contains a series of complex ions and polymers thereof. The predominant species present depend both on the molar ratio of acidic oxide (network former) to basic oxide (network breaker) and on the chemical nature of each oxide.

A number of molten oxide mixtures have been studied by high temperature calorimetry at 600-900°C. These include the systems PbO-V_2O_5 (Yokokawa and Kleppa, 1964a), PbO-B_2O_3 (Holm and Kleppa, 1967), PbO-SiO_2 (Østvold and Kleppa, 1969), PbO-GeO_2 (Müller and Kleppa, 1973), Na_2O-MoO_3 (Navrotsky and Kleppa, 1967c), Li_2O-B_2O_3,

Na_2O-B_2O_3 (Østvold and Kleppa, 1970), and binary mixtures of alkali metaphosphates (Ko and Kleppa, 1971). These studies were generally performed by dissolving small amounts of component oxides into a relatively large amount of melt of a given composition; thus the partial molar enthalpies of solution of each component were obtained as a function of composition.

The structure and thermodynamics of naturally occurring silicate melts is one of the unsolved problems in petrology. Clearly, calorimetric study of silicate and aluminosilicate melts along the lines described above would be very useful. However, the temperatures at which these mixtures are molten and have relatively low viscosities (1100-1300°C) are at the upper limit of calorimetric capability. Recently, both Hong and Kleppa (1974) and Warner, Roye and Jeffes (1973) have reported the construction of calorimeters having all-alumina blocks and capable of precise operation to at least 1300°C. Such instruments could be used for the direct study of heats of mixing in silicate melts.

Another approach is the study of silicate glasses by solution calorimetry in oxide melts near 700°C. Although a glass is not fully representative of the liquid state, it is clearly more so at 700°C (at or near the glass transition temperature) than near room temperature where HF calorimetry is practiced. We are currently working on selected silicate glasses using this approach.

The oxide melts used as solvents in high temperature solution calorimetry must meet several requirements. They must dissolve both acidic oxides (such as SiO_2) and basic oxides (such as MgO, CaO) readily and with reproducible heats of solution. Furthermore, the measured enthalpy of solution of a given oxide should be independent of its concentration, of the presence of other solutes, and of small fluctuations in solvent composition. These conditions are best fulfilled in a composition region where the melt acts as a buffer; i.e., where the addition of small amounts of acidic or basic oxide merely changes slightly the proportions of two (or more) complex anions present in comparable concentrations. The solvents used most extensively thus far, $2PbO.B_2O_3$ and $3Na_2O.4MoO_3$, both lie within buffer regions in their respective binary systems.

Lead borate is the solvent of choice for most oxides, but cannot be used for oxides containing TiO_2, SnO_2, or Mn_2O_3 because of the exothermic formation of precipitates of ternary lead titanates, stannates, or manganites. The sodium molybdate melt works well for those three oxides but dissolves acidic oxides such as SiO_2 too slowly. Neither of the above melts can be used under conditions of low oxygen fugacity, as are necessary to control ferrous/ferric ratios in silicates, because of the easy reduction of Pb^{2+} and Mo^{6+}. To overcome this difficulty, we are developing an alkali borate or borosilicate solvent which can be used with $CO-CO_2$ gas mixtures. To date, the most satisfactory composition seems to be one in the system $Na_2O-B_2O-SiO_2$, which is near a ternary eutectic and of low viscosity near 800°C. However, the water content of the solvent presents a problem which requires careful control. This is an area of future development in calorimetry.

3. APPLICATIONS OF HIGH TEMPERATURE CALORIMETRY TO MINERAL THERMODYNAMICS

3.1 Enthalpies of formation of minerals

By measuring the heat of solution of a mineral and of its component oxides, the enthalpy of formation of the mineral from the oxides can be obtained directly, without recourse to the rather complex reaction schemes necessary in HF calorimetry. The following systems have been studied by high temperature solution calorimetry, using lead-cadmium-borate, lead borate, and/or sodium molybdate solvents. Navrotsky and Kleppa (1968) measured the enthalpies of formation of a series of spinels containing Mg^{2+}, Co^{2+}, Ni^{2+}, Zn^{2+}, Cd^{2+}, Al^{3+}, Fe^{3+}, Ga^{3+}, Mn^{3+}, Ti^{4+} and Ge^{4+}. A series of chromite spinels (MCr_2O_4) were studied (Muller and Kleppa, 1973). The tungstates MWO_4 (M = Mg, Co, Ni, Cu, Zn, Cd), were studied by Navrotsky and Kleppa (1969). Heats of formation of silicates and germanates of Mg, Co, Ni, Cu, and Zn have been determined (Navrotsky, 1971a) as well as those of $MgCaSi_2O_6$, $CoCaSi_2O_6$, $NiCaSi_2O_6$, Mn_2SiO_4 and $MnSiO_3$ (Navrotsky and Coons, 1976). Recently, very precise determinations of the enthalpies of formation of compounds in the system $MgO-Al_2O_3-SiO_2$ were made by Charlu, Newton, and Kleppa (1975). Phases studied were enstatite and an aluminous enstatite, sapphirine, sillimanite, forsterite, spinel, cordierite, and pyrope. Calorimetric precisions were significantly improved in these experiments by making efforts to control the (low) water content of the lead borate melt used.

Future work on enthalpies of formation of minerals may well concentrate in three areas. Firstly, additional precise determination of enthalpies of formation of phases, for example in the system $MgO-CaO-SiO_2-Al_2O_3$, along the lines of the study by Charlu, Newton and Kleppa (1975) will be made. Secondly, the further development of calorimetry under atmospheres of controlled low oxygen fugacity will permit systematic study of iron-bearing minerals. Thirdly, although water- and hydroxyl-containing phases cannot be studied by solution in molten oxides at atmospheric pressure because the H_2O degasses from the melt at variable rates, it appears that the fluorine end-member of substitutional

(F^-, OH^-) solid solution series (micas, amphiboles, apatite, topaz) can be readily studied. We are currently working on certain of these systems, and have data on the heats of formation of fluoropargasite and fluoroapatite.

3.2 Polymorphic transition and petrologic phase relations

High temperature solution calorimetry is well suited to the study of the enthalpy of phase transitions. When both phases or phase assemblages can be prepared (one being metastable under ambient conditions), the difference in their heats of solution gives the enthalpy of transformation (at calorimetric temperature) directly. The enthalpy of fusion (difference in heat of solution of crystalline and glassy samples) has been measured for SiO_2 (Holm, Kleppa, and Westrum, 1967), GeO_2 (Navrotsky, 1971b), cordierite (Navrotsky and Kleppa, 1973) and diopside (Navrotsky and Coons, 1976). The polymorphs of SiO_2 (quartz, coesite and stishovite) and of GeO_2 (quartz and rutile types) have been studied also (Holm, Kleppa and Westrum, 1967; Navrotsky, 1971b), as have the metastable modifications of Al_2O_3 (Yokokawa and Kleppa, 1964b) and those of TiO_2 (Navrotsky and Kleppa, 1967a; Navrotsky, Jamieson and Kleppa, 1967). The enthalpies of the olivine-spinel transition in Ni_2SiO_4 (Navrotsky, 1973a) and Mg_2GeO_4 (Navrotsky, 1973b), and of the ilmenite-perovskite transition in $CdTiO_3$ (Neil, Navrotsky and Kleppa, 1971) have been measured. Navrotsky and Kasper (1976) have determined the enthalpies of disproportionation of the stannate spinels, Mg_2SnO_4 and Co_2SnO_4, to the oxides. These spinel reactions are models for silicate reactions in the lower mantle. Polymorphism in the aluminum silicates has been studied (Holm and Kleppa, 1966; Anderson and Kleppa, 1969; Navrotsky, Newton and Kleppa, 1973). The enthalpy of the breakdown of albite to jadeite plus quartz has been determined (Hlabse and Kleppa, 1968). In the system $MgO-Al_2O_3-SiO_2$ (Newton, Charlu and Kleppa, 1974; Charlu, Newton and Kleppa, 1975) calorimetric data have permitted the calculation of the slopes of phase boundaries for the reaction, cordierite $\rightarrow$ sapphirine + quartz and of equilibria involving pyrope garnet.

3.3 Problems of Order-Disorder

In their study of spinels, Navrotsky and Kleppa (1967b, 1968) paid particular attention to the variation of the distribution of cations between octahedral and tetrahedral sites. They observed a change in enthalpy of solution upon heating a natural $MgAl_2O_4$ spinel and proposed a simple model to describe the thermodynamics of disordering. Navrotsky (1975) has applied a similar model to compounds having the pseudobrookite structure. An important consequence of massive cation disorder is the stabilization of such disorder phases at high temperature.

In silicate minerals, the theme of Al-Si order-disorder on tetrahedral sites recurs frequently, and several calorimetric studies have been made to measure the enthalpies associated with these processes. In addition to the extensive work on the alkali feldspars using HF calorimetry (Hovis, 1971), the disordering process in albite has been studied by oxide melt calorimetry (Holm and Kleppa, 1968). Al-Si disordering in sillimanite, Al_2SiO_5, has been studied by oxide melt calorimetry, the calorimetric data correlated with crystallographic data, and a disordering model proposed (Navrotsky, Newton and Kleppa, 1973). Less detailed calorimetric studies have been made of Al-Si disorder in sapphirine (Charlu, Newton and Kleppa, 1975), and of possible Al-Si and Mg-Al disorder in cordierite (Navrotsky and Kleppa, 1973; Newton, Charlu and Kleppa, 1974).

Future work should emphasize calorimetric and structural studies on the same samples, these materials being well-characterized in terms of chemistry and temperature of equilibration. Problems of cation ordering in pyroxenes are ripe for such study.

3.4 Direct Reaction Calorimetry at High Temperature

This includes two types of studies. The first is the direct measurement of enthalpies of rapidly occurring phase transitions such as the transition from stishovite to glass in SiO_2 (Holm, Kleppa and Westrum, 1967), and the transition from the α-PbO_2 structure to rutile in TiO_2 (Navrotsky, Jamieson and Kleppa, 1967).

The second type of experiment is direct reaction in the calorimeter at high temperature. It can be applied when the reaction of interest occurs rapidly, goes to completion, and leads to a well-defined product. Thus, while generally not applicable to silicates, the method has been used to study reactions between a liquid and a solid oxide when these differ considerably in acidity, such as, for example, in the formation of solid lead vanadates from liquid V_2O_5 and solid PbO (Yokokawa and Kleppa, 1964a). This method is very useful in the study of gas-metal reactions. Of some interest to geologists are studies of the partial molal enthalpy of solution of oxygen in nonstoichiometric wüstite (Marucco, Gerdanian and Dode, 1970) and UO_{2+x} (Gerdanian, 1974). Further application to sulfide equilibria, to alloy phases in meteorites, and to Fe^{2+}/Fe^{3+} redox equilibria could be made in the future.

4. CALORIMETRY OF GEOTHERMAL FLUIDS AND WATER-CONTAINING SYSTEMS AT PRESSURES UP TO 2 KB

Thermochemical data for aqueous electrolyte solutions at high pressures and temperatures are necessary for understanding and modelling geothermal systems but experimental data are virtually non-existent above 300°C. We are currently measuring enthalpies of aqueous solutions up to pressures of 2 kbar and temperatures of 800°C by a method which combines standard cold-seal pressure vessel techniques with a Calvet-type twin microcalorimeter. We have constructed a new calorimeter for use with larger pressure vessels in the 200-500°C range, which is the temperature range of greatest interest for applications to geothermal fluids and ore deposits.

We are measuring the heat capacities and heat contents of aqueous solutions in the temperature range 200-800°C and at pressures up to 2kb. Our technique is not restricted to the liquid-vapour coexistence curve and we can work easily in the super-critical region. We plan to develop *in situ* isothermal mixing devices for further study of water-containing systems at high pressure and temperature. Application to equilibria involving hydrous solid minerals and water-containing melts are possible.

REFERENCES

Anderson, P.A.M. and Kleppa, O.J. Am. Jour. Sci., 267, 285-290 (1969).

Calvet, E. and Prat, H. "Microcalorimetrie". (Masson et Cie, Paris, France). (1954).

Charlu, T.V., Newton, R.C. and Kleppa, O.J. Geochim. et Cosmochim. Acta, 39, 1487-1498 (1975).

Gerdanian, P. J. Phys. Chem. Solids, 35, 163-170 (1974).

Hlabse, T. and Kleppa, O.J. Amer. Mineral., 53, 1281-1292 (1968).

Holm, J.L. and Kleppa, O.J. Amer. Mineral., 51, 1608-1622 (1966).

Holm, J.L. and Kleppa, O.J. Inorg. Chem., 6, 645-648 (1967).

Holm, J.L., Kleppa, O.J. and Westrum, E.F. Jr. Geochim. et Cosmochim. Acta, 31, 2289-2307 (1967).

Holm, J.L. and Kleppa, O.J. Amer. Mineral., 53, 123-133 (1968).

Hovis, G.L. Ph.D. Thesis, Geology Dept., Harvard Univ. (1971).

King, E.G. Jour. Amer. Chem. Soc., 74, 4446-4448 (1952).

Kleppa, O.J. Thermodynamics-International Atomic Energy Agency, I, 383-407 (1966).

Kleppa, O.J. CNRS No. 201 - Thermochimie, 119-127 (1972).

Ko, H.C. and Kleppa, O.J. Inorg. Chem., 10, 771-775 (1970).

Marucco, J., Gerdanian, P. and Dode, M. J. Chim. Phys., 67, 906-913 (1970).

Müller, F. and Kleppa, O.J. Z. Anorg. Allg. Chem., 397, 171-178 (1973).

Navrotsky, A. J. Inorg. Nucl. Chem., 33, 4035-4050 (1971a).
Navrotsky, A. J. Inorg. Nucl. Chem., 33, 1119-1124 (1971b).
Navrotsky, A. Earth Planet. Sci. Lett., 19, 474-475 (1973a).
Navrotsky, A. Material Science, Pergamon Press, Oxford, 383-398 (1973b).
Navrotsky, A. Amer. Mineral., 60, 249-256 (1975).
Navrotsky, A. Physics and Chemistry of Minerals (a new Springer Verlag journal; in press).
Navrotsky, A. and Coons, W.E. Geochim. et Cosmochim. Acta (in press).
Navrotsky, A., Jamieson, J.C. and Kleppa, O.J. Science, 158, 388-389 (1967).
Navrotsky, A. and Kasper, R.B. Earth Planet. Sci. Lett., 31, 247-254 (1976).
Navrotsky, A. and Kleppa, O.J. J. Am. Ceram. Soc., 50, 626 (1967a)
Navrotsky, A. and Kleppa, O.J. J. Inorg. Nucl. Chem., 29, 2701-2714 (1967b).
Navrotsky, A. and Kleppa, O.J. Inorg. Chem., 6, 2119-2121 (1967c).
Navrotsky, A. and Kleppa, O.J. J. Inorg. Nucl. Chem., 30, 479-498 (1968).
Navrotsky, A. and Kleppa, O.J. Inorg. Chem., 8, 756-758 (1969).
Navrotsky, A. and Kleppa, O.J. J. Am. Ceram. Soc., 56, 198-199 (1973).
Navrotsky, A., Newton, R.C. and Kleppa, O.J. Geochim. et Cosmochim. Acta, 37, 2497-2508 (1973).
Neil, J.M., Navrotsky, A. and Kleppa, O.J. Inorg. Chem., 10, 2076-2077 (1971).
Neuvonen, K.J. Bull. Comm. Geol. Finlande, No. 158, 1-50 (1952).
Newton, R.C., Charlu, T.V. and Kleppa, O.J. Contr. Mineral. Petrol., 44, 295-311 (1974).
Østvold, T. and Kleppa, O.J. Inorg. Chem., 8, 78-82 (1969).
Østvold, T. and Kleppa, O.J. Inorg. Chem., 9, 1395-1400 (1970).
Shearer, J.A. and Kleppa, O.J. J. Inorg. Nucl. Chem., 35, 1073-1078 (1973).
Torgeson, O. and Sahama, T. Jour. Amer. Chem. Soc., 70, 2156-2160 (1948).
Warner, A.E.M., Roye, M.P. and Jeffes, J.H.E. Trans. Inst. Mining and Metall., 82, C246-C248 (1973).
Yokokawa, T. and Kleppa, O.J. Inorg. Chem., 3, 954-957 (1964a).
Yokokawa, T. and Kleppa, O.J. Jour. Phys. Chem., 68, 3246-3249 (1964b).

STUDY PROBLEMS

(1) (a) In recent work (Navrotsky and Kasper, in press), we measured enthalpies of solution in molten sodium molybdate at 713°C as follows:

MgO	- 8.56 ± 0.24 kcal/mol.
CoO	- 5.04 ± 0.06
SnO_2	- 0.21 ± 0.05
Mg_2SnO_4	-18.47 ± 0.35
Co_2SnO_4	- 7.98 ± 0.26

Calculate the enthalpies of formation from the oxides of Mg_2SnO_4 and Co_2SnO_4 spinels. Would both these compounds be stable to low temperatures at atmospheric pressure?

(b) Jackson, Ringwood and Liebermann (1974) reported that Co_2SnO_4 "disproportionates" to the component oxides at 12 kb and 1000°C, with $\partial P/\partial T = 0 \pm 7$ bar/deg, while Mg_2SnO_4 disproportionates at 26 kb and 1000°C, with $\partial P/\partial T = 40 \pm 10$ bar/deg. Do the calorimetric data support these rather different P-T slopes for the two reactions? The volume changes (at room temperatures and atmospheric pressure) for disproportionation are - 3.80 and - 3.85 cc/mol for Co_2SnO_4 and Mg_2SnO_4, respectively.

(c) Would you expect heat capacity measurements alone to provide accurate values for the entropies of formation of Co_2SnO_4 and Mg_2SnO_4?

(d) Obviously, the problems of direct geologic interest are the olivine-spinel transition and possible disproportionation of further phase transitions in $(Mg,Fe)_2SiO_4$ silicates. What useful information about such reactions can be gotten by calorimetry? Propose some possible calorimetric experiments and discuss their feasibility in terms of existing technology (limitations of calorimetry and high pressure technology, sample size and purity, etc.).

(2) The enthalpies of formation of some olivines and clinopyroxenes from the oxides are given below. Using these data and simple assumptions about entropies and activity-composition relations, formulate an expression for the exchange of M and Mg (M = Fe,Ni, Co) between olivine and a clinopyroxene approaching diopside in compositions and calculate values of K_D.

	ΔH_f^o (kcal/mol) near 1000 K
Mg_2SiO_4	- 14.7
Fe_2SiO_4	- 7.0
Co_2SiO_4	- 4.2
Ni_2SiO_4	- 3.3
$MgCaSi_2O_6$	- 34.3
$FeCaSi_2O_6$	- 26.5
$CoCaSi_2O_6$	- 26.7
$NiCaSi_2O_6$	- 27.1

These K_D values can be compared to experimental values obtained by Leeman and Weill, Lindstrom and Weill, and others.

EXPERIMENTAL DETERMINATION OF THE MIXING PROPERTIES OF SOLID SOLUTIONS WITH PARTICULAR REFERENCE TO GARNET AND CLINOPYROXENE SOLUTIONS

Bernard J. Wood

Department of Geology, University of Manchester,
Manchester M13 9PL,

INTRODUCTION

It is, in general, only possible to apply thermodynamic data to rocks in a quantitative way if the mixing properties of the natural, multi-component minerals, are known. For this reason there have been, in recent years, a considerable number of experimental and theoretical studies of the thermodynamic properties of mineral solid solutions (e.g. Thompson and Waldbaum, 1968; Orville, 1972; Nafziger, 1973).

The first object of this paper is to review the general approach to determining activity-composition relationships for binary and ternary solid solutions. The latter part of the paper will be devoted to some new and recently published results on garnet and aluminous clinopyroxene solutions.

Experimental determinations of activity-composition relationships may be broadly grouped into those in which the activity of i, a_i is varied by varying the bulk composition of the system at constant P and T and those in which this variation is produced by varying one of the intensive parameters at constant bulk composition.

(a) The first type of experiment involves studying the partitioning of one or more elements between the phase of interest and some other phase whose thermodynamic properties are known. Nafziger and Muan (1967) used this approach to determine the mixing properties of (Fe-Mg) olivine solid solutions by studying the exchange equilibrium:

D. G. Fraser (ed.), Thermodynamics in Geology, 11-27. All Rights Reserved.

$$\underset{\text{oxide}}{FeO} + \underset{\text{olivine}}{MgSi_{0.5}O_2} \rightleftarrows \underset{\text{oxide}}{MgO} + \underset{\text{olivine}}{Fe_{0.5}SiO_2} \qquad (1)$$

Knowledge of the thermodynamic properties of (Fe,Mg)O oxide solution and of the compositions of coexisting (Fe,Mg) oxide and (Fe, Mg) olivine solid solutions enable $a^{ol}_{FeSi_{0.5}O_2}$ and $a^{ol}_{MgSi_{0.5}O_2}$ to be determined (see below).

(b) The second method involves fixing the activity of one of the components of the solid solution at a known value by using an appropriate mineral and/or gas assemblage. Determination of the equilibrium composition of the solid solution, X_i, at fixed value of a_i yields the desired mixing properties. The activity of i, a_i, is varied over the required range by making appropriate adjustments to one of the intensive parameters, P, T, a_{O_2}, etc.

Hahn and Muan (1962) used this method to determine the thermodynamic properties of MgO-"FeO" oxide solutions by studying the assemblage (Mg,Fe)O solid solution, metallic iron, gas. Equilibrium coexistence of the oxide$_{ss}$, pure iron and gas of known a_{O_2} fixes the activity of FeO in the oxide solid solution through the relationship:

$$\underset{\text{metal}}{Fe} + \tfrac{1}{2}O_2 \rightleftarrows \underset{\text{oxide}}{FeO} \qquad (2)$$

Similarly, equilibrium between anorthite, quartz (both pure) and a clinopyroxene solid solution fixes at known pressure and temperature, the activity of $CaAl_2SiO_6$ component in the pyroxene:

$$\underset{\text{anorthite}}{CaAl_2Si_2O_8} \rightleftarrows \underset{\text{clinopyroxene}}{CaAl_2SiO_6} + \underset{\text{quartz}}{SiO_2} \qquad (3)$$

Before discussing equilibria (2) and (3) any further let us return to a more detailed description of method (a) using the examples of feldspar solid solutions and (Fe,Mg) olivine solid solutions.

$NaAlSi_3O_8$ - $KAlSi_3O_8$ FELDSPAR SOLUTIONS

Orville (1963) has investigated the partitioning of sodium and potassium between high temperature alkali feldspar solid solution and a chloride-bearing fluid phase. The exchange of Na and K between these two phases can be represented by the reaction:

$$\underset{\text{feldspar}}{NaAlSi_3O_8} + \underset{\text{fluid}}{KCl} \rightleftarrows \underset{\text{feldspar}}{KAlSi_3O_8} + \underset{\text{fluid}}{NaCl} \tag{4}$$

for which the equilibrium constant K_4 is defined as follows:

$$K_4 = \frac{a^{fsp}_{KAlSi_3O_8} \cdot a^{fluid}_{NaCl}}{a^{fsp}_{NaAlSi_3O_8} \cdot a^{fluid}_{KCl}} \tag{5}$$

In equation (5) $a^{fsp}_{KAlSi_3O_8}$ refers to the activity of $KAlSi_3O_8$ component in the feldspar phase, a^{fluid}_{NaCl}, the activity of NaCl component in the fluid phase and so on. Orville determined the compositions, (K/Na + K ratios), of coexisting feldspar and fluid phases as a function of bulk K/K + Na ratio at a pressure of 2000 atmospheres and for fixed total chloride concentration of the fluid (2 molal).

The equilibrium constant K_4, may be related to the measured compositions of phases, by introducing activity coefficients γ^{fluid}_{NaCl}, $\gamma^{fsp}_{KAlSi_3O_8}$ and so on, defined in the following way:

$$\begin{aligned} a^{fsp}_{KAlSi_3O_8} &= X^{fsp}_{KAlSi_3O_8} \quad \gamma^{fsp}_{KAlSi_3O_8} \\ a^{fsp}_{NaAlSi_3O_8} &= X^{fsp}_{NaAlSi_3O_8} \quad \gamma^{fsp}_{KAlSi_3O_8} \end{aligned} \tag{6}$$

$$\begin{aligned} a^{fluid}_{NaCl} &= M^{fluid}_{NaCl} \quad \gamma^{fluid}_{NaCl} \\ a^{fluid}_{KCl} &= M^{fluid}_{KCl} \quad \gamma^{fluid}_{KCl} \end{aligned} \tag{7}$$

In equations (6) and (7), $X^{fsp}_{NaAlSi_3O_8}$ refers to the mole fraction of $NaAlSi_3O_8$ component in the feldspar phase and M^{fluid}_{KCl} to the number of moles of KCl per 1000 gms of water. Substituting (6) and (7) into (5) and taking logarithms gives:

$$\ln K_4 = \ln \frac{X^{fsp}_{KAlSi_3O_8} \cdot M^{fluid}_{NaCl}}{X^{fsp}_{NaAlSi_3O_8} \cdot M^{fluid}_{KCl}} + \ln \frac{\gamma^{fluid}_{NaCl}}{\gamma^{fluid}_{KCl}} +$$

$$+ \ln \gamma^{fsp}_{KAlSi_3O_8} - \ln \gamma^{fsp}_{NaAlSi_3O_8} \qquad (8)$$

$$\ln K_4 = \ln K_D + \ln \frac{\gamma^{fl}_{NaCl}}{\gamma^{fl}_{KCl}} + \ln \gamma^{fsp}_{KAlSi_3O_8} - \ln \gamma^{fsp}_{NaAlSi_3O_8} \qquad (9)$$

The quotient in brackets on the right hand side of equation (8)

$$\frac{X^{fsp}_{K} \cdot M^{fl}_{NaCl}}{X^{fsp}_{Na} \cdot M^{fl}_{KCl}}$$

generally referred to as K_D, is what Orville measured as a function of bulk composition.

Values of K_D determined at any one temperature and pressure for a wide range of bulk compositions enable $\gamma^{fsp}_{NaAlSi_3O_8}$ and $\gamma^{fsp}_{KAlSi_3O_8}$ to be determined provided the mixing properties of the fluid phase are known. The first step is to differentiate equation (9) at fixed pressure and temperature:

$$- d \ln K_D = d \ln \frac{\gamma^{fl}_{NaCl}}{\gamma^{fl}_{KCl}} + d \ln \gamma^{fsp}_{KAlSi_3O_8} - d \ln \gamma^{fsp}_{NaAlSi_3O_8} \qquad (10)$$

The Gibbs-Duhem equation for the binary $NaAlSi_3O_8$ solution gives the following condition:

$$d \ln \gamma^{fsp}_{KAlSi_3O_8} = - \frac{X^{fsp}_{NaAlSi_3O_8}}{X^{fsp}_{KAlSi_3O_8}} d \ln \gamma^{fsp}_{NaAlSi_3O_8} \qquad (11)$$

Substituting (11) into (10) and making the assumption that $\gamma^{fl}_{NaCl}/\gamma^{fl}_{KCl}$ is constant, independent of K/Na ratio at fixed total chloride concentration yields:

$$-d \ln K_D = - \frac{1}{X^{fsp}_{KAlSi_3O_8}} d \ln \gamma^{fsp}_{NaAlSi_3O_8} \qquad (12)$$

or

$$\ln \gamma^{fsp}_{NaAlSi_3O_8} = \int_0^{X_{KAlSi_3O_8}} X_{KAlSi_3O_8}\, d \ln K_D \qquad (13)$$

The analogous equation for $\ln \gamma^{fsp}_{KAlSi_3O_8}$ may be derived in a similar way:

$$\ln \gamma^{fsp}_{KAlSi_3O_8} = \int_0^{X^{fsp}_{NaAlSi_3O_8}} X_{KAlSi_3O_8}\, d \ln K_D \qquad (14)$$

The right hand sides of equations (13) and (14) may be evaluated graphically by plotting ln K_D vs $X^{fsp}_{NaAlSi_3O_8}$ and determining the area under the curve. This procedure is illustrated in Fig. 1 for experiments performed at 700°C. The assumption that the ratio of activity coefficients in the fluid is independent of K/Na ratio is generally adopted for dilute alkali chloride solutions. It may be tested by decreasing the total chloride content of the solution and determining the change, if any, in K_D. Two experiments performed by Orville at 0.2M total chloride had K_D values almost identical to those for 2M chlorides; these data imply a constant activity coefficient ratio.

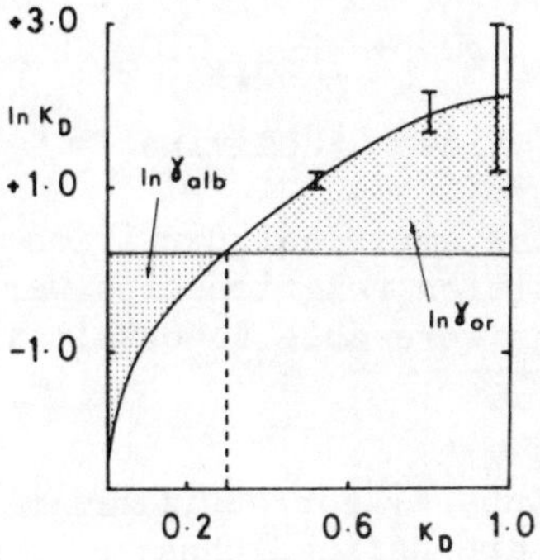

Fig. 1. Graphical integration of equation (13) and (14) for the experimental data of Orville (1963).

From Fig. 1 it may be seen that the measured values of K_D have considerable uncertainties attached to them, and that these uncertainties produce a wide range of possible values of the areas under the $\ln K_D - X_{NaAlSi_3O_8}$ curve. Approximate error bars are shown on the activity-composition diagram for $NaAlSi_3O_8 - KAlSi_3O_8$ (Fig. 2) which has been constructed by the graphical method shown in Fig. 1.

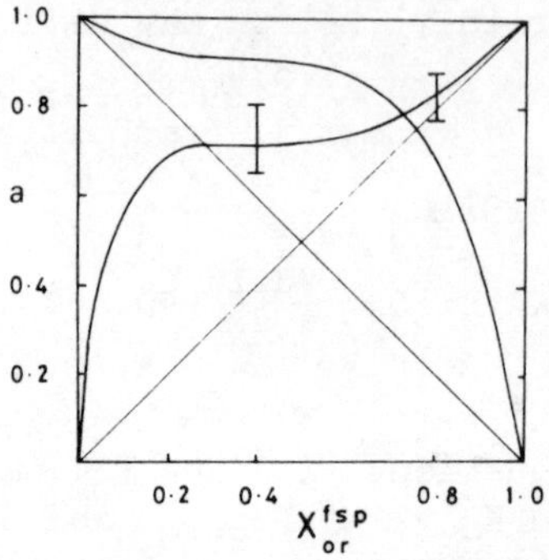

Fig. 2. Activity-composition relationships for sanidine-high albite solid solutions at 700°C.

$MgSi_{0.5}O_2 - FeSi_{0.5}O_2$ OLIVINE SOLUTIONS

Nafziger and Muan (1967) used a similar method to determine the thermodynamic properties of Mg-Fe olivine solid solutions using equilibrium (1):

$$\underset{\text{oxide}}{FeO} + \underset{\text{olivine}}{MgSi_{0.5}O_2} \rightleftarrows \underset{\text{oxide}}{MgO} + \underset{\text{olivine}}{Fe_{0.5}SiO_2}$$

These authors determined the empirical partition coefficient K_{D_1}, and, since activity-composition relationships were available for MgO-FeO oxide solutions were able to obtain $\gamma^{ol}_{MgSi_{0.5}O_2}$ and $\gamma^{ol}_{FeSi_{0.5}O_2}$ as follows:

The equilibrium constant K_1 for equilibrium (1) and partition coefficient K_{D_1} are defined thus:

$$K_1 = \frac{a^{ox}_{MgO} \cdot a^{ol}_{FeSi_{0.5}O_2}}{a^{ox}_{FeO} \cdot a^{ol}_{MgSi_{0.5}O_2}} \qquad (15)$$

$$K_{D_1} = \frac{X^{ox}_{MgO} \cdot X^{ol}_{FeSi_{0.5}O_2}}{X^{ox}_{FeO} \cdot X^{ol}_{MgSi_{0.5}O_2}} \qquad (16)$$

where a^{ox}_{MgO} and X^{ox}_{MgO} are the activity and mole fraction of MgO component in the oxide phase. Since activity-composition relationships for the oxide phase are known, K_1 can be expressed in terms of the parameter C_1 (known if compositions of both phases are known), $\gamma^{ol}_{MgSi_{0.5}O_2}$ and $\gamma^{ol}_{FeSi_{0.5}O_2}$:

$$\ln K_1 = \ln C_1 + \ln \gamma^{ol}_{FeSi_{0.5}O_2} - \ln \gamma^{ol}_{MgSi_{0.5}O_2} \qquad (17)$$

where C_1 is given by:

$$C_1 = \frac{a^{ox}_{MgO} \cdot X^{ol}_{FeSi_{0.5}O_2}}{a^{ox}_{FeO} \cdot X^{ol}_{FeSi_{0.5}O_2}} \qquad (18)$$

Using the Gibbs-Duhem equation as before we obtain:

$$\ln \gamma^{ol}_{MgSi_{0.5}O_2} = \int_0^{X^{ol}_{FeSi_{0.5}O_2}} X_{FeSi_{0.5}O_2} \, d \ln C_1 \qquad (19)$$

$$\ln \gamma^{ol}_{FeSi_{0.5}O_2} = \int_0^{X^{ol}_{MgSi_{0.5}O_2}} X_{MgSi_{0.5}O_2} \, d \ln C_1$$

The results of Näfziger and Muan's experiments at 1200°C, with attendant uncertainties are shown in the activity-composition diagram of Fig. 3. A slight complication in their experiments arises from the fact that the oxide solid solution is not a simple binary (MgO-FeO) since it may contain variable amounts of oxygen. Similar behaviour is exhibited by the iron "end-member" FeO. Hahn and Muan (1962) determined the mixing properties of MgO-FeO solutions by equilibrating them with metallic iron at known a_{O_2} and T (equilibrium (2)):

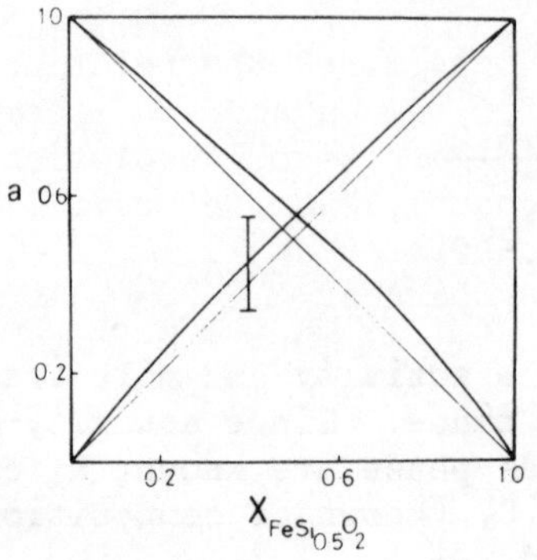

Fig. 3. Activities of $FeSi_{0.5}O_2$ and $MgSi_{0.5}O_2$ components in olivine solid solutions at 1200°C.

$$\underset{\text{metal}}{Fe} + \underset{\text{gas}}{\tfrac{1}{2}O_2} \rightleftarrows \underset{\text{oxide}}{FeO}$$

The activity of FeO in the oxide solid solution is thus fixed relative to a standard state of "FeO" in equilibrium with iron metal at the same temperature. The activity of FeO in the oxide solution is varied by varying a_{O_2}, keeping iron metal present. In order, therefore, for Nafziger and Muan to use the activity-composition data for the oxide solid solution it was simply necessary for them to ensure that all their experiments were saturated in metallic iron and that the oxygen contents of the oxides were the same as in the experiments of Hahn and Muan. The presence of iron metal would of course be unnecessary if the mixing properties of MgO-FeO-O solid solutions were known. A graphical method of solving the Gibbs-Duhem equation for such a case (ternary) has been described by Schuhmann (1955) and was used by Schwerdtfeger and Muan (1967) to deduce the properties of the analogous solid solutions MnO-FeO-O.

ALUMINOSILICATE GARNET SOLID SOLUTIONS

The principal problem associated with the determination of the mixing properties of aluminosilicate garnets is that most of the solutions involved are stable only at high pressures. The apparatus used in experimental studies at these pressures does not, of course, enable as accurate a control of the intensive variables as is possible at 1 atmosphere. Nevertheless, activity-composition relationships determined at high pressures need not necessarily have extremely large uncertainties provided that the equilibria used to fix activities are carefully chosen.

Hensen et al. (1975) determined the relationship between activity and composition for the $Ca_3Al_2Si_3O_{12}$ component in $Mg_3Al_2Si_3O_{12}$ - $Ca_3Al_2Si_3O_{12}$ garnet solid solutions using an assemblage of anorthite, quartz, kyanite or sillimanite and garnet. At known pressure and temperature, $a^{gt}_{Ca_3Al_2Si_3O_{12}}$ is fixed in this assemblage, (with all other phases pure $a_i = 1$) by the equilibrium:

$$\underset{\text{anorthite}}{3CaAl_2Si_2O_8} \rightleftarrows \underset{\text{garnet}}{Ca_3Al_2Si_3O_{12}} + \underset{\text{kyanite/ sillimanite}}{2Al_2SiO_5} + \underset{\text{quartz}}{SiO_2} \quad (20)$$

Given the pressures and temperatures of equilibrium co-existence of pure $Ca_3Al_2Si_3O_{12}$ garnet with the other phases the activities of $Ca_3Al_2Si_3O_{12}$ may be calculated in the following manner. Let us suppose that the equilibrium pressure for the pure phases with grossular garnet ($a^{gt}_{Ca_3Al_2Si_3O_{12}} = 1$) at some temperature T is P^0. The activity of $Ca_3Al_2Si_3O_{12}$ component in the garnet at some other pressure P^1 and the same temperature T is given by:

$$\int_{P^1}^{P^0} \Delta V^0_{20} = RT \ln \frac{a^{gt}_{Ca_3Al_2Si_3O_{12}}}{1} \quad (21)$$

where ΔV^0 is the standard state volume change for the reaction. As a good approximation, ΔV^0 may be regarded as constant over most of the P-T range of interest so that the left hand side of (21) can be integrated thus:

$$(P^0 - P^1)\, \Delta V^0_{20} = RT \ln a^{gt}_{Ca_3Al_2Si_3O_{12}} \quad (22)$$

Since ΔV^0_{20} is large (-1.58 cal bar^{-1} in the kyanite field) a wide range of values of $a^{gt}_{Ca_3Al_2Si_3O_{12}}$ may in principle be obtained by relatively small variations in pressure P^1. The unfavourable aspect of these large activity variations is that small uncertainties in P^1 or P^0 will, of course, produce fairly large uncertainties in activity.

The stability of the assemblage garnet $-Al_2SiO_5$ polymorph-anorthite-quartz is fairly small in the system CaO-MgO-Al_2O_3-SiO_2 (14-21 kb at 1300°C) so that the range of accessible activity values $(0.10)^3$ to $(0.26)^3$ is not as large as one would like. Note that Ca-Mg garnets involve mixing on three sites per formula unit so that the relationship between $a_{Ca_3Al_2Si_3O_{12}}$ and $X_{Ca_3Al_2Si_3O_{12}}$

is:

$$a^{gt}_{Ca_3Al_2Si_3O_{12}} = (X^{gt}_{Ca_3Al_2Si_3O_{12}} \ \gamma^{gt}_{Ca_3Al_2Si_3O_{12}})^3 \quad (23)$$

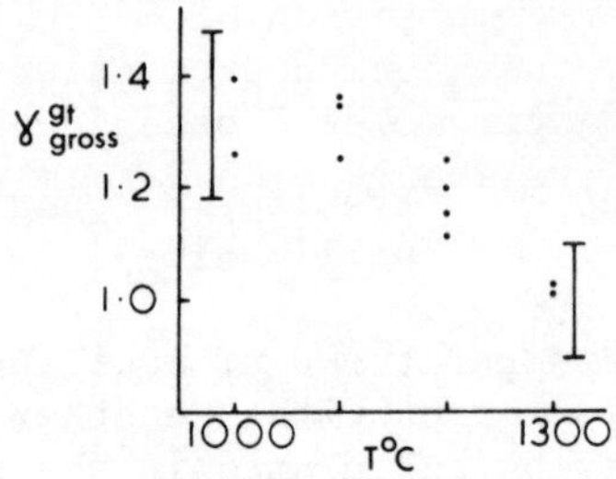

Fig. 4. Activity coefficient of the $Ca_3Al_2Si_3O_{12}$ component in $Mg_3Al_2Si_3O_{12}$-$Ca_3Al_2Si_3O_{12}$ solid solutions as a function of temperature (applies to the composition range 0.10 - 0.22 Ca/Ca+Mg only).

The accessible range of activity values corresponds to a composition range of 0.10 to 0.22 in $X^{gt}_{Ca_3Al_2Si_3O_{12}}$ and activity coefficients (see Fig. 4) which are somewhat greater than 1.0. It should be noted that the activity coefficients shown in this figure are slightly smaller than those given in the original paper by Hensen et al. This is because of a small upward revision of the values of P^O for the experimental apparatus and techniques used in Manchester by Hensen and by the author.

The activity coefficients shown in Fig. 4 may be represented (in the composition-temperature range studied) by the equation:

$$\ln \gamma^{gt}_{Ca_3Al_2Si_3O_{12}} = \frac{(1-X_{Ca_3Al_2Si_3O_{12}})^2 (5780 - 3.52\ T)}{RT} \quad (24)$$

Equation (24) has the form of lnγ for a regular solution. Its use here is not, however, intended to imply that $Ca_3Al_2Si_3O_{12}$ - $Mg_3Al_2Si_3O_{12}$ garnets behave as regular solutions over their entire composition range.

The properties of $Ca_3Al_2Si_3O_{12}$-$Fe_3Al_2Si_3O_{12}$ garnets are more readily amenable to study using equilibrium (20) than are those of the grossular-pyrope series. In the system CaO-FeO-Al_2O_3-SiO_2 the assemblage garnet-anorthite-quartz-Al_2SiO_5 polymorph is stable over a wide range in pressure (11-23 kb at 1100°C) which corresponds to $X^{gt}_{Ca_3Al_2Si_3O_{12}}$ of between 0.1 and 1.0. The relationship between $a_{Ca_3Al_2Si_3O_{12}}$ mole fraction $X^{gt}_{Ca_3Al_2Si_3O_{12}}$ and is currently under study by G. Cressey (personal communication). Preliminary results at 900°C and 1100°C are shown in Fig. 5 and show that the grossular-almandine solid solution is close to ideal ($\gamma_{Ca_3Al_2Si_3O_{12}} = 1$) in this temperature range and for composition of 0 to 0.7 $X_{Ca_3Al_2Si_3O_{12}}$. Note that the uncertainties in activity become large as $a_{Ca_3Al_2Si_3O_{12}}$ approaches 1 because ΔP ($P^0 - P^1$) approaches zero but its absolute uncertainty does not diminish.

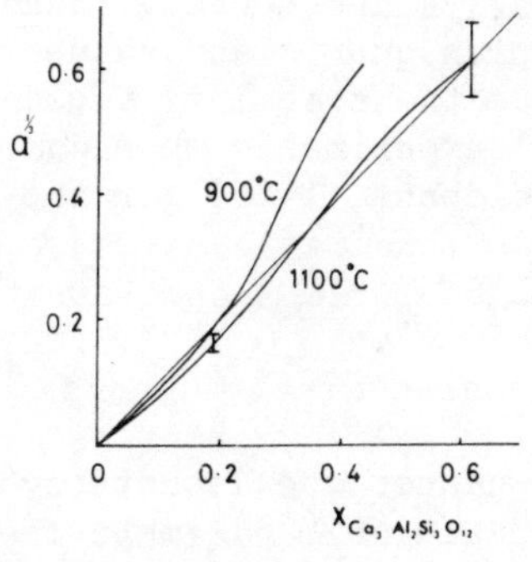

Fig. 5. Activity of the $Ca_3Al_2Si_3O_{12}$ component in $Fe_3Al_2Si_3O_{12}$ - $Ca_3Al_2Si_3O_{12}$ solid solution as a function of composition (G. Cressey, personal communication).

CLINOPYROXENES CONTAINING $CaAl_2SiO_6$ COMPONENT

The relationship between activity and mole fraction of the $CaAl_2SiO_6$ component in clinopyroxenes has been studied by the author (see Wood, 1976 for a preliminary report) using a similar method to that just described for garnet solutions. If clinopyroxene coexists with anorthite and quartz under equilibrium conditions at known P and T then $a^{cpx}_{CaAl_2SiO_6}$ is defined by the relationship:

$$\underset{\text{anorthite}}{CaAl_2Si_2O_8} \rightleftarrows \underset{\text{clinopyroxene}}{CaAl_2SiO_6} + \underset{\text{quartz}}{SiO_2} \qquad (25)$$

The compositions of clinopyroxenes coexisting with anorthite and quartz have been determined over a wide range of pressure and temperature for bulk compositions in the systems $CaO-MgO-Al_2O_3-SiO_2$, $CaO-FeO-Al_2O_3-SiO_2$, and $CaO-MgO-FeO-Al_2O_3-SiO_2$.

In principle, equilibrium (25) should enable more accurate determination of $a^{cpx}_{CaAl_2SiO_8}$ than is possible for $a_{Ca_3Al_2Si_3O_{12}}$ using equilibrium (20). This is because ΔV^0_{24} is considerably smaller than ΔV^0_{20} (- 0.349 cal bar^{-1} as opposed to - 1.58 cal bar^{-1}) and a given error in P produces a much smaller error in activity.

Although uncertainties in $a^{cpx}_{CaAl_2SiO_8}$ due to uncertainties in the pressures of specific experiments are quite small, a major problem has arisen from the inconsistencies between the experimental results of Hays (1966) and Hariya and Kennedy (1968) for the system $CaO-Al_2O_3-SiO_2$. Hariya and Kennedy found a stability field for $CaAl_2SiO_6$ pyroxene plus quartz and values of P^0 for reaction (25) which lie outside the stability field of $CaAl_2SiO_6$ pyroxene given by Hays. Hays' experiments when combined with free energy data (from Robie and Waldbaum, 1968) for the reaction:

$$\underset{\text{corundum}}{Al_2O_3} + \underset{\text{quartz}}{SiO_2} \rightleftarrows \underset{\text{kyanite}}{Al_2SiO_5} \qquad (26)$$

imply that $CaAl_2SiO_6$ pyroxene cannot coexist stably with anorthite and quartz. The position of reaction (25) (metastable) for the $CaO-Al_2O_3-SiO_2$ system has been deduced from Hays' data to lie over 3 kb above that experimentally determined as stable by Hariya and Kennedy. The possible errors in standard-state thermodynamic data (or P^0) for (25) result in large uncertainties in the values of $a^{cpx}_{CaAl_2SiO_6}$ which are adopted. Although both sets of $a^{cpx}_{CaAl_2SiO_6}$ will be used in this paper it should be stated that a new study of the $CaO-Al_2O_3-SiO_2$ system by the author supports the results of Hays rather than those of Hariya and Kennedy.

Fig. 6 is a diagram of activity versus composition for the $CaAl_2SiO_6$ component in $CaMgSi_2O_6-CaAl_2SiO_6$ clinopyroxenes at temperatures of 1100 and 1300°C. The error bars shown on each point refer to an approximate uncertainty of ± 0.5 kb for each experiment and to ± 1σ for the pyroxene compositions determined by the electron microscope microanalyser (EMMA IV). The set of data points clustering around the line of $a^{cpx}_{CaAl_2SiO_6}$ equal to $X^{cpx}_{CaAl_2SiO_6}$ were calculated using Hariya and Kennedy's standard-state data, combined with entropy and volume changes of reaction (25) from Robie and Waldbaum (1968). These data yield the following equation for P^0_{25} with T in °C.

$$P^0 = 30\,800 - 11.3\,(1400 - T)\text{ bars} \qquad (27)$$

(Hariya and Kennedy)

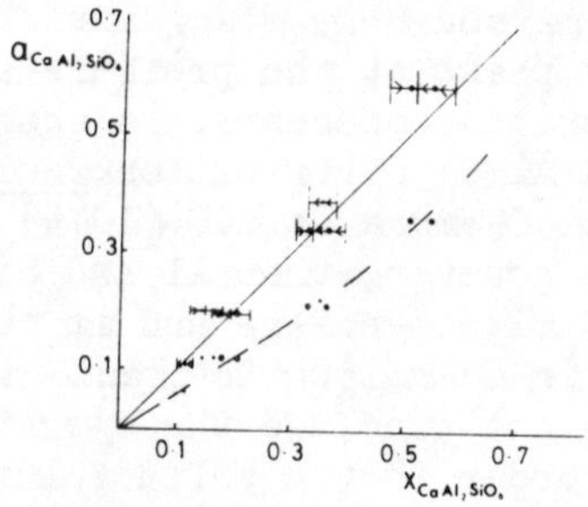

Fig. 6. Two possible sets of activity-composition relationships for $CaAl_2SiO_6$ component in $CaMgSi_2O_6$-$CaAl_2SiO_6$ clinopyroxene solution. Upper set of points based on Hariya and Kennedy's data for (25). Lower set based on Hays' data.

The lower group of points in Fig. 6 are the same experimental data displaced to lower values of $a_{CaAl_2SiO_6}$ through adoption of Hays' data for P^0:

$$P^0 = 35\,500 - 11.3\,(1500 - T) \text{ bars} \quad \text{(Hays)} \tag{28}$$

Given that equation (28) is consistent with experiments performed in Manchester it is apparent from Fig. 6 that $a^{cpx}_{CaAl_2SiO_6}$ is rather less than $X^{cpx}_{CaAl_2SiO_6}$, a situation which is in marked contrast to all the other solid solutions discussed so far in this paper. This does not necessarily imply, however that $\gamma^{cpx}_{CaAl_2SiO_6}$ is less than 1.0 as may be deduced from the form of the free energy of $CaMgSi_2O_6$-$CaAl_2SiO_6$ solid solution.

If $CaMgSi_2O_6$ and $CaAl_2SiO_6$ were to mix as charge-balanced units in the clinopyroxene structure the molar free energy of a solid solution of mole fraction $X_{CaAl_2SiO_6}$ would be:

$$\begin{aligned} G_{ss} = {} & X_{CaAl_2SiO_6}\, \mu^0_{CaAl_2SiO_6} + (1-X_{CaAl_2SiO_6})\, \mu^0_{CaMgSi_2O_6} \\ & + RT\, X_{CaAl_2SiO_6} \ln X_{CaAl_2SiO_6} \\ & + (1-X_{CaAl_2SiO_6}) \ln (1-X_{CaAl_2SiO_6}) \\ & + X_{CaAl_2SiO_6}\, RT \ln \gamma_{CaAl_2SiO_6} \\ & + (1-X_{CaAl_2SiO_6})\, RT \ln \gamma_{CaMgSi_2O_6} \end{aligned} \tag{29}$$

In equation (29) $\mu^0_{CaAl_2SiO_6}$ and $\mu^0_{CaMgSi_2O_6}$ are the standard state chemical potentials (pure phase at the pressure and temperature of interest) of the subscript components. A charge-balanced model of this type with ordered Al-Si distribution would lead to $\gamma_{CaAl_2SiO_6}$ less than 1.0. Okamura et al. (1974) have found from crystallographic studies, however, that Al and Si atoms are disordered in pure $CaAl_2SiO_6$ clinopyroxene and it seems probable that such disorder is present in $CaMgSi_2O_6$-$CaAl_2SiO_6$ solid solutions. Although this result does not preclude the operation of some short-range "charge-balancing" order in the solid solutions let us make the simplest assumption of complete Al-Si disorder over the entire structure (some other possibilities are briefly discussed in Wood, 1976). Disorder of this type yields a large entropy of mixing of $CaMgSi_2O_6$-$CaAl_2SiO_6$ solid solutions and the following result for G_{ss}:

$$\begin{aligned} G_{ss} = {} & X_{CaAl_2SiO_6}\,\mu^0_{CaAl_2SiO_6} + (1-X_{CaAl_2SiO_6})\,\mu^0_{CaMgSi_2O_6} \\ & + RT\left[X_{CaAl_2SiO_6}\ln X_{CaAl_2SiO_6} + (1-X_{CaAl_2SiO_6})\ln(1-X_{CaAl_2SiO_6})\right] \\ & + RT\left[X_{CaAl_2SiO_6}\ln\frac{X_{CaAl_2SiO_6}}{2} + (2-X_{CaAl_2SiO_6})\ln\frac{2-X_{CaAl_2SiO_6}}{2}\right] \\ & + X_{CaAl_2SiO_6}\,RT\ln\gamma_{CaAl_2SiO_6} + (1-X_{CaAl_2SiO_6})\,RT\ln\gamma_{(1-X_{CaAl_2SiO_6})} \end{aligned} \quad (30)$$

Differentiation of G_{ss} with respect to $X_{CaAl_2SiO_6}$ yields the following result for activity of the $CaAl_2SiO_6$ component:

$$a^{cpx}_{CaAl_2SiO_6} = X^2_{CaAl_2SiO_6}(2-X_{CaAl_2SiO_6})\,\gamma_{CaAl_2SiO_8} \quad (31)$$

In this case $\gamma_{CaAl_2SiO_6}$ and $\gamma_{CaMgSi_2O_6}$ (obtained from Gibbs-Duhem integration) are greater than 1.0 for most compositions of the $CaMgSi_2O_6$-$CaAl_2SiO_6$ series even using Hays' standard state data.

The molar excess free energy of mixing of $CaMgSi_2O_6$-$CaAl_2SiO_6$ solid solution is defined as follows:

$$G^{xs} = X_{CaAl_2SiO_6}\, RT \ln \gamma_{CaAl_2SiO_6} + (1-X_{CaAl_2SiO_6})\, RT \ln \gamma_{CaMgSi_2O_6} \qquad (32)$$

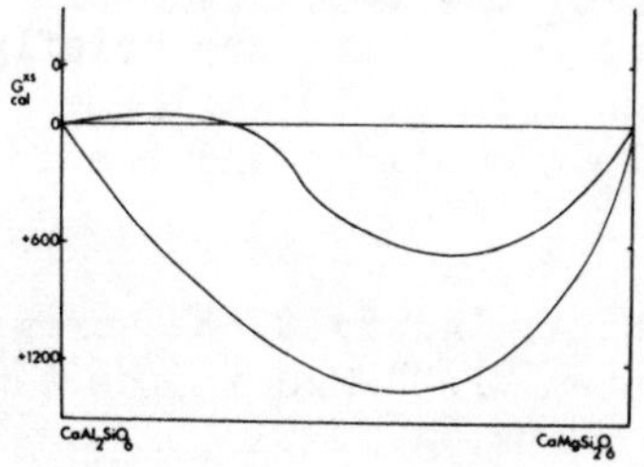

Fig. 7. G^{xs} for $CaMgSi_2O_6$-$CaAl_2SiO_6$ solid solution. Upper curve, based on Hays' data, lower curve on Hariya and Kennedy data.

Fig. 7 shows G^{xs} as a function of composition at 1200°C for the disordered model of $CaMgSi_2O_6$-$CaAl_2SiO_6$ solid solution discussed above. Both sets of standard state data discussed here lead to positive values of G^{xs} with a maximum difference between them of only 800 cals.

Newton (accompanying paper) has found positive values of H^{xs} for this solid solution series, confirming the general form of the activity-composition relationships discussed here. It is doubtful, however, whether calorimetric results on the solid solutions provide a good constraint on the standard state data because of lack of knowledge of excess entropy contributions to G^{xs}. The author's results on CaO-Al_2O_3-SiO_2 lead him to prefer the upper curve for G^{xs} in Fig. 7 and lower activity curves in Fig. 6.

CONCLUSIONS

The silicate solid solutions discussed in this paper all exhibit deviations from ideality. It is apparent, however, that the techniques used to characterise these thermodynamic properties do not allow activity coefficients to be determined much more accurately than ± 10%. This lack of accuracy does not preclude the application of experimentally determined activity coefficients to natural systems provided that the equilibria considered are carefully chosen (see, for example, Wood and Fraser, 1976, Chapter 4). It should, however, lead the reader to view with caution any attempts to apply complex solution models to poorly constrained experimental results.

ACKNOWLEDGEMENT

The experimental work on garnet and clinopyroxene solid solutions was supported by the Natural Environment Research Council.

REFERENCES

Hahn, W.C. Jr. and Muan, A. Activity measurements in oxide solid solutions: The system "FeO"-MgO in the temperature interval 1100° to 1300°C. Trans. A.I.M.E., 224, 416-420 (1962).

Hariya, Y. and Kennedy, G.C. Equilibrium study of anorthite under high pressure and high temperature. Amer. J. Sci., 266, 193-203 (1968).

Hays, J.F. Lime-alumina-silica. Carnegie Inst. Washington Yearbook 65, 234-239 (1966).

Hensen, B.J., Schmid, R. and Wood, B.J. Activity-composition relationships for pyrope-grossular garnet. Contrib. Mineral. Petrol., 51, 161-166 (1975).

Nafziger, R.H. High-temperature activity-composition relations of equilibrium spinels, olivines and pyroxenes in the system Mg-Fe-O-SiO_2. Amer. Mineral., 58, 457-465 (1973).

Nafziger, R.H. and Muan, A. Equilibrium phase compositions and thermodynamic properties of olivines and pyroxenes in the system MgO-"FeO"-SiO_2. Amer. Mineral., 52, 1364-1385 (1967).

Okamura, F.P., Ghose, S. and Ohashi, H. Structure and crystal chemistry of calcium Tschermak's pyroxene $CaAl_2SiO_6$. Amer. Mineral., 59, 549-577 (1974).

Orville, P.M. Alkali ion exchange between vapour and feldspar phases. Amer. J. Sci., 261, 201-237 (1963).

Orville, P.M. Plagioclase cation exchange equilibria with aqueous chloride solution: Results at 700°C and 2000 bars in the presence of quartz. Amer. J. Sci., 272, 234-272 (1972).

Robie, R.A. and Waldbaum, D.R. Thermodynamic properties of minerals and related substances at 298.15°K (25.0°C) and one atmosphere (1.013 bars) pressure and at higher temperatures. U.S. Geol. Survey Bull., 1259, 256 pp. (1968).

Schuhmann, R. Jr. Application of Gibbs-Duhem equations to ternary systems. Acta Met., 3, 219-226 (1955).

Schwerdtfeger, K. and Muan, A. Phase equilibria in the system Fe-Mn-O involving "(Fe,Mn)O" and $(Fe,Mn)_3O_4$ solid solutions.

Trans. A.I.M.E., 239, 1114-1119 (1967).
Thompson, J.B. Jr. and Waldbaum, D.R. Mixing properties of sanidine crystalline solutions: I. Calculations based on ion-exchange data. Amer. Mineral., 53, 1965-1999 (1968).
Wood, B.J. Mixing properties of tschermakitic clinopyroxenes. Amer. Mineral., 61, 599-602 (1976).
Wood, B.J. and Fraser, D.G. Elementary Thermodynamics for Geologists. O.U.P. (in press).

THERMOCHEMISTRY OF GARNETS AND ALUMINOUS PYROXENES IN THE CMAS SYSTEM

Robert C. Newton

Department of the Geophysical Sciences
University of Chicago
Chicago, Illinois 60637

IMPORTANCE OF THE CMAS SYSTEM

The bulk chemistry of the earth's upper mantle, as deduced from analysis of exotic fragments from explosive igneous pipes and from geochemical arguments, can be "modelled" to quite a good approximation by the simple system $CaO-MgO-Al_2O_3-SiO_2$. The only significant departures from this quaternary system in Green and Ringwood's (1967) "pyrolite III", which is their favored mantle composition, are 8.0 percent FeO and 0.4 percent Cr_2O_3. FeO plays a role in the ferromagnesian minerals very similar to that of MgO, which dominates it by a factor of ten in molar amount, hence its influence should be minor. Cr_2O_3 is closely similar to Al_2O_3 in its behavior and again greatly inferior in amount. The quaternary system yields all of the major phases which coexist in peridotites, namely olivine, orthopyroxene, clinopyroxene, spinel, garnet and plagioclase. For these reasons the simple system has been very useful in geophysical phase-equilibrium investigations.

The most important reaction in the 4-component system for geophysical purposes can be approximately written:

$$\text{A)}\quad \underset{\text{orthopyrox.}}{2MgSiO_3} + \underset{\text{clinopyrox.}}{CaMgSi_2O_6} + \underset{\text{spinel}}{MgAl_2O_4} \rightleftharpoons \underset{\text{garnet}}{Mg_2CaAl_2Si_3O_{12}} + \underset{\text{olivine}}{Mg_2SiO_4}$$

where the Ca/Mg ratio of garnet actually varies along the reaction boundary as do the Al_2O_3 contents of the pyroxenes, although the reaction remains univariant. This simple equilibrium provides a quantitative criterion for the major phase composition of upper mantle material.

D. G. Fraser (ed.), Thermodynamics in Geology, 29-55.

EXPERIMENTAL DIFFICULTIES

The simple univariant reaction described above, although it is of great importance to geophysical interpretations, is not well understood quantitatively, in spite of several experimental determinations. The reasons are simple and two-fold. First, the kinetics of reaction in dry silicate systems become impossibly slow at temperatures below about 1200°C, so that convincing demonstration of chemical equilibrium in the form of "reversals", or growth of one assemblage at the expense of an isochemical one across the reaction boundary, is often impossible. Second, the compositions and ordering states of the synthetic phases may be quite non-equilibrium. This is particularly true of the Al content of the pyroxenes, equilibrium with respect to which is extremely difficult to achieve experimentally.

As a consequence of these difficulties the four experimental determinations of reaction A) are in considerable disagreement. The actual equilibrium investigated is the same in each case, with the exception of the Green and Ringwood (1967) "pyrolite III" determination. In the high-temperature portions of the determinations there is substantial accord. The lower-temperature and extrapolated portions show a fundamental discrepancy, with opinions falling into two groups. According to Green and Ringwood (1967) and O'Hara et al (1971), strong curvature develops in the low-temperature portions, so that the field of garnet peridotite is confined to high, probably subcrustal pressures. According to the alternative opinion (MacGregor, 1965; Kushiro and Yoder, 1966), the univariant boundary drives into a low-pressure region, such that garnet peridotite could, under some circumstances, be stable in dry deep crustal (that is, granulite-grade) metamorphism (Kushiro and Yoder, 1966, p. 361), rather than require a subcrustal origin. According to the MacGregor (1965) and Kushiro and Yoder (1966) diagrams, the upper mantle would lie almost everywhere in the field of garnet peridotite, except under the active spreading centers, whereas the O'Hara et al (1971) and Green and Ringwood (1967) diagrams call for an extensive field of spinel peridotite stable under much of the earth's crust.

THE ROLE OF THERMOCHEMISTRY

The powerful agency of thermodynamics ought to be the natural arbiter of controversies of this sort. Starting from the high temperature region where the equilibria may be experimentally reversed, the univariant curves may be produced to lower temperatures by means of the Clausius-Clapeyron equation. This powerful method has not yet realized its potential in application to reactions in refractory mineral systems because of lack of requisite data.

The form of the Clausius-Clapeyron equation most commonly used is:

$$\frac{dP}{dT} = \frac{\Delta S}{\Delta V} \qquad (1)$$

where dP/dT is the slope of the univariant equilibrium, ΔS is the entropy change, and ΔV is the volume change. The ratio $\Delta S/\Delta V$ is quite insensitive to pressure, so that for most purposes, $\Delta S/\Delta V$ may be replaced by $\Delta S^o/\Delta V^o$, where the superscript zeroes refer to one-atmosphere quantities. The volume change is known very well, even for reactants which are solid solutions, and thermal expansion and compressibility corrections may be applied. Evaluation of ΔS presents a bigger problem. Heat capacities have been measured for most of the solid solution end members and sometimes can be estimated closely for intermediate solid solutions. Thus, the "Third Law entropy" can be evaluated by:

$$S(T) = \int_0^T \frac{C_p}{T}\,dT \qquad (2)$$

where C_p is the heat capacity. Extensive tables exist for the Third Law entropies of many end-member minerals (Robie and Waldbaum, 1968). In addition to the Third Law entropies there are usually important entropy contributions arising from "configurational" sources, that is, from entropy of mixing and of disorder. These can sometimes be elucidated by careful X-ray diffraction or other measurements of atomic site populations in mix-crystals, but in many cases there is an irreducible ambiguity which leaves one with a few or several "models" of atomic mixing among which it is difficult to choose. The configurational entropies of the various models for a given solid solution series will be different, and the numerator of expression 1) is thus uncertain to a considerable degree. In some cases it may be possible to choose among various atomic mixing models by a judicious combination of phase-equilibrium measurements and enthalpy of solution measurements. The example of the aluminous clinopyroxenes below illustrates this combined approach.

The alternative form of the Clausius-Clapeyron equation:

$$\frac{dP}{dT} = \frac{\Delta H}{T\Delta V} \qquad (3)$$

offers another approach. Here ΔH is the enthalpy change of the univariant reaction, and can be evaluated by heat of solution measurements:

$$\Delta H(T,P) = \Delta H_{sol'n}(\text{react}) - \Delta H_{sol'n}(\text{prod}) + \int_1^P [\Delta V - T\Delta(V\alpha)]\,dP \qquad (4)$$

where the temperature of the calculation is the same at which the heat of solution measurements are made. Correction to other temperatures is made from heat capacity measurements. The third term on the right, the pressure correction to ΔH, is important. For most purposes, the integrand term involving the thermal expansions, α, can be ignored. This use of the Clausius-Clapeyron equation has one very important advantage: the resort to mixing "models" is obviated. Heat of solution data are rapidly accumulating for the major rock-forming mineral series, due largely to the advent of new methods for high-temperature solution calorimetry, as described by A. Navrotsky in this symposium. Application of the data to the calculation of the peridotite univariant equilibrium A) will be made below.

$CaMgSi_2O_6$-$CaAl_2SiO_6$ CLINOPYROXENE

Aluminous clinopyroxene is an important member of the minerals comprising spinel and garnet lherzolites. The intake of Al_2O_3 into the pyroxenes is an increasing function of temperature and pressure, to the point where, according to Green and Ringwood (1967), the aluminous spinel of model mantle material is entirely absorbed into the pyroxenes at temperatures above 1300°C. The mixing entropies of Al in the pyroxenes and the departures from ideality of Al substitution will have a large effect on the calculated dP/dT of the spinel to garnet lherzolite transition.

Aluminous diopside may be regarded as a solid solution of two end-members, $CaMgSi_2O_6$ (diopside) and $CaAl_2SiO_6$ (Ca-Tschermak pyroxene, or CaTs). X-ray crystallography has shown that there are three distinct kinds of crystallographic sites which are occupied by the cations. Calcium occupies the larger octahedral (six-coordinated) sites M_2, magnesium and aluminum share the smaller octahedral sites M_1, and silicon and aluminum share the tetrahedral (four-coordinated) sites. The large amount of cation mixing in the solid solutions stabilizes them by virtue of the configurational entropy, if concommitant heat effects are not large; that is, if the mixing is nearly ideal.

The ideality of mixing, or departure from it, may be investigated by measuring the activities of the components in solid solution as functions of composition. This was done by Wood (1975) in the following manner. Compositions in the system $CaMgSi_2O_6$-$CaAl_2Si_2O_8$ (anorthite) were experimentally equilibrated at temperatures in the range 900°-1300°C and pressures of 10-25 kb in the solid-pressure-medium piston-cylinder apparatus. The resulting assemblages were pyroxene-anorthite-quartz. The activity of CaTs in the pyroxene is given to a close approximation by:

$$RT\ln a_{CaTs} \cong \Delta V^{\circ}(P-P^*) \quad (5)$$

where ΔV^o is the one-atmosphere volume change of the end-member reaction:

$$\text{B)} \quad \underset{\text{Cpx}}{CaAl_2SiO_6} + \underset{\text{quartz}}{SiO_2} \rightleftharpoons \underset{\text{anorthite}}{CaAl_2Si_2O_8}$$

P* is the experimentally-determined pressure of the equilibrium end-member reaction, as worked out by Hariya and Kennedy (1968). The assumptions are made of independence of the partial molal volume of CaTs of composition, and independence of ΔV^o of pressure. The reference state of the activity of CaTs is the pure phase at the T and P of equilibration. Data could be obtained only over a restricted pressure range because higher pressures result in garnet synthesis.

Wood's data give the very simple relationship that the activity of CaTs is nearly equal to its mole fraction in diopside-CaTs solid solutions. This would be true of an ideal solution of CaTs and diopside end members; Al and Si are ordered on the tetrahedral sites and Mg and Al are ordered on the corresponding M_1 sites. The effective mixing units for ordered pyroxenes are (MgSi) and (AlAl). Because of the simplicity of this substitution scheme and because it maintains strict nearest-neighbor charge balance, one is tempted to favor it strongly. However, as Wood (1975) pointed out, the X-ray crystallographic study of Okamura et al (1974) showed that synthetic CaTs pyroxene has a high degree of (Si,Al) tetrahedral disorder. The ideal model activity in the case of complete cation randomness on M_1 and T sites would be:

$$a^{id}_{CaTs} = X^2_{CaTs}(2-X_{CaTs}) \qquad (6)$$

and X_{CaTs} would be considerably larger than a_{CaTs} for an ideal solution (zero enthalpy of mixing). Thus, the finding of activity nearly equal to mole fraction implies a significant positive enthalpy of mixing.

It is quite possible that the state of (Al,Si) order in diopside-CaTs solid solutions is greater than completely random. Short-range order might be present which would not be observable in an X-ray diffraction experiment. Various models are possible. A simple model was developed by Wood (1975 and personal communication) in which electrical charge balance is maintained on a unit-cell scale. In this model, each unit cell contains four "molecules", which may be all $CaMgSi_2O_6$, all CaTs, or any combination of these. The unit cell contents are internally randomly mixed with regard to the tetrahedral and M_1 sites. Each of the five molecular combination possibilities corresponds to a different configurational entropy of the unit cell. Different compositions of the pyroxene correspond to different probabilities of encountering any of the five cell types. The configurational entropies of the five cell types are each multiplied by the probability of encountering that cell type in a random selection, and the sum is multiplied by Avogadro's number to give the molar configurational entropy of four pyroxene formula units. Fig. 1 shows the molar configurational entropy of the short-range-order model and of the completely disordered model. It is seen that the configurational entropy of the former is only about half that of the latter, even though the two disordering types could not be distinguished in an X-ray diffraction experiment.

The activity coefficient, γ, in its broadest usage, is a number which multiplies an ideal activity to get a real activity. Thus, if the actual activities of components in a solid solution have been measured, the activity coefficient will depend on the mixing model. For instance, for the completely disordered model, γ is defined by:

$$a_{CaTs} = \gamma_{CaTs}[X^2_{CaTs}(2-X_{CaTs})] \quad (7)$$

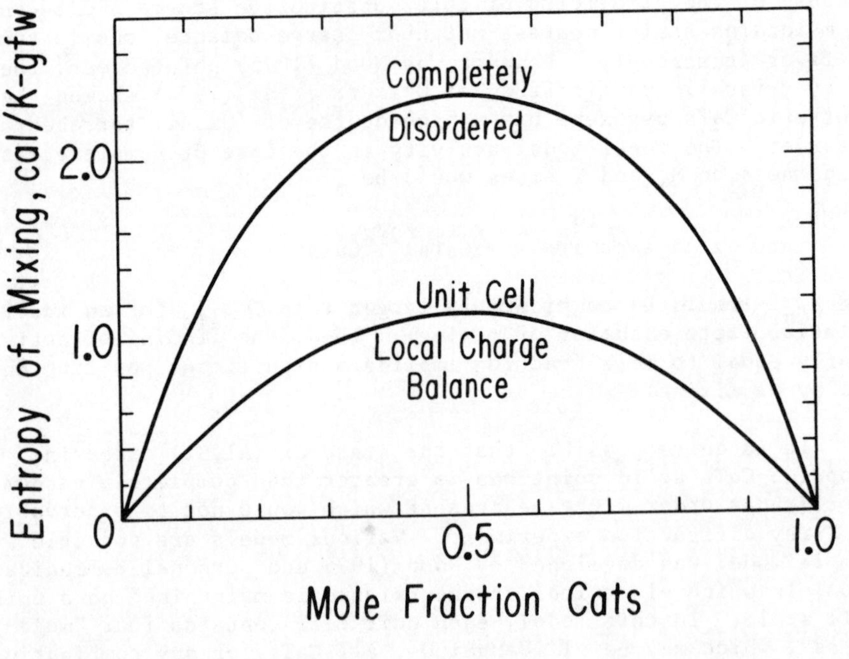

Fig. 1 Configurational entropy of mixing per mole of $CaMgSi_2O_6$ (diopside) - $CaAlSi_2O_6$ (CaTs) pyroxenes, for completely disordered and unit cell charge balanced models.

TABLE I

Activity coefficients of $CaMgSi_2O_6$ (Di) and $CaAlSi_2O_6$ (CaTs) in clinopyroxene, after Wood (1975), according to two models of cation mixing.

X_{CaTs}	Complete Disorder		Unit Cell Charge Balance	
	γ_{CaTs}	γ_{Di}	γ_{CaTs}	γ_{Di}
0.07	10.6	1.10	3.1	1.01
0.16	4.0	1.25	2.3	1.06
0.33	2.0	1.58	1.6	1.19
0.55	1.35	2.11	1.26	1.42

Table I gives the activity coefficients of CaTs for various pyroxene compositions calculated by Wood (1975) from his experimental data, assuming the completely disordered and the unit-cell charge-balance models. Given also are the activity coefficients of the $CaMgSi_2O_6$ component, calculated from the CaTs activity coefficients by means of the Gibbs-Duhem equation.

The <u>excess Gibbs energy of mixing</u> is that portion of the Gibbs energy of mixing different from the ideal Gibbs energy of mixing:

$$\Delta G^{ex} = \Delta G + T\Delta S^{conf} \qquad (8)$$

where ΔG is the Gibbs energy difference between the solid solution and the end-members in the unmixed condition, and ΔS^{conf} is an analogous configurational entropy quantity. The quantity $-T\Delta S^{conf}$ is the ideal Gibbs energy of mixing. According to the simplest kinds of non-ideal mixing theory, ΔG^{ex} is independent of temperature, and it is expected that this will approximately hold for real solutions over temperature ranges of a few hundreds of degrees, as seems to be borne out by Wood's experimental data. Thus, we have to a good approximation:

$$\Delta G^{ex} = RT(X_{CaTs} \ln \gamma_{CaTs} + X_{Di} \ln \gamma_{Di}) \cong \Delta H^{ex} \qquad (9)$$

where ΔH^{ex} is the <u>excess enthalpy of mixing</u>. This quantity is a function of composition, and can be obtained from enthalpy of solution measurements on the solid solutions and the end-members. Thus, solution calorimetry allows us to discriminate among the various mixing models. Relations 8) and 9) predict ΔH^{ex} values which are model-dependent. These may be then compared with the results of the calorimetry.

Fig. 2 shows the plotted calorimetric measurements on diopside-CaTs solid solutions obtained by Newton, Charlu and Kleppa (1976) at 970 K in a lead borate calorimeter solvent. The heat of solution trends predicted by the completely disordered, charge-balance and completely ordered models of Wood (1975) are shown also.

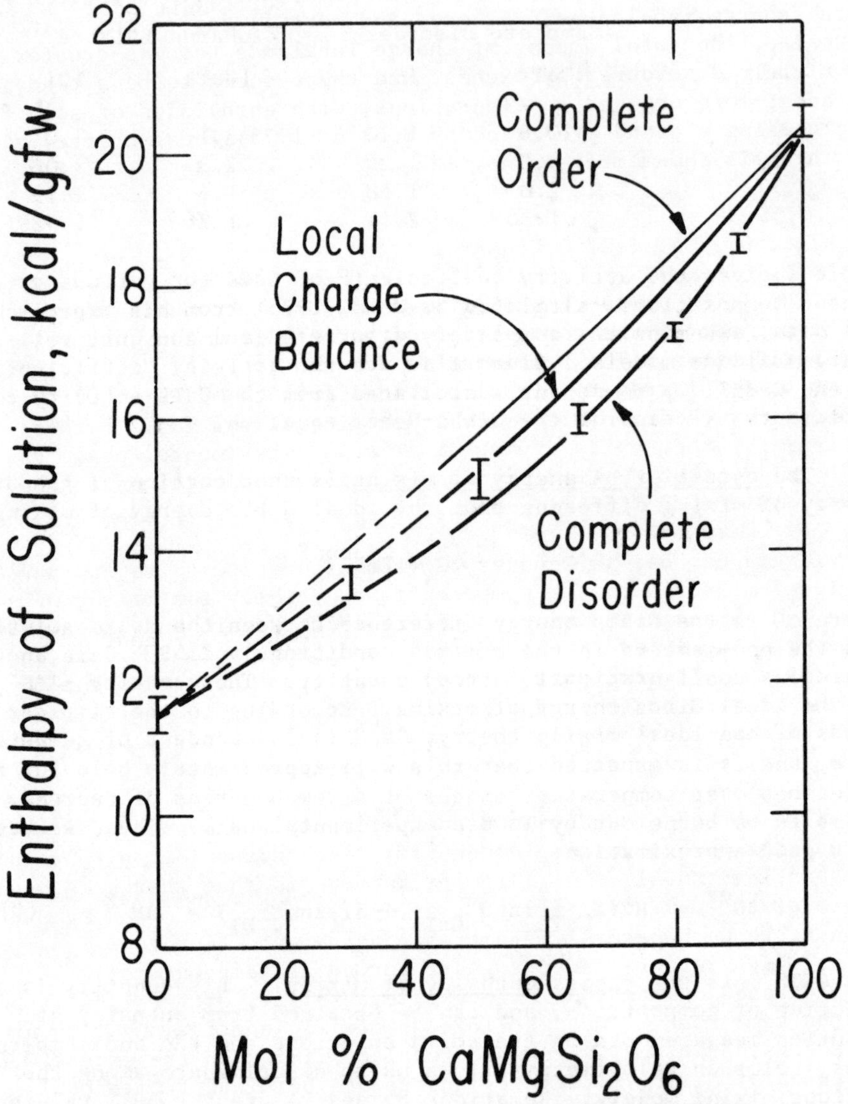

Fig. 2 Enthalpy of solution of synthetic diopside-CaTs pyroxenes in a lead borate melt at 970 K (Newton et al, 1976), compared with trends predicted by completely ordered, unit cell charge balanced, and completely disordered models of Wood (1975).

It is evident on inspection that the heat of solution data are in best agreement with the completely disordered model for pyroxenes in the diopside-rich range. Apparently, the greater mixing entropy of this configuration stabilizes it in spite of local charge inbalances. As more and more Al enters the clinopyroxene, the total amount of charge inbalance becomes greater for completely disordered pyroxenes, and there seems to be a tendency to adopt more ordered configurations, with enthalpies of solution approaching a trend predicted by Wood's (1975) local charge model as the CaTs end-member is neared.

$Mg_3Al_2Si_3O_{12}$-$Ca_3Al_2Si_3O_{12}$ GARNET

The join pyrope-grossular is a key one, accounting for as much as 85 percent of the composition of garnets from many natural garnet peridotites. The remainder is dominantly $Fe_3Al_2Si_3O_{12}$, which might be considered to mix almost ideally with $Mg_3Al_2Si_3O_{12}$, if analogy may be drawn with many other mineral systems. Pyrope-grossular garnets play a pivotal role in experimental phase relations of synthetic peridotite and eclogite compositions in the model system CaO-MgO-Al_2O_3-SiO_2.

There has been a tendency in the literature to regard garnets as ideal mixtures of the components, especially for mixing of pyrope and almandine (Mysen and Heier, 1972; Råheim and Green, 1974). However, Ganguly and Kennedy (1974) and Ganguly (1976) have shown theoretically that important deviations from ideality occur in mixing of most of the garnet end-members. These excess Gibbs energies of mixing are mostly positive and lead to fairly high temperature solvi in many of the binary and ternary joins.

The method of analysis used by Ganguly and Kennedy (1974) to deduce the ΔG^{ex} of mixing among almandine, grossular, pyrope and spessartine ($Mn_3Al_2Si_3O_{12}$) is instructive in that they used arguments of three kinds. First, they analyzed the Fe^{2+} and Mg partitioning between natural metamorphic garnet and biotite pairs from rocks which were believed to have formed at temperatures of 630° ± 40°C based on the coexistence of staurolite, either locally or in the same rock. The distribution coefficients, along with atomic size considerations, were used to place limits on the excess Gibbs energies of mixing. Further constraints obtained from experimental syntheses of pyrope-grossular garnets at temperatures down to 750°C, which experiments failed to locate a binary solvus. The reader is referred to their paper for the interesting details of the analysis.

To a first approximation, the deviations from ideality may be expressed by <u>regular solution theory</u> (Guggenheim, 1952, p. 29), in which the activity coefficient of a component in a binary system is given by the simple expression:

$$RT\ln \gamma = W(1-X)^2 \tag{10}$$

where X is the mole fraction of the component and W, the "interaction parameter", is a constant, independent of temperature, pressure and composition, in the simplest case. From the petrologic, crystal chemical, and experimental evidence, Ganguly and Kennedy (1974) deduced the magnitude of the pyrope-grossular interaction parameter as, W = 3.83 ± 0.22 kcal for the 1/3 formula. The value most appropriately applies to a temperature near 650°C but, if the regular solution formulation is valid, it should be independent of temperature.

Hensen et al (1975) attacked the problem of pyrope-grossular mixing by a method identical in principle to that used by Wood (1975) on the aluminous clinopyroxenes. They experimentally equilibrated bulk compositions on the pyrope-anorthite join at temperatures from 1000°-1300°C and pressures in the range 15-21 kilobars. The resulting assemblages were garnet-Al_2SiO_5-quartz-anorthite. The run pressures were considerably lower than those for which grossular is stable with Al_2SiO_5 (kyanite or sillimanite) and quartz relative to anorthite (Hariya and Kennedy, 1968). This solid solution of $Ca_3Al_2Si_3O_{12}$ into a pyrope-rich host at reduced pressures takes place because the activity is low for the $Ca_3Al_2Si_3O_{12}$ component, being directly related to the mole fraction.

From the calculated activity coefficients a regular solution parameter for pyrope-grossular garnets is derived:

$$W = \frac{RT\ln \gamma_{CaAl_{2/3}SiO_4}}{(1-X_{CaAl_{2/3}SiO_4})^2} \tag{11}$$

It should be recognized here that the regular solution theory is to be regarded as only a useful approximation which applies over a restricted range of composition. The garnets synthesized by Hensen et al (1975) ranged from 10 to 22 mole percent grossular, which is at the upper limit of pyrope mole fractions in the natural garnets considered by Ganguly and Kennedy (1974). A linear least squares fit to 11 data points gave a temperature dependence of W = 7460 - 4.3T in calories. Thus, it would seem that the garnets are getting more ideal with increasing temperature. At 630°C, the temperature assumed by Ganguly and Kennedy (1974), W would be 3560 calories, which is in very good agreement with Ganguly and Kennedy's value.

Strictly speaking, the regular solution formulation does not allow for a temperature variation of W because the entropy is considered to be entirely ideal configurational. In Hensen et al's (1975) approximation, however, we must have:

$$\left.\frac{\partial G^{ex}}{\partial T}\right)_{p,X} = - \Delta S^{ex} = 4.3\, X_{Gr} X_{Py} \qquad (12)$$

For $X_{Gr} = 0.15$ we have that $\Delta S^{ex} = 0.55$ cal/K for the 1/3 formula, which is comparable to the configurational entropy of 0.84 cal/K. This feature makes the regular solution formulation somewhat less than satisfactory.

Newton et al (1976) measured the enthalpies of solution of synthetic pyrope-grossular garnets at 970 K in a calorimeter solvent of composition $2PbO \cdot BrO_3$. Their results are shown in Fig. 3 (brackets). The trend predicted by the results of Ganguly and Kennedy (1974) with the regular solution assumption in the strict sense, which includes the assumption that $\Delta S^{ex} = 0$, are also shown. It is seen that, in the range zero to thirty percent grossular, which includes the range of Ganguly and Kennedy's natural samples and Hensen et al's (1975) syntheses, the calorimetry results are in good agreement with the regular solution theory and a W of about 3.8 kcal. At higher grossular contents the regular solution

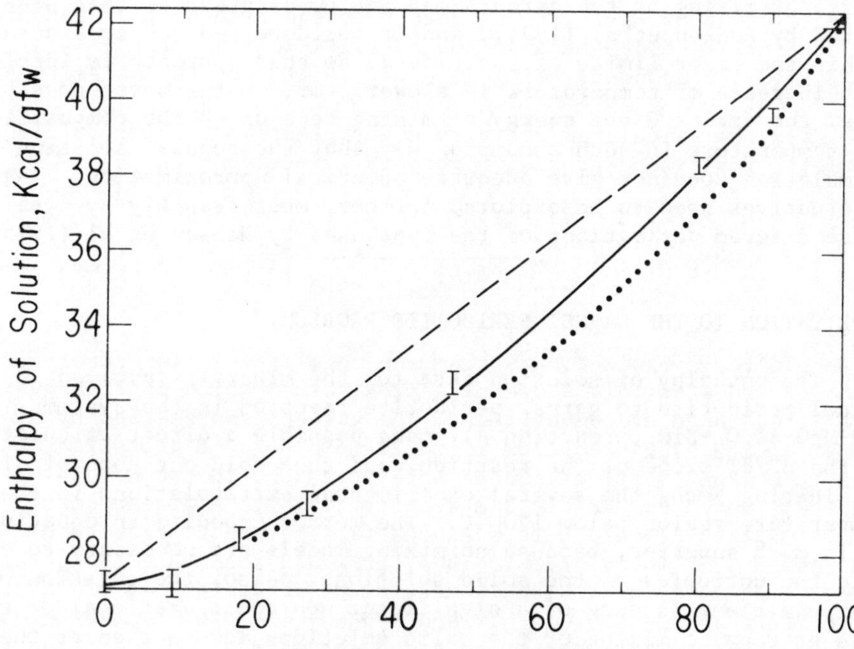

Fig. 3 Enthalpy of solution of pyrope-grossular garnets in a lead borate melt at 970 K (Newton et al, 1976). Dashed line is trend for ideal mixing; dotted line is trend for regular solution with W = 3.82 kcal (Ganguly and Kennedy, 1974).

formulation is not adequate, and a sub-regular formulation (Thompson, 1967) may be substituted. Here, the excess enthalpy of mixing per three moles of divalent cations is given by:

$$\Delta H^{ex} = 3(W_{H_1} X_{Py} X_{Gr}^2 + W_{H_2} X_{Py}^2 X_{Gr}) \tag{13}$$

The enthalpy of mixing data are well fit across the entire range of composition by the parameters W_{H_1} = 2.00 kcal, W_{H_2} = 3.82 kcal.

In spite of the apparent good agreement of the calorimetry results with the regular solution theory of Ganguly and Kennedy (1974), the marked temperature dependence of W found by Hensen et al (1975) would require a different enthalpy of mixing. According to their results we should have, at 970 K and X_{Gr} = 0.15, ΔH^{ex} = $\Delta G^{ex} + T\Delta S^{ex}$ = 3(0.42 + 0.53) = 2.85 kcal. This is nearly twice the value indicated by calorimetry, so that the agreement among the three methods cannot be considered quantitative at present.

Various alternatives are possible to rectify the three sets of observations. The temperature dependence of the excess Gibbs energy of mixing of the garnets may not be as great as that preferred by Hensen et al (1975), and it could be reduced considerably within the error limits of their data, so that approach to ideality with increase of temperature is slower. Or, in the worst possible case, the excess Gibbs energy of mixing depends on the composition and temperature in such a complex way that the regular solution formulation does not give adequate numerical approximation. These alternatives need to be explored further, most feasibly by detailed phase diagram derivations of the type used by Hensen et al (1975).

APPLICATION TO THE GARNET PERIDOTITE PROBLEM

The enthalpy of solution data for the minerals involved in the spinel peridotite to garnet peridotite reaction in the system $CaO-MgO-Al_2O_3-SiO_2$, reaction A), make possible a direct calculation of the dP/dT slope of the reaction, and thus hold out hope of discriminating among the several experimental extrapolations in the temperature region below 1200°C. The method embodied in equation 3) is much superior, because no mixing models are necessary to estimate the entropies of the solid solutions. Also, the question of the possible existence of a significant non-configurational or excess entropy of mixing of the solid solutions does not enter the discussion. The dP/dT slope may be calculated at any pressure using the one-atmosphere enthalpies of solution and equation 4). The variation of ΔH of transition with temperature at constant composition and pressure will be very small for temperatures within a few hundred degrees of the calorimeter temperature (970 K), because it depends on a small difference between the heat capacities of products and reactants, and will be ignored here.

At the outset the compositions of the coexisting pyroxenes and garnets must be known. Experimental determinations in a higher temperature range are those of Akella (1975) and Herzberg and Chapman (1975). The data of Akella (1975) extend near to the conditions of present interest, $1000^{o}C$ and 15-20 kilobars, and are based on reversed equilibration. A short extrapolation of Akella's $1000^{o}C$ orthopyroxene composition isotherm yields very nearly 6.0 wt. percent Al_2O_3. His microprobe analyses of orthopyroxene and clinopyroxene equilibrated with garnet in the range 1000°-$1200^{o}C$ indicate equal weight percents of Al_2O_3 for both. This was concluded also by Obata (1975) from a thermodynamic analysis of the garnet to spinel lherzolite reaction using Boyd's (1970) analytic data of synthetic pyroxenes equilibrated with garnet in the system $CaSiO_3$-$MgSiO_3$-Al_2O_8 at 30 kb and $1200^{o}C$. The clinopyroxene further departs from diopside composition in having about 8 mole percent excess $Mg_2Si_2O_6$ at $1000^{o}C$, according to the experiments of Davis and Boyd (1966) and Mori and Green (1975) on the enstatite-diopside join, and the small amount of Al_2O_3 present should not affect this value significantly. From these considerations we arrive at the formula $(CaMgSi_2O_6{}_{.793}\ CaAl_2SiO_6{}_{.127}\ Mg_2Si_2O_6{}_{.080})$ for clinopyroxene and $(Mg_2Si_2O_6{}_{.882}\ MgAl_2SiO_6{}_{.118})$ for orthopyroxene. The Ca/Ca+Mg ratio of garnet is taken as $0.16 \pm .02$ following Akella (1975) and Obata (1975).

Using the foregoing compositions, the univariant equilibrium garnet to spinel lherzolite in the quaternary system has the following coefficients:

A') $$0.522\ Cpx + 1.478\ Opx + 0.759\ Sp \rightleftharpoons 1.000\ Gt + 0.759\ Fo$$

where Cpx denotes clinopyroxene, Opx orthopyroxene, Sp spinel, Gt garnet, and Fo forsterite.

The enthalpy of solution data may now be compiled to give ΔH^{o} for the reaction. The measurements of Charlu et al (1975) in the lead borate melt at 970 K gave 17.56 ± 0.26 kcal for $Mg_2Si_2O_6$, 14.29 ± 0.14 for $MgAl_2O_4$ and $16.11 \pm .24$ for Mg_2SiO_4. If the $Mg_2Si_2O_6$ content of the clinopyroxene is ignored, the CaTs mole fraction becomes 0.138, and a binary solution of this composition has a heat of solution in lead borate at 970 K of 18.10 kcal/gfw with an uncertainty of about 0.15, referring to Fig. 2. A small correction for the non-ideality of the $Mg_2Si_2O_8$ component in clinopyroxene can be made from the equilibrium data of Warner and Luth (1974) on the diopside-enstatite join in the pressure range 0-10 kb. Using T = 1273 K and P = 18 kb we get ΔG^{ex} = 632 cal/gfw for 8 mole percent $Mg_2Si_2O_6$ from their equations to determine ΔG^{ex}. If we make further the plausible assumptions that the contribution to the excess Gibbs energy of the excess entropy is small and that the enthalpy of interaction of $Mg_2Si_2O_6$ with a pyroxene of composition $Di_{.86}CaTs_{.14}$ is the same for pure diopside, we arrive at an "effective heat of solution" of the clinopyroxene of 17.43 kcal/gfw. The appropriate orthopyroxene has a heat of solution of 8.16 kcal/gfw (Charlu et al, 1975). The garnet value is 28.05, as seen from Fig. 3. From the coefficients of equation A') we get:

$$\begin{aligned}\Delta H^{o} &= .522x17.43+1.48x16.32+.759x14.29-28.05-.759x16.11 \\ &= 3.79 \text{ kcal/gfw}\end{aligned} \tag{14}$$

at 970 and 1 atm, with an uncertainty of 0.41, taken as the square root of the sums of the squares of the uncertainties of the individual quantities each multiplied by its stoichiometric coefficient. Because of the small dependence of ΔH on temperature for solid-solid reaction at constant composition, this value may be applied with confidence to a temperature range extending a few hundred degrees away from 970 K.

To make the important pressure correction to ΔH of the reaction, we must know the volume change of the reaction and its dependence on temperature. Using the volume data of Skinner and Boyd (1964) for aluminous orthopyroxene, of Clark et al (1962) and Warner and Luth (1974) for the clinopyroxenes, unpublished data of the author for pyrope-grossular garnets, and data for spinel and forsterite from Robie and Waldbaum (1968) we have V_{Cpx} = 65.48 cm^3/gfw, V_{Opx} = 62.14, V_{Sp} = 39.71, V_{Gt} = 115.34 and V_{Fo} = 43.78 at 298 K. This gives a volume change of reaction of -7.57 cm^3. With the use of the thermal expansion data of Skinner (1966) for diopside, enstatite, spinel, pyrope and forsterite, the volume change at 1000°C is -7.61 cm^3. Thus, fortunately, the differential thermal expansion may be ignored.

The enthalpy change of the reaction at 18 kb and 1000°C is given by equation 4) as:

$$\Delta H = 3.79 - \frac{7.61x18}{41.81} = +0.51 \text{ kcal}$$

ignoring the differential thermal expansion term. At 15 kb we get 1.06 kcal and at 12 kb we get 1.61 kcal. The dP/dT slopes of the reaction are now calculated from the Clausius-Clapeyron relation, equation 3), and are -6.96 bar/°C at 12 kb, -4.57 at 15 kb and -2.2 bar/°C at 18 kb, each with a maximum uncertainty from all sources of only $\pm$ 2 bar/°C.

Fig. 4 shows the calculated slopes at 1000°C and the experimental determinations of Kushiro and Yoder (1966) and of O'Hara et al (1971). The two sets of experimental data are mutually supportive in the actual temperature range of the investigations, as shown by the solid lines. The main discrepancies in the actual experimental range probably result from pressure calibration in the solid medium piston-cylinder apparatus. The major difference is in the ways the authors chose to extrapolate their data to lower temperatures. A simple straight line extension is quite compatible with the data points of Kushiro and Yoder (1966). On the other hand, the curve of O'Hara et al (1971) expresses the anticipated curvature which is owing to reduced Al_2O_3 in the pyroxenes at lower temperatures.

The present thermodynamic analysis shows immediately some features of interest. First, the straight-line extrapolation of Kushiro and Yoder's (1966) data is inconsistent. The reaction must curve strongly and, to be compatible with their experiments, it would seem that the curve cannot go much below 16 kb at 1000°.

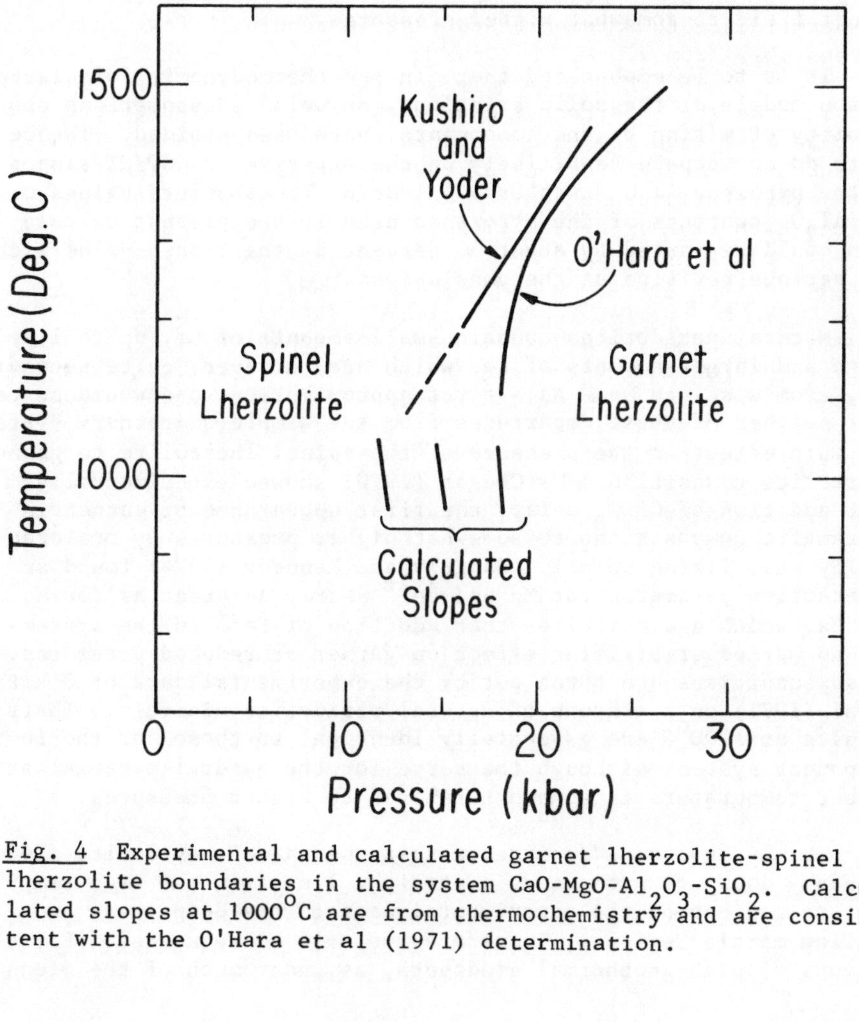

Fig. 4 Experimental and calculated garnet lherzolite-spinel lherzolite boundaries in the system $CaO\text{-}MgO\text{-}Al_2O_3\text{-}SiO_2$. Calculated slopes at 1000°C are from thermochemistry and are consistent with the O'Hara et al (1971) determination.

The calculated dP/dT slope at 18 kb is compatible with extrapolation to the same temperature range favored by O'Hara et al (1971).

The large curvature which is necessary to satisfy both the experimental and the calorimetric data is a consequence of the continual reduction of the Al_2O_3 content of the pyroxenes with decreasing temperature and the attendant reduction of the entropy of mixing in the high temperature assemblage, which greatly affects dP/dT. The slope of the curve will become more negative at still lower temperatures for this reason. The stability field of garnet + forsterite is very likely, therefore, to have an absolute minimum of about 18 kilobars in the neighborhood of 800°C in the system CaO-MgO-Al_2O_3-SiO_2. This conclusion is quite in line with that of Obata (1975), except that his minimum is at about 12 kilobars, owing mainly to his assumption of ideal mixing in the garnet phase. The effect of garnet non-ideality will be to remove the garnet field to somewhat higher pressures.

It is to be emphasized that, in the thermodynamic formulation, cation models of the solid solutions, as well as assumptions about ideality of mixing of the components, have been avoided. The results do not depend sensitively on the experimental dP/dT slopes of the pyroxene Al_2O_3 isopleths. Indeed, the absolute values of the Al_2O_3 contents of the pyroxenes used in the present calculation could be varied by about 30 percent of the chosen value without serious revision of the conclusions.

Natural peridotites contain small amounts of Cr, up to 1 percent, and larger amounts of Fe, which are, however, quite secondary, atom-wise, to Mg. As a first approximation, one would suspect that neither of these departures from the simple quaternary system has much effect on the pressure of the spinel lherzolite to garnet lherzolite transition. MacGregor (1970) showed experimentally that addition of Cr_2O_3 delays the first appearance of garnet in ultramafic compositions to somewhat higher pressures by preferentially stabilizing spinel. Ganguly and Kennedy (1974) found an interaction parameter for Mg and Fe^{2+} nearly as great as for Mg and Ca, which again implies that addition of Fe^{2+} to the system has no marked stabilizing effect on garnet at reduced pressures. These suspicions are borne out by the experimental data of O'Hara et al (1971) on a mixture of natural peridotite minerals. Their results at 1200°C are essentially identical to those for the four-component system, although the curve for the natural material at higher temperature is somewhat removed to higher pressures.

We may have some confidence that the garnet peridotite field boundary does not project as a straight line from the high temperature regions of the experimental data to the deep crust and shallow mantle regions. There are two main consequences. First, regions of high geothermal gradients, as under much of the ocean basins, probably have an uppermost mantle of spinel lherzolite, as advocated by Green and Ringwood (1967), which gives way at a certain depth, perhaps near 60 km, to a garnet lherzolite mantle. Old shield areas of the continental interiors may be underlain by garnet lherzolite, if the geothermal gradients are low enough and the crust is thick enough. Second, the crustal garnet peridotites of the Norwegian type could only have been formed at the deepest levels of the crust, or in the upper mantle, and must have been transported to regions of shallower crustal metamorphism by tectonic processes, as advocated by O'Hara and Mercy (1963) on textural and field relations. A possibility which may explain some occurrences, suggested by Dickey (1970) for rare garnet in the peridotite of the Er Rif, Morocco, is that the garnet may have been mechanically reworked from adjacent garnet pyroxenite lenses and incorporated into olivine-bearing units without further reaction.

ACKNOWLEDGEMENTS

Thermochemical data presented here were obtained through support of a Petroleum Research Fund grant, American Chemical Society, No. 7051-AC2, and National Science Foundation grant No. NSF DES74-22951. Additional support was provided by the Materials Research Laboratory, NSF, at Chicago.

REFERENCES

Akella, J. Extended Abstr., Int. Conf. on Geotherm. and Geobarom., Penn. St. Univ., Oct. 5-10, 1975.
Boyd, F.R. Mineral. Soc. Amer. Spec. Pap., 3, 63 (1970).
Charlu, T.V., Newton, R.C. and Kleppa, O.J. Geochim. Cosmochim. Acta, 39, 1487 (1975).
Clark, S.P., Schairer, F.J. and de Neufville, J. Carnegie Inst. of Wash. Yrbk., 61, 59 (1962).
Davis, B.T.C. and Boyd, F.R. J. Geophys. Res., 71, 3507 (1966).
Dickey, J.M. Mineral. Soc. Amer. Spec. Pap., 3, 33 (1970).
Ganguly, J. Contr. Mineral. Petrol., 55, 81 (1976).
Ganguly, J. and Kennedy, G.C. Contr. Mineral. Petrol., 48, 137 (1974).
Green, D.H. and Ringwood, A.E. Earth Plan. Sci. Lett., 3, 151 (1967).
Guggenheim, E. "Mixtures", Oxford University Press, 270 pp. (1952).
Hariya, Y. and Kennedy, G.C. Amer. J. Sci., 266, 193 (1968).
Hensen, B.J., Schmid, R. and Wood, B.J. Contr. Mineral. Petrol., 51, 161 (1975).
Herzberg, C.T. and Chapman, N.A. Int. Conf. on Geotherm. and Geobarom., Penn. St. Univ., Oct. 5-10, 1975.
Kushiro, I. and Yoder, H.S. J. Petrol., 7, 337 (1966).
MacGregor, I.D. Carnegie Inst. of Wash. Yrbk., 64, 126 (1965).

MacGregor, I.D. Phys. Earth Plan. Int., 3 372 (1970).
Mori, T. and Green, D.H. Earth Plan. Sci. Lett., 26, 277 (1975).
Mysen, B.O. and Heier, K.S. Contr. Mineral. Petrol., 36, 73 (1972).
Newton, R.C., Charlu, T.V. and Kleppa, O.J. Geochim. Cosmochim. Acta, 1976 (in press).
Obata, M. Int. Conf. on Geotherm. and Geobarom., Penn. St. Univ., Oct. 5-10, 1975.
O'Hara, M.J., Richardson, S.W. and Wilson, G. Contr. Mineral. Petrol., 32, 48 (1971).
O'Hara, M.J. and Mercy, E.L.P. Trans. Roy. Soc. Edin., 65, 251 (1963).
Okamura, F.P., Ghose, S. and Ohashi, H. Amer. Mineral., 59, 549 (1974).
Råheim, A. and Green, D.H. Contr. Mineral. Petrol., 48, 179 (1974).
Robie, R.A. and Waldbaum, D.R. U.S. Geol. Surv. Bull., 1259, 256 pp (1968).
Skinner, B.J. in S.P. Clark (Ed.), "Handbook of Physical Constants", Geol. Soc. Amer. Mem., 97, 75 (1966).
Skinner, B.J. and Boyd, F.R. Carnegie Inst. of Wash. Yrbk., 63, 163 (1964).
Thompson, J.B. in Ph. H. Abelson, (Ed.), "Researches in Geochemistry", vol.II, Wiley, 340 (1967).
Warner, R.D. and Luth, W.C. Amer. Mineral., 59, 98 (1974).
Wood, B.J. Int. Conf. on Geotherm. and Geobarom., Penn. St. Univ., Oct. 5-10, 1975.

STUDY PROBLEMS

Mixing Properties of Aluminous Clinopyroxene

It is known from X-ray crystallography that aluminous diopside has essentially three distinct kinds of crystallographic sites which are occupied by the cations. Calcium occupies the larger octahedral (six-coordinated) sites M_2, magnesium and aluminum share the smaller octahedral sites M_1, and silicon and aluminum share the tetrahedral (four-coordinated) sites. The occupancies for the end members of the aluminous diopside series may be diagrammed as follows:

	M_2	M_1	T
$CaMgSi_2O_6$ (Diop.)	Ca	Mg	Si_2
$CaAl_2SiO_6$ (CaTs)	Ca	Al	(AlSi)

The activity of $CaAl_2SiO_6$ (Ca-Tschermak's molecule) in $CaMgSi_2O_6$ solid solutions was measured over a range of temperatures and compositions by Wood (1975) in the following manner. Compositions in the system $CaAl_2Si_2O_8$-$CaMgSi_2O_6$ were equilibrated at pressures in the range 10-25 kilobars and temperatures of 900° - 1300°C in the solid-pressure-medium piston-cylinder apparatus. The resulting assemblages were pyroxene-anorthite-quartz. The activity of CaTs, α_{CaTs}, in the pyroxene is given to a close approximation by:

$$RT\ln\alpha_{CaTs} \simeq \Delta V^o (P-P^*)$$

where ΔV^o is the one-atmosphere volume change of the end member reaction:

$$\underset{\text{cpx}}{CaAl_2SiO_6} + \underset{\text{qtz}}{SiO_2} = \underset{\text{anorthite}}{CaAl_2Si_2O_8},$$

P^* is the experimentally-determined pressure of the equilibrium end member reaction (Hariya and Kennedy, 1968) and the assumptions are made of independence of the partial molal volume of CaTs of composition and independence of ΔV^o of pressure. The reference state of CaTs is the pure phase at the T and P of solid solution equilibration.

Wood's data give the very simple relationship that activity of CaTs is nearly equal to its mole fraction in diopside-CaTs solid solutions. This would be true of an ideal solution of completely ordered CaTs and diopside end members. However, as Wood pointed out, the recent X-ray crystallographic study of

Okamura, Ghose and Ohashi (1974) showed that synthetic CaTs pyroxene has a high degree of (Si,Al) disorder. The ideal activity in the case of complete cation randomization would be:

$$\alpha^{id}_{CaTs} = X^2_{CaTs}(2-X_{CaTs})$$

and X_{CaTs} would be considerably larger than αCaTs for an ideal solution (zero enthalpy of mixing). Thus, the finding of activity nearly equal to mole fraction implies a significant positive enthalpy of mixing. Using the relationship:

$$\alpha_{CaTs} = \gamma_{CaTs}\ X^2_{CaTs}(2-X_{CaTs})$$

for a completely disordered pyroxene, where γ is the activity coefficient, Wood derived the following table of activity coefficients from his experimental data:

X_{CaTs}	$\gamma^{disordered}_{CaTs}$
0.07	10.7
0.16	4.0
0.33	2.0
0.55	1.35

From the values of γ_{CaTs} it is possible to obtain the activity coefficient of diopside γ_{Di}, over the same compositions range by use of the Gibbs-Duhem relation:

$$\ln\gamma_{Di} = -\int_0^{X_{CaTs}} \frac{X_{CaTs}}{X_{Di}}\, d\ln X_{CaTs}$$

The activity coefficients are considered to vary only slowly as functions of pressure and temperature over the ranges of interest. The excess free energy (ΔG^{ex}) of mixing, given by $-RT(\ln\gamma_{CaTs} + \ln\gamma_{Di})$ is expected, by analogy with many other solutions, to be nearly independent of temperature.

Thus, we can set up, on the basis of the phase equilibrium reductions, various "models" for the mixing of aluminous clino-pyroxene components. For a completely (Si,Al) ordered pyroxene the activity coefficients would be nearly unit, and the predicted excess enthalpy of mixing, ΔH^{ex}, is small. For a completely (Si, Al) disordered pyroxene the activity coefficients are large and the predicted ΔH^{ex} is large. Pyroxenes intermediate between these two extremes of ordering yield intermediate values of γ and ΔH^{ex}.

Solution calorimetry serves to discriminate among these models. Newton, Charlu and Kleppa (recent work) measured the enthalpies of solution of synthetic CaTs-diopside clinopyroxenes in a melt of composition $2PbO.B_2O_3$ at 1000 K. Their heats of solution are tabulated:

Composition mol per cent	$\Delta H^{sol'n}$ kcal/gfw
Diopside 100	20.53 ± .25
$Di_{90}CaTs_{10}$	18.66 ± .15
$Di_{80}CaTs_{20}$	17.33 ± .16
$Di_{65}CaTs_{35}$	15.97 ± .22
$Di_{50}CaTs_{50}$	15.08 ± .30
$Di_{30}CaTs_{70}$	13.50 ± .24
CaTs 100	11.55 ± .23

Problems

1) Derive the relationship:

$$\alpha_{CaTs} = X^2_{CaTs}\ (2-X_{CaTs})$$

for ideal solution of completely disordered diopside-CaTs pyroxenes (that is, for completely independent random mixing of Si and Al on tetrahedral sites and Mg and Al on the M_1 octahedral site).

2) Show that, if the function ΔG^{ex} is nearly independent of temperature, the excess enthalpy of mixing is given by:

$$\Delta H^{ex} = X_{CaTs} RT\ln\gamma_{CaTs} + X_{Di} RT\ln\gamma_{Di}.$$

3) From the data of Wood for γ^{Dis}_{CaTs} derive γ^{Dis}_{Di} as a function of composition and ΔH^{ex} as a function of composition.

4) Show that ΔH^{ex} is given directly from the heat of solution data as the departure, at a given composition, from the value read from a straight line between the heats of solution of the end members. A negative departure corresponds to a positive heat of mixing.

5) Use the heat of solution data of Newton, Charlu and Kleppa in conjunction with the ΔH^{ex} derived from the data of Wood for the two models, to elucidate the ordering state of CaTs-diopside mix-crystals.

References

Hariya, Y. and Kennedy, G.C. Amer. J. Sci., 266, 193 (1968).
Okamura, F.P., Ghose, S. and Ohashi, H. Amer. Mineral., 59, 549 (1974).
Wood, B.J. Extended Abstr., Int. Conf. on Geotherm., Geobarom., Penn. St. Univ., Oct. 5-10, 1975.

SOLUTIONS TO PROBLEMS

1) First we consider a completely disordered model of mixing on the join diopside-CaTs ($CaMgSi_2O_6$-$CaAl_2SiO_6$). The activities of diopside (D) and CaTs (C) are related to the molar Gibbs energy of mixing by:

$$\Delta G = \Delta H - T\ \Delta S \tag{1}$$

$$= X_C RT\ln a_C + (1-X_C)RT\ln a_D \tag{2}$$

where ΔH and ΔS are, respectively, the enthalpy of mixing and entropy of mixing, X_C is the mole fraction of CaTs, and the a's are the activities. If the mixing is ideal, we have:

$$\Delta S = S^{conf} - X_C S_C^{conf}\ (1-X_C)S_D^{conf} \tag{3}$$

where S^{conf} is the configurational entropy of the mix-crystal per mole and S_C^{conf} and S_D^{conf} are the configurational entropies of the pure end members at the same physical conditions. The molar configurational entropy, S^1, of mixing of two kinds of atoms on a structural site is given by:

$$S^1 = -R[X^1 \ln X^1 + (1-X^1)\ln(1-X^1)] \tag{4}$$

(Denbigh, 1971, p.237) where X^1 is the mole fraction of one kind of atom on the site.

The total configurational entropy of the pyroxene will be the sum of the configurational entropies for all structural sites where mixing may occur. For completely disordered pure CaTs the configurational entropy is given by random mixing of equal numbers of Al and Si on the two tetrahedral sites. The X-ray crystallographic data show that Ca and Al atoms are each ordered on two different kinds of octahedral sites. Thus the quantity S_C^{conf} is given by $-2R(\frac{1}{2}\ln\frac{1}{2} + \frac{1}{2}\ln\frac{1}{2}) = 2R\ln 2$. The corresponding quantity for pure diopside is zero.

The molar configurational entropy of the mix-crystal is given by:

$$S^{conf} = -R[X_C \ln X_C + (1-X_C)\ln(1-X_C)]$$

$$-2R \left[\frac{X_C}{2} \ln \frac{X_C}{2} + \frac{X_C + 2(1-X_C)}{2} \ln \frac{X_C + 2(1-X_C)}{2} \right] \quad (5)$$

The first bracketed term accounts for mixing of Mg and Al on the M_1 octahedral sites and the second bracketed term accounts for mixing on the two tetrahedral sites. Expression (5) can be rearranged:

$$S^{conf} = -RX \{\ln[X_C^2(2-X_C)] + 2\ln\tfrac{1}{2}\}$$

$$-R(1-X_C) \{ \ln \frac{(2-X_C)^2(1-X_C)}{4} \}$$

The molar configurational entropy of mixing follows from equation (3):

$$\Delta S = -RX_C \ln[X_C^2(2-X_C)] - R(1-X_C)\ln \frac{(2-X_C)^2(1-X_C)}{4}$$

which, by comparison with (2), gives:

$$a_C = X_C^2 (2-X_C)$$

$$a_D = \frac{(2-X_C)^2(1-X_C)}{4}$$

for the ideal activities.

It is quite possible that the state of (Al,Si) ordering in diopside-CaTs solid solutions is greater than completely random. Short-range order might be present which would not be observable in an X-ray diffraction experiment. Various models are possible. A simple model was developed by Wood (1975 and personal communication) in which electrical charge balance is maintained on a unit-cell scale. In this model, each unit cell contains four "molecules", which may be all diopside, all CaTs, or any combination of these. The unit-cell contents are internally randomly mixed with regard to the tetrahedral sites and the M_1 sites. Each of the five molecular combination possibilities corresponds to a different configurational entropy of the unit cell. Different compositions of the pyroxene correspond to different probabilities of encountering any of the five cell types. The configurational entropies of the five cell types are each multiplied by the probability of encountering that cell type in a random selection, and the sum is multiplied by Avogadro's number to give the configurational entropy of four pyroxene formula units. Fig. 1 shows the molar configurational entropy of the

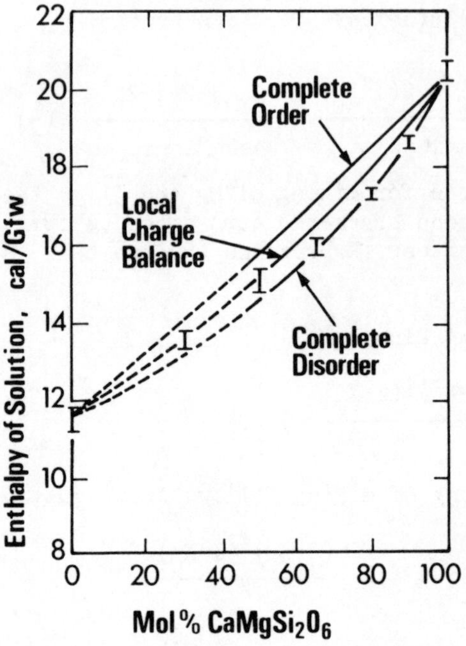

Fig. 1. Configurational entropy of mixing per mole of $CaMgSi_2O_6$-$CaAl_2SiO_6$ clinopyroxene as a function of composition for the "completely disordered" and "unit cell local charge balance" models.

short-range-order model and of the completely disordered model. It is seen that the configurational entropy is only half that of the completely disordered model, even though the two disordering types cannot be distinguished from one another by X-ray diffraction.

2) The ideal Gibbs energy of mixing, ΔG^{id}, is given by:

$$\Delta G^{id} = RT(X_C \ln a_c^{id} + X_D \ln a_D^{id}) \tag{6}$$

The excess Gibbs energy of mixing is defined by:

$$\Delta G^{ex} = \Delta G - \Delta G^{id} \tag{7}$$

It is expected from simple kinds of solution theories that ΔG^{ex} will be reasonably independent of temperature over a range of a few hundred to several hundred degrees. If this is so, as the Wood (1975) data suggest, then the analogous molar excess entropy of mixing, ΔS^{ex}, given by $-\frac{\Delta G^{ex}}{T}$, will be nearly zero. We can thus, to a good approximation, replace ΔG^{ex} by ΔH^{ex}, the excess molar enthalpy of mixing. Thus, we may hope to discriminate among the order-disorder models proposed for CaTs-diopside mixing by making calorimetric measurements of the enthalpy of mixing of the solid solutions.

3) The activity coefficient γ_C for CaTs is given by:

$$a_C = X_C^2\,(2-X_C)\gamma_C \qquad (8)$$

for the completely disordered model. a_C has been measured over a range of P and T by Wood (1975) by determining the compositions of diopside-CaTs clinopyroxenes in equilibrium with anorthite and quartz. The activity coefficients for the unit-cell charge balance model may also be derived given activity-composition data.

Activity coefficients for the $CaMgSi_2O_6$ component may be gotten from those of the $CaAl_2SiO_6$ component by use of the Gibbs-Duhem equation:

$$X_C d\ln\gamma_C + X_D d\ln\gamma_D = 0 \qquad (9)$$

In integrated form this becomes:

$$\gamma_D = \int_{X_D}^{1} \frac{X_C}{X_D}\,\frac{d\ln\gamma_C}{dX_D}\,dX_D$$

This may be approximated numerically to any desired degree of accuracy by:

$$\gamma_D \simeq \Sigma\,(\overline{\frac{X_C}{X_D}})\;\Delta\ln\gamma_C$$

where $(\overline{\frac{X_C}{X_D}})$ is the mean value of the component ratio over an interval ΔX_D and $\Delta\ln\gamma_C$ is taken over the same interval in the sense of increasing X_D. Table 1 gives the activity coefficients for $CaAl_2SiO_6$ and $CaMgSi_2O_6$ over the range of composition investigated by Wood (1975), for the completely disordered and charge-balance models.

4) The most effective method of determining the enthalpy of mixing in solid solutions is to measure the enthalpy of solution of various compositions in a suitable solvent. It can be shown easily that the molar enthalpy of mixing for a given composition is given by the difference between a proportional combination of the molar enthalpies of solution of the end-members and the heat of solution per mole of the solid solution. Consider the experimental results of R.C. Newton, T.V. Charlu and O.J. Kleppa (recent work) on the enthalpy of solution of a synthetic clinopyroxene of compositions $CaTs_{.5}$ $diopside_{.5}$ in a $Pb_2B_2O_6$ melt at 970 K:

a) $.5CaTs + .5diop. = .5CaTs^{(sol'n)} + .5diop^{(sol'n)}, \Delta H_a = 16.04$ kcal

b) $CaTs_{.5}\, diop_{.5} = .5CaTs^{(sol'n)} + .5diop^{(sol'n)}, \Delta H_b = 15.08$ kcal

c) $.5CaTs + .5diop = CaTs_{.5} diop_{.5}, \Delta H_c = \Delta H_a - \Delta H_b = 0.96$ kcal

In step a) the end-member pyroxenes are dissolved separately and in step b) the solid solution is dissolved. The assumption is made that the partial molal enthalpies of the components in the $Pb_2B_2O_5$ melt are the same in steps a) and b); that is, they are independent of the way in which the components are introduced. This will be valid to a high degree so long as the solutions are very dilute in $CaAl_2SiO_6$ and $CaMgSi_2O_6$. Thus it is seen that negative departures from linearity in the heat of solution versus composition curve correspond to positive enthalpies of mixing.

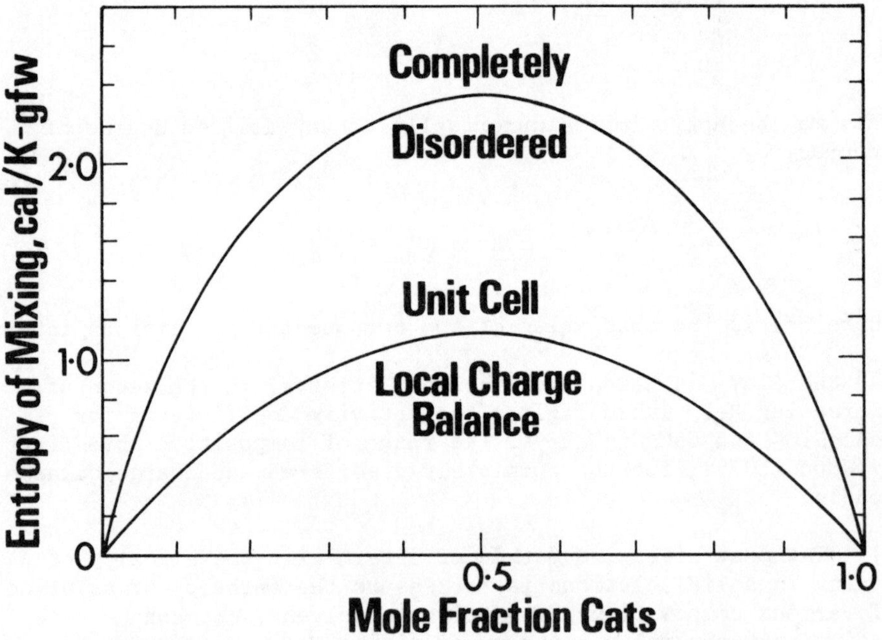

Fig. 2. Enthalpy of solution of $CaMgSi_2O_6$-$CaAl_2SiO_6$ clinopyroxenes in $Pb_2B_2O_5$ melt at 970 K. Brackets show experimental determination of Newton, Charlu and Kleppa (recent work). Lines show trends predicted by the "ordered", "completely disordered" and "local charge balance" models of Wood (1975), given the heats of solution of the pure end-members.

5) The heat of solution data of Newton, Charlu and Kleppa is plotted in Fig. 2. Shown also are the heats of solution predicted by the Wood (1975) completely disordered and unit-cell local charge balance models. The most straightforward interpretation of Fig. 2 is that, for solid solutions close to the diopside end there is nearly complete (Al,Si) disorder. However, as the pyroxenes become more Al-rich, the structure is unable to tolerate the increasing amounts of charge in-balance and so the ordering state approaches that of the local charge-balance configuration.

ACTIVITY-COMPOSITION RELATIONSHIPS FOR CRYSTALLINE SOLUTIONS

R. Powell

Department of Earth Sciences,
University of Leeds, Leeds 2

INTRODUCTION

Equilibrium thermodynamic calculations on mineral assemblages usually involve the equilibrium relation for a balanced chemical reaction:

$$\Sigma i\nu_i = 0$$

Thus:

$$\Sigma \nu_i \mu_i = 0 \qquad (1)$$

where μ_i = chemical potential of component i
ν_i = reaction coefficient of component i

For a standard state of pure i at the temperature and pressure of interest:

$$\mu_i = \mu_i^o + RT\ln a_i = \mu_i^o + RT\ln X_i\gamma_i$$

and (1) can be rewritten in its more familiar form:

$$-\Delta G^o = RT\ln K$$

$$\text{where } \Delta G^o = \Sigma\nu_i\mu_i^o$$

$$K = \Pi a_i^{\nu_i} = \Pi(X_i\gamma_i)^{\nu_i} \qquad (2)$$

where a_i = activity of component i

D. G. Fraser (ed.), Thermodynamics in Geology, 57-65. All Rights Reserved.

X_i = thermodynamic mole fraction of i

γ_i = activity coefficient of i, $\gamma_i = 1$ for ideal solutions

μ_i^o = standard chemical potential of i

ΔG^o = standard Gibbs energy of reaction

K = equilibrium constant

Thus, to perform calculations on mineral assemblages it is necessary to know standard Gibbs energies of reactions and the activity-composition (a-x) relations for the components in the minerals involved. This paper is addressed to the latter problem.

How well we need to know the a-x relationships depends on the temperature dependence of the Gibbs energy of reaction. For example, consider two reactions (a and b) for which the equilibrium relations (2) are:

For (a) $50000 - 50T = RT\ln a_i$

If $a_i = 0.5 \pm 0.1$; this corresponds to $T = 1028 \pm 10K$

For (b) $5000 - 5T = RT\ln a_i$

If $a_i = 0.5 \pm 0.1$; this corresponds to $T = 1380 \pm 180K$

Clearly for (a), which corresponds to many dehydration and decarbonation reactions, detailed knowledge of the a-x relationships is not important, and the assumption of ideal mixing may well suffice. However for (b), which corresponds to many solid-solid reactions, calculations will only produce useful results if a-x relations are known in some detail. This latter case is the reason for worrying about a-x relationships.

Several terms must be introduced. The first is long range order (lro) which refers to the fractionation of elements between different sites in the structure, for example Fe and Mg between the M1 and M2 octahedral sites in pyroxenes. The second term is short range order (sro) which refers to departures from random mixing, with elements showing a preference for nearest neighbours in a structure. One factor which might be responsible for sro is local charge balance in the structure. For example, octahedral Al is inferred to prefer to be adjacent to tetrahedral Al in CaTs- and MgTs-rich pyroxenes (Newton, this volume). Sro effects are a considerable problem not only in that they are difficult to measure, but because they seriously complicate the formulation of a-x relationships.

This paper is divided into two parts, the first considering thermodynamic mole fractions, the second activity coefficients.

MOLE FRACTIONS

In a complicated solution with many elements and many sites, separating the mole fraction and activity coefficient expressions is arbitrary although a choice can usually be made which simplifies the resulting expressions (cf. Grover, 1974). There is no problem in defining activity for a standard state of pure component i at P and T:

$$RT\ln a_i = \left(\frac{\partial nG^{mix}}{\partial n_i}\right)_{P,T,n_j(j\neq i)}$$

where G^{mix} = molar Gibbs energy of mixing
n_i = number of moles of i
$n = \Sigma n_i$ = number of moles of phase

A useful choice of defining expressions for mole fraction and activity coefficient is:

$$RT\ln X_i = \left(\frac{\partial nG^{id}}{\partial n_i}\right)_{P,T,n_j(j\neq i)} \quad (3)$$

$$RT\ln \gamma_i = \left(\frac{\partial nG^{ex}}{\partial n_i}\right)_{P,T,n_j(j\neq i)} \quad (4)$$

where G^{id} = molar Gibbs energy of random mixing
$G^{ex} = G^{mix} - G^{id}$

The Gibbs energy of random mixing, G^{id}, is obtained from the combinatorial expression for the number of ways of mixing n things, some of which are distinguishable:

$$\frac{n!}{\Pi(x_i n)!}$$

After applying Stirling's approximation, the contribution to G^{id} is given (for example, Guggenheim, 1952) by:

$$RT\Sigma x_i \ln x_i \quad (5)$$

Before continuing, it is important to consider what this mixing refers to. If sro effects are small or non-existent, then terms (5), for the mixing on each site in the structure contribute to G^{id}; thus:

$$G^{id} = RT\Sigma m_j \Sigma x_{ij} \ln x_{ij} \quad (6)$$

where m_j = number of j sites in the formula unit

x_{ij} = mole (or site) fractions of i in site j

However if sro effects are substantial then it is not mixing on sites that is involved, but mixing of "molecular" species. Although in simple cases it is easy to express G^{id} using (5) (where x_i refer to the molecular species); G^{id} has not yet been formulated for the case where only some of the elements in the structure are involved in strong sro. Thus the development in the rest of this paper applies only to crystalline solutions which show little or no sro.

The application of (3) to (6) gives the expression for the thermodynamic mole fraction of a component in the phase. However, in the multi-site case, this presents a problem. Consider a phase in which elements A and B mix randomly on site 1 and elements C and D mix randomly on site 2. A phase of some particular composition can be represented as a block diagram:

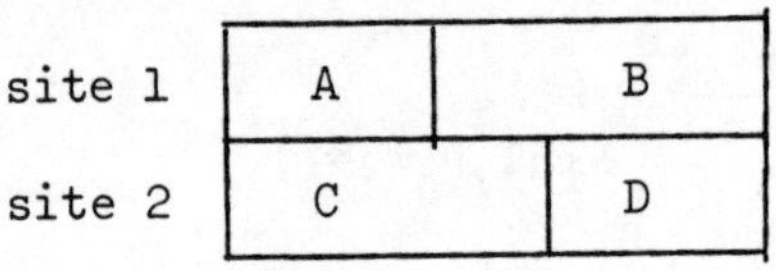

From (6):

$$G^{id} = RT\,(x_{A1}\ln x_{A1} + x_{B1}\ln x_{B1} + x_{C2}\ln x_{C2} + x_{D2}\ln x_{D2}) \quad (7)$$

To find X_{AC} (7) must be differentiated with respect to n_{AC} and so these site fractions must be expressed in terms of some bulk composition parameters including n_{AC}. Although n_{AC} is dependent on the choice of a set of independent components, the result of the differentiation is independent of this choice (Powell, 1975). Taking AC, BC and BD as the independent set of components:

$$x_{A1} = \frac{n_{AC}}{n} \quad x_{B1} = \frac{n_{BC}+n_{BD}}{n} \quad x_{C2} = \frac{n_{AC}+n_{BC}}{n} \quad x_{D2} = \frac{n_{BD}}{n} \quad (8)$$

where $n = n_{AC} + n_{BC} + n_{BD}$

These are substituted into (7) and then by applying (3)

$$RT\ln X_{AC} = RT\ln \frac{n_{AC}(n_{AC} + n_{BC})}{n.n} = RT\ln x_{A1}x_{C2}$$

This procedure can be generalised for a multi-component multi-site phase:

$$X_a = \prod\prod x_{ij}^{c_{aij}} \tag{9}$$

where X_a = thermodynamic mole fraction of component a

c_{aij} = number of atoms of i in site j in the formula unit of component a

For example, for the thermodynamic mole fraction of $CaMgSi_2O_6$ in pyroxene:

$$X_{CaMgSi_2O_6,px} = x_{Ca,M2}x_{Mg,M1}x_{Si,T}^2$$

A normalisation constant must be introduced when more than one element occupies a site in pure a, to ensure that $X_a = 1$ for pure a. Thus for a disordered standard state for CaTs ($CaAl_2SiO_6$) in pyroxene:

$$X_{CaAl_2SiO_6,px} = 4x_{Ca,M2}x_{Al,M1}x_{Al,T}x_{Si,T}$$

This derivation of a general expression for the ideal mixing activity (9), here called the thermodynamic mole fraction, avoids the complications achieved by Grover (1974).

ACTIVITY COEFFICIENTS

Calorimetric measurements (Newton, this volume), phase equilibrium experiments (Wood, this volume), and the occurrence of solvi in many systems of geological interest indicate that many crystalline solutions are non-ideal. Whereas metallurgists use expressions for activity coefficients to fit experimentally measured a-x relationships so that empirical methods like the Margules equations are adequate, in geology most a-x relationships must be calculated from experimentally determined phase equilibria. Solution models based on some physical approximation of the system are more likely to be successful than empirical methods. Many solution models, mostly empirical, have been devised and a comparison of a few of the most commonly used binary models is given by Powell (1974). The main observation was that almost any model involving more than one adjustable parameter could describe the binary phase equilibrium data, in this case asymmetric solvi. A further observation could have been made that even one parameter models suffice if thermodynamic information is required only in the region of one of the limbs of the solvus, as the effect of the asymmetry of the system is small in the Henry's Law/Raoult's Law regions.

As most crystalline silicate solutions are multi-site and multi-component it is essential to have a mixing model that is not only easy to express and manipulate but also has few adjustable parameters which have to be found. Fortunately we can achieve this with models that include a naive physical representation of interactions in the phase. The basis of this physical representation is that nearest-neighbour interactions are considered to be so dominant that other interactions can be neglected. If the energy of a nearest-neighbour AB pair is not the same as the average of the energy of an AA and a BB pair (the case for non-ideality) then it would be expected that A and B would be distributed non-randomly. If this non-randomness is included in the derivation, then the quasi-chemical model results (Guggenheim, 1952). Unfortunately it is not possible to derive closed expressions for this model even for ternary systems (Hagemark, 1968). However if this non-randomness is ignored then simple expressions for the a-x relations can be derived for the single-site and multi-site cases.

Single-site case

Consider A and B mixing randomly on one site with the nearest-neighbour approximation for the energy. In this case:

$$G = RT\,(x_A \ln x_A + x_B \ln x_B) + x_A^2 \varepsilon_{AA} + 2x_A x_B \varepsilon_{AB} + x_B^2 \varepsilon_{BB} \qquad (10)$$

where ε_{ij} refers to the energy of an ij nearest-neighbour pair; each such energy parameter is multiplied by the probability of finding an ij nearest-neighbour. The 2 is required in the second energy term to account for AB and BA nearest-neighbours. Differentiating (10) with respect to n_A:

$$\mu_A = \varepsilon_{AA} + RT\ln x_A + x_B^2 w_{AB}$$

where $w_{AB} = \varepsilon_{AA} + \varepsilon_{BB} - 2\varepsilon_{AB}$

This is the regular solution model, Guggenheim's zeroth order model. The interchange energy, w_{AB}, can be a function of temperature and pressure (for example, Guggenheim, 1966, p.82). This is a symmetric mixing model. For example it predicts a symmetrical solvus whose critical temperature, T_c, is given by $T_c = \frac{w}{2R}$ (if $w \neq f(T)$), for large positive deviations from ideality in the system.

The multi-component regular solution equations can be derived in the same way as the binary equation. Thus:

$$G^{id} = RT\Sigma x_i \ln x_i + \Sigma\Sigma x_i x_j \varepsilon_{ij}$$

From which:

$$RT\ln \gamma_k = \sum_{i\neq k} x_i(1 - x_k)w_{ik} - \sum_{\substack{i\ j\\ i,j\neq k}}\sum x_i x_j w_{ij} \quad (11)$$

where $w_{ij} = \varepsilon_{ii} + \varepsilon_{jj} - 2\varepsilon_{ij}$

There is one interchange energy parameter between each pair of elements for the multi-component regular solution model.

It will not be sensible to use (11) for each site in a multi-site phase because nearest-neighbours are no longer on that site (cf. Powell, 1975) given that the starting point for the derivation of (11) is the nearest-neighbour energy approximation. For multi-site phases equation (11) for each site may reflect next nearest-neighbour interactions.

Multi-site case

Nearest-neighbour interactions in the multi-site case will involve elements on different sites (as well as possibly some on the same site). This is investigated in the same way as above for the $(A,B)_1(C,D)_2$ phase used earlier. When each site only has nearest-neighbours of the other site, then:

$$G = RT(x_{A1}\ln x_{A1} + x_{B1}\ln x_{B1} + x_{C2}\ln x_{C2} + x_{D2}\ln x_{D2})$$
$$+ x_{A1}x_{C1}\varepsilon_{AC} + x_{A1}x_{D2}\varepsilon_{AD} + x_{B1}x_{C2}\varepsilon_{BC} + x_{B1}x_{D2}\varepsilon_{BD} \quad (12)$$

Substituting (8) and differentiating with respect to n_{AC}:

$$\mu_{AC} = \varepsilon_{AC} + RT\ln x_{A1}x_{C2} + x_{B1}x_{D2}w_{ACBD}$$

where $w_{ACBD} = \varepsilon_{AD} + \varepsilon_{BC} - \varepsilon_{AC} - \varepsilon_{BD}$

This is the Bragg-Williams formulation for the activity coefficient, originally devised to describe long range order in brass and other alloys (see e.g. Guggenheim, 1952), and since used by molten salt chemists (e.g. Lumsden, 1966). The multi-component equations can be derived by generalising (12):

$$G = RT(\Sigma x_{i1}\ln x_{i1} + \Sigma x_{j2}\ln x_{j2} + \Sigma\Sigma x_{i1}x_{j2}w_{ij})$$

from which:

$$\mu_{k\ell} = \varepsilon_{k\ell} + RT\ln x_{k1}x_{\ell 2} + \sum_{\substack{i\neq k \\ j\neq \ell}}\sum x_{i1}x_{j2}w_{ijk\ell} \tag{13}$$

where $w_{ijk\ell} = \varepsilon_{i\ell} + \varepsilon_{kj} - \varepsilon_{ij} - \varepsilon_{k\ell}$

One problem that arises here which doesn't arise in the one-site formulation is that not all the w's are independent. If there are n species on site 1 and m species on site 2 then there are $\binom{n}{2}\binom{m}{2}$ w's only (n - 1) (m - 1) of which are independent. The general expression for the activity coefficients using an independent set of w's whose first two subscripts are arbitrarily fixed (r and s) is:

$$RT \ln \gamma_{k\ell} = \sum_{\substack{i\neq r \\ j\neq s}}\sum w_{rsij} (\delta_{ki} + x_{i1}(-1)^{\delta_{ki}})(\delta_{\ell j} + x_{j2}(-1)^{\delta_{\ell j}}) \tag{14}$$

where $\delta_{ab} = 0$ when $a=b$
$= 1$ $a\neq b$

Non-ideality in multi-site phases should involve terms of the form (11) and (14), and in many cases the 'cross-site' terms (14) may be most important because there are more inter-site than intra-site nearest-neighbours. Using both (11) and (14) should account for nearest and next nearest-neighbour interactions.

CONCLUSIONS

For cases involving no substantial short range order, equations (6), (11) and (14) provide a starting point for formulating a-x relationships for multi-site multi-component crystalline solutions. Clearly much needs to be done. In the end there is no substitute for calorimetric measurements of mixing properties (for example, Newton, this volume) and phase equilibrium experiments devised to determine a-x relationships unambiguously (e.g. Wood, this volume). There is also room for much more theoretical work on the form of a-x relationships, particularly including short range order effects, to allow the interpolation and extrapolation of the measurements of a-x relationships as they become available. This is a plea to abandon empirical methods for methods based on a physical approximation of the interactions actually occurring in mineral structures.

REFERENCES

Grover, J. Geochim. Cosmochim. Acta, 38, 1527 (1974).
Guggenheim, E.A. Mixtures. Oxford University Press, Oxford (1952).
Guggenheim, E.A. Applications of statistical mechanics. Oxford University Press, Oxford (1966).
Hagemark, K. J. Phys. Chem., 72, 2316 (1968).
Lumsden, J. Thermodynamics of molten salt mixtures. Academic Press, New York (1966).
Powell, R. Contrib. Mineral. Petrol., 46, 274 (1974).
Powell, R. Contrib. Mineral. Petrol., 51, 29 (1974).

CHEMICAL MIXING IN MULTICOMPONENT SOLUTIONS: AN INTRODUCTION TO THE USE OF MARGULES AND OTHER THERMODYNAMIC EXCESS FUNCTIONS TO REPRESENT NON-IDEAL BEHAVIOUR

John Grover

Department of Geology,
University of Cincinnati, Cincinnati, Ohio 45221.

INTRODUCTION

Our principal goal in devising thermodynamic models for the chemical behaviour of crystalline solutions is to provide a realistic basis for the calculation of reaction equilibria, phase diagrams or reaction kinetics. We recognise in this that the activities of the chemical components of crystalline solutions are usually unequal to the mole fractions of those components, and we attempt in forming our models to express chemical activity as a function of composition and of some minimum number of additional variables, preferably not a function of composition.

The activity, a_i, of component i in a solution is related to the chemical potential of i, $\mu_i(T,P,X)$, and a standard-state (or ideal <u>reference</u>) chemical potential, $\mu_i^o(T,P)$ by

$$RT \ln a_i = \mu_i(T,P,X_i,X_j) - \mu_i^o(T,P) \quad (1)$$

(Prigogine and Defay, 1954, p.88; parentheses here denote functional dependence). If we define the rational activity coefficient, γ_i, as

$$\gamma_i \equiv (a_i/X_i), \quad (2)$$

then (1) may be rewritten in the form:

$$\mu_i = \mu_i^o(T,P) + RT \ln X_i \gamma_i . \quad (3)$$

D. G. Fraser (ed.), Thermodynamics in Geology, 67-97. All Rights Reserved.

In general, γ_i is a function of temperature, pressure and composition. The total molar Gibbs energy of a solution, $\bar{G}_T$ is therefore given by:

$$\bar{G}_T = \sum_i \mu_i^o X_i + RT\sum_i X_i \ln X_i + RT\sum_i X_i \ln \gamma_i(T,P,X). \tag{4}$$

This may be written alternatively as the sum of an ideal and a non-ideal (or excess) contribution:

$$\bar{G}_T = \bar{G}_{ID} + \bar{G}_{EX} \tag{5a}$$

$\bar{G}_{ID}$ can be decomposed into terms representing the free energy of mechanical mixing and the free energy of ideal chemical mixing:

$$\bar{G}_{ID} = \sum_i \mu_i^o(T,P)\, X_i \;+\; RT \sum_i X_i \ln X_i \;, \tag{5b}$$

respectively. The excess Gibbs energy may then be associated with that part of equation (4) that involves activity coefficients:

$$\bar{G}_{EX} = RT \sum_i X_i \ln \gamma_i(T,P,X) \;. \tag{5c}$$

THE SOLUTION PROPERTIES OF A BINARY MIXTURE

In the case of a binary solution (i = 1,2), the interpretation of equation (4) is straightforward. $\bar{G}_{ID}$ has the form

$$G_{ID} = \mu_1^o X_1 + \mu_i^o X_2 + RT\,(X_1 \ln X_1 + X_2 \ln X_2); \tag{6a}$$

and, because $X_1 + X_2 = 1$, this can be written as

$$\bar{G}_{ID} = [\mu_1^o + (\mu_2^o - \mu_1^o)X_2\;] + RT\;[(1-X_2)\ln(1-X_2) + X_2 \ln X_2\;] \tag{6b}$$

The first term in brackets on the right side of (6b) represents mechanical mixing, and has the form of a straight line, $\mu_1^o - \mu_2^o$, in $\bar{G}_T - X_2$ space (P, T fixed). The slope of this line is $(\mu_2^o - \mu_1^o)$ and the G intercept (at $X_2 = 0$) is μ_1^o (see Fig. 1). If there were no chemical mixing in the system 1-2, the total free energy for any specified bulk composition X_2 would be given on this line. Insoluble substances mix according to this linear law.

The second term in brackets, equation (6b), represents the contribution of ideal chemical mixing to $\bar{G}_T$. Because $0 \leqslant X_i \leqslant 1$,

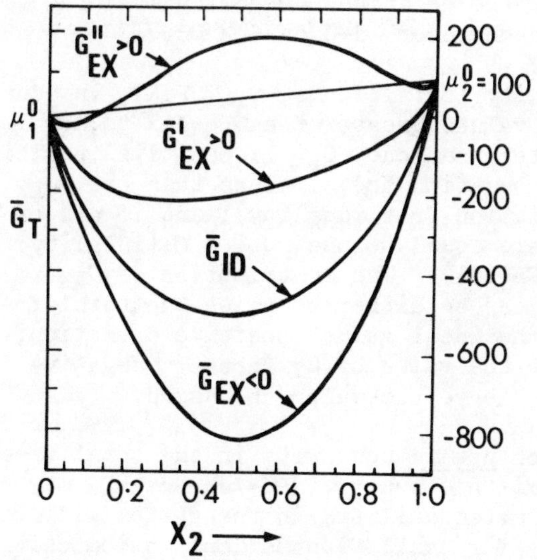

Fig. 1. Variation of the total molar Gibbs energy of a binary solution with composition (see text). The parameters used to calculate these curves are (cf. eqn. 24):

μ_1^o = 0 cal. mol^{-1}; μ_2^o = 100 cal. mol^{-1}

$\bar{G}_{EX}$ < 0 : W_{G1} = -1500 cal. mol^{-1}; W_{G2} = -1000 cal. mol^{-1}

$\bar{G}'_{EX}$ > 0 : W_{G1} = 1500 cal. mol^{-1}; W_{G2} = 1000 cal. mol^{-1}

$\bar{G}''_{EX}$ > 0 : W_{G1} = 3000 cal. mol^{-1}; W_{G2} = 2500 cal. mol^{-1}

$\bar{G}_{EX}$ = 0 ($\bar{G}_{ID}$) : $W_{G1} = W_{G2} = 0$

R = 1.98726 cal. K^{-1} mol^{-1}; T = 400 K

this term is always negative. Its addition to the first term gives rise to a negative departure from the straight line $\mu_1^o - \mu_2^o$, and the sum, $\bar{G}_{ID}$, has the form of a smooth curve that is everywhere concave upward. (The first derivative or slope of the curve is given by

$$\frac{d\bar{G}_{ID}}{dX_2} = (\mu_2^o - \mu_1^o) + RT \ln \frac{X_2}{X_1} \quad ; \tag{7}$$

the second derivative is

$$\frac{d^2\bar{G}_{ID}}{dX_2{}^2} = \frac{RT}{X_1X_2} \quad , \tag{8}$$

which can never be negative. Recall that a function $\bar{G} = f(X_2)$ is concave upward wherever $d^2G/dX_2^2 > 0$).

The activity coefficients, $\gamma_i(T,P,X)$, in equation (4) can, however, have values greater than, equal to, or less than one. If they are greater than one, $\bar{G}_{EX}$ is positive (positive deviation from ideality; cf. equation 5c); if less than one, $\bar{G}_{EX}$ is negative (negative deviation from ideality); and if all of the activity coefficients are equal to one, $\bar{G}_{EX} = 0$ (ideality: $\bar{G}_T = \bar{G}_{ID}$; cf. equation 2) (Fig. 1). The contribution of $\bar{G}_{EX}$ to $\bar{G}_T$ at any composition, then, can be either to raise the total free energy of mixing above the ideal value (positive deviation; $\bar{G}'_{EX}$ and $\bar{G}''_{EX}$, Fig. 1), to depress the value of $\bar{G}_T$ further (negative deviation), or, if $\bar{G}_{EX} = 0$, to leave the value unchanged.

Regions of upward concavity in the total free energy function for a binary mixture, where $(d^2\bar{G}_T/dX_2^2) > 0$, may be associated with stability (or metastability) in the system with respect to unmixing and a tendency for small fluctuations in composition not to survive. Regions of downward concavity in the free energy of mixing curve are inherently unstable with respect to small compositional perturbations in an initially homogeneous material. Thus, with time, compositional heterogeneities in the solution will appear wherever $(d^2\bar{G}_T/dX_2^2) < 0$.

Whether, in the case of positive deviation from ideality, the equation of state is concave downward for a particular range of compositions depends upon the functional nature of the excess molar free energy of mixing, $\bar{G}_{EX}$. From equations (5a) and (8) we have

$$\frac{d^2\bar{G}_T}{dX_2^2} = \left(\frac{RT}{X_1X_2}\right) + \left(\frac{d^2\bar{G}_{EX}}{dX_2^2}\right) \tag{9}$$

Thus, if $(d^2\bar{G}_{EX}/dX_2^2)$ becomes sufficiently negative, it may override the positive contribution from the RT/X_1X_2 term to produce downward concavity in $\bar{G}_T(X_2)$. Unless $\bar{G}_{EX}$ is highly asymmetric with composition, this effect should be most pronounced near the middle of the composition range because the function $d^2\bar{G}_{ID}/dX_2^2 = RT/X_1X_2$ has a singular positive minimum at $X_1 = (1-X_2) = X_2 = \frac{1}{2}$. Regardless of the magnitude and direction of the effect induced by $\bar{G}_{EX}$, however, there will always be a tendency toward negative deviations at compositions sufficiently close to $X_2 = 0$ and $X_2 = 1$, because $\ln X_i$ approaches $-\infty$ as X_i approaches 0. In the context of equation (7) this tendency guarantees that the slope of $\bar{G}_T(X_2)$ will be negative at X_2 near 0; that the slope for X_2 near 1 will be positive; and that as X_2 approaches either 0 or 1, the magnitude of the slope will approach infinity (i.e. the curve $\bar{G}_T(X_2)$ will become vertical). If the free energy of mixing, $\bar{G}_T$, is concave downward over some part of the composition range, the points of inflection between regions of upward and downward concavity will occur wherever $(d^2\bar{G}_T/dX_2^2) = 0$, or

$$\frac{d^2\bar{G}_{EX}}{dX_2^2} = -\left(\frac{RT}{X_1X_2}\right) \tag{10}$$

Equation (10) represents the binary spinode, and serves to delimit regions of thermodynamic stability and instability with respect to compositional unmixing.

TERNARY AND HIGHER-ORDER CHEMICAL MIXTURES

The interpretation of equation (4) for a ternary or higher-order solution is no more difficult in principle than the interpretation already given for a binary solution. However, every additional component requires the addition of one compositional variable to the free energy of mixing function, and therefore one additional dimension in free energy - composition space. The graphical depiction of a binary system thus requires two dimensions $(\bar{G}_T, X_2)$; that of a ternary system, three dimensions $(\bar{G}_T, X_2, X_3)$, and so on. The graphical representation of multicomponent solutions with $c > 3$ is virtually impossible, except in projection, and it becomes necessary to rely on strictly functional treatments of the properties of state.

In a ternary system $(i = 1,2,3)$, the ideal molar free energy of mixing is

$$\bar{G}_{ID} = (X_1\mu_1^o + X_2\mu_2^o + X_3\mu_3^o) + RT\sum_{i=1}^{3} X_i \ln X_i \quad , \tag{11a}$$

which may also be written

$$\bar{G}_{ID} = [\,\mu_1^o + (\mu_2^o - \mu_1^o)X_2 + (\mu_3^o - \mu_1^o)X_3\,] +$$

$$+ RT\left(\ln X_1 + \sum_{i=2}^{3} X_i \ln \frac{X_i}{X_1}\right) \tag{11b}$$

if $(1-X_2-X_3)$ is substituted for X_1. The expression in brackets, equation (11b), represents the free energy of mechanical mixing in a ternary system and plots as a plane in $\bar{G}_T$-X_2-X_3 space. This plane intersects the free energy axes $X_1=1$; $X_2=1$; and $X_3=1$ at μ_1^o, μ_2^o and μ_3^o, respectively. The RT term in (11b) represents ideal chemical mixing, and contributes to a negative departure from the plane of mechanical mixing because $0 \leqslant X_i \leqslant 1$. For a binary system, the maximum departure in $\bar{G}_T$ produced by the ideal chemical mixing term is equal to RT ln $\frac{1}{2}$ ($\simeq$ - 0.6931 RT), whereas it is RT ln 1/3 ($\simeq$ -1.0986 RT) for a ternary system and RT ln 1/c for a c-component system.

Departures from ideality in any ternary or higher-order solution can be considered in terms of the molar excess free energy of mixing, $\bar{G}_{EX}$ (equations 5a, 5c), but the analysis of its effect on the form of the total free energy surface (or hypersurface) in terms of the generation of positive or negative departures from ideality (with associated tendencies toward downward, or accentuated upward, concavity) may be complex. The analysis is, nonetheless, important because the functional form of $\bar{G}_{EX}$ can often be determined only on the basis of the compositions of coexisting phases in a system, and these depend upon the position of free energy minima. For the geologist or chemist concerned with a specific multicomponent system, an iterative rather than explicit approach to the location of free energy minima may be more productive (Brown and Skinner, 1974).

REPRESENTATION OF THE EXCESS GIBBS FREE ENERGY OF MIXING

In order to generate the total free energy surface for a system by means of equations (5), it is necessary to specify a value for the activity coefficient of at least one component for each bulk composition of the solution at every temperature and pressure of interest. While this is possible in theory, given a large number of experimental data and the Gibbs-Duhem equation to relate γ_1, γ_2, .., and γ_c in each phase (see Darken, 1950, and Lewis and Randall, 1961, pp.551-567), it may be an arduous chore in practice. Various authors have attempted to overcome the awkward form of (5c) by devising polynomial or other series-form expansions in composition for $\bar{G}_{EX}$ or ln γ_i. In these the coefficients of the composition terms are taken to be functions only of temperature and pressure. These coefficients are usually evaluated empirically from data on the compositions of coexisting phases at fixed temperature and pressure. It is convenient, therefore, to express the condition for chemical equilibrium in a system containing several phases in terms of the series coefficients. This presents no difficulty in principle because the series coefficients may always be related to the chemical potentials and activity coefficients of individual components.

The choice of a specific expression that best represents $\bar{G}_{EX}(T,P,X_i)$ for a given system depends upon: (1) the extent to which the function can predict the solution properties of the system at temperatures, pressures and compositions other than those for which the adjustable parameters (e.g. series coefficients) of the function were initially determined; and (2) the ease and accuracy with which the function can be used to calculate reaction equilibria and kinetics or to generate phase diagrams. No model can be said to be better than another, *a priori*, until it has been tested against reliable experimental data. Because it is always easier to accommodate a limited number of data in a model with a large number of variables, it is usually assumed that the model that accounts for a given collection of data with the fewest fitting parameters is the most theoretically rigorous and therefore to be preferred.

Among expressions for $\bar{G}_{EX}$ (or $\bar{H}_{mix}$) that have been found useful in describing the solution behaviour of specific non-ideal systems are:

(1) *The regular solution model* (Hildebrand, 1929; Hildebrand and Scott, 1962; Denbigh, 1966, pp.432-435; Prigogine and Defay, 1954, pp.256-261, 394-408);

(2) *Margules's equations* (Margules, 1895; Carlson and Colburn, 1942; Wohl, 1946 and 1953; Thompson, 1967; King, 1969, p.316 *et seq.*), and the mathematically equivalent expressions that comprise:

(3) *The subregular solution model* (Hardy, 1953);

(4) *The subregular model with a factor correcting for non-random association of species* (Sharkey, Pool and Hoch, 1971; see also Guggenheim, 1935, and Rushbrooke, 1938);

(5) *The extended subregular solution model* (Currie and Curtis, 1976; Wohl, 1946 and 1953);

(6) *The van Laar equations* (Carlson and Colburn, 1942; King, 1969, p.317 *et seq.*);

(7) *The Guggenheim and Redlich-Kister equations* (Guggenheim, 1937; Redlich and Kister, 1948a, b; King, 1969, pp.325-327);

(8) *The Wilson equation* (Wilson, 1964; King, 1969, pp.557-560).

(9) *The quasi-chemical solution model* (Guggenheim, 1935; 1952, especially chapters 4, 10 and 11; Rushbrooke, 1938; 1949, pp.287-310; Blander and Braunstein, 1960; Green, 1970b);

(10) *Wohl's equations of the second, third and higher orders in effective volume fraction, q* (Wohl, 1946 and 1953; King, 1969, pp.327-336, 555-556);

(11) *The theory of conformal ionic solutions* (Blander and Yosim, 1963; Blander and Topol, 1966); and

(12) The multiple-site model for homogeneous solutions (Thompson, 1969 and 1970; Grover, 1974).

THE MARGULES EQUATIONS

The Margules formulation for $\bar{G}_{EX}$ has been especially useful in dealing with systems in which phase separation (positive deviation from ideality) occurs or which exhibit extrema in the variation of the activity coefficients with composition. The remainder of this paper will be devoted to a brief discussion of the properties of the Margules equations.

Margules' original formulation was intended to represent the partial vapour pressure for either component of a binary solution in terms of the mole fraction of the component X_i (i=1,2), the vapour pressure of the pure end-member P_i^o, and an exponential series in composition, for which the coefficients are constant at specified conditions of temperature and pressure (Margules, 1895, pp.1265-1268; see also Denbigh, 1966, pp.240-242). However, the method can be extended rigorously to any of the other intensive variables conforming to the Gibbs-Duhem equation or one of its modifications (e.g. the Duhem-Margules equation). Margules' expression for the partial pressure of component i is:

$$p_i = P_i^o X_i^{\alpha_o^{(i)}} \exp\{\alpha_1^{(i)} (1-X_i) + \frac{\alpha_2^{(i)}}{2} (1-X_i)^2 + \frac{\alpha_3^{(i)}}{3} (1-X_i)^3 + \ldots\} \tag{12}$$

where, because of Raoult's Law $\alpha_o^{(i)} = 1$ and $\alpha_1^{(i)} = 0$. The analogous expression for activity is:

$$a_i = (a_i^o)\, X_i \exp\{ \frac{K_2^{(i)}}{2} (1-X_i)^2 + \frac{K_3^{(i)}}{3} (1-X_i{}^3 + \ldots \} \tag{13}$$

where the usual definition of standard state for a solid mixture requires $a_i^o = 1$. Thus, with equation (2),

$$\ln \gamma_i \equiv \ln(\frac{a_i}{X_i}) = \sum_{j=2}^{n} \frac{1}{j} K_j^{(i)} (1-X_i)^j \tag{14}$$

(cf. Guggenheim, 1937). Using equation (14), Hardy (1953) writes the chemical potential of a binary mixture in terms of the third power in composition:

$$\mu_1 - \mu_1^o = RT \ln X_1 + \tfrac{1}{2}\alpha_2 X_2^2 + \frac{1}{3}\alpha_3 X_2^3 \tag{15}$$

and

$$\mu_2 - \mu_2^o = RT \ln X_2 + \tfrac{1}{2}\beta_2 X_1^2 + \frac{1}{3}\beta_3 X_1^3 \tag{16}$$

It follows that for a binary mixture:

$$\mu_1^{EX} = RT \ln \gamma_1 = \tfrac{1}{2}\alpha_2 X_2^2 + \frac{1}{3}\alpha_3 X_2^3 \tag{17a}$$

and

$$\mu_2^{EX} = RT \ln \gamma_2 = \tfrac{1}{2}\beta_2 X_1^2 + \frac{1}{3}\beta_3 X_1^3 \tag{17b}$$

Hardy (1953, equation 5) also presents an expression for the excess free energy of a binary mixture:

$$\bar{G}_{EX} = X_1 X_2 (W_{G1} X_2 + W_{G2} X_1) , \tag{18}$$

which he calls the "subregular" approximation. The parameters W_{G1} and W_{G2} are functions only of temperature and pressure, not of composition. This equation can be considered a linear combination of the excess components of two <u>regular</u> (single-coefficient) solutions (Hildebrand and Scott, 1962; Denbigh, 1966, pp.432-435). Thus, if for solution 1

$$\bar{G}_{EX}^{(1)} = W_{G1} X_1 X_2 , \tag{19}$$

and for solution 2

$$\bar{G}_{EX}^{(2)} = W_{G2} X_1 X_2 , \tag{20}$$

the subregular approximation has X_1 moles of solution 2 mixing with X_2 moles of solution 1. By the lever rule,

$$\bar{G}_{EX}^{(1)+(2)} \equiv \bar{G}_{EX} = X_1 (W_{G2} X_1 X_2) + X_2 (W_{G1} X_1 X_2) , \tag{21}$$

or

$$\bar{G}_{EX} = W_{G2} (X_1 X_2) + ([W_{G1} - W_{G2}]\, X_2)(X_1 X_2) ; \tag{22}$$

and these expressions are clearly equivalent to (18). The form of equation (22) emphasizes the linear character of the combination. Dividing both sides of (22) by $(X_1 X_2)$ we have

$$\frac{\bar{G}_{EX}}{X_1X_2} = W_{G2} + (W_{G1} - W_{G2})\ X_2\ ; \tag{23}$$

and it is clear that this model requires $(\bar{G}_{EX}/X_1X_2)$[1] to vary linearly with X_2 if W_{G1} and W_{G2} are to be independent of composition. When $W_{G1} = W_{G2}$ the slope of the straight line (23) is zero. (Compare Sharkey, Pool and Hoch, 1971).

The expression for the total free energy of binary mixing corresponding to equations (5) and (18) is:

$$\begin{aligned}\bar{G}_T = {} & \mu_1^o + (\mu_2^o - \mu_1^o)\ X_2 + \\ & RT\ (X_1 \ln X_1 + X_2 \ln X_2) + \\ & X_1X_2\ (W_{G1}X_2 + W_{G2}X_1)\end{aligned} \tag{24}$$

and the chemical potentials of the individual components, calculated from (24) are given by

$$\mu_1 = \mu_1^o + RT \ln X_1 + X_2^2\ [W_{G1} + 2(W_{G2} - W_{G1})X_1] \tag{25a}$$

and

$$\mu_2 = \mu_2^o + RT \ln X_2 + X_1^2\ [W_{G2} + 2(W_{G1} - W_{G2})X_2]\ . \tag{25b}$$

By substituting X_1 as $(1 - X_2)$ in (25a) and X_2, conversely, in (25b), we find that

$$\mu_1^{EX} = RT \ln \gamma_1 = (2W_{G2} - W_{G1})X_2^2 + 2(W_{G1} - W_{G2})X_2^3 \tag{26a}$$

and

$$\mu_2^{EX} = RT \ln \gamma_2 = (2W_{G1} - W_{G2})X_1^2 + 2(W_{G2} - W_{G1})X_1^3 \tag{26b}$$

Comparison of equations (26) and (17) shows immediately that the functional properties of the model described by (18) and those deriving from (14), expanded to j=3, are identical. These equations comprise <u>the two-parameter (asymmetric) Margules model for a binary solution</u>. To complete the analogy it is necessary only to equate the coefficients in the expressions for chemical potential, (17) and (26):

---

[1] or $\bar{H}_{mix}/X_1X_2$, a quantity that may be measurable by calorimetric or spectroscopic means; the equivalent to $\bar{G}_{EX}/X_1X_2$ if $\bar{S}_{EX} = 0$.

$$\tfrac{1}{2}\alpha_2 = (2W_{G2} - W_{G1});\ \frac{1}{3}\alpha_3 = 2(W_{G1} - W_{G2}) \tag{27}$$

$$\tfrac{1}{2}\beta_2 = (2W_{G1} - W_{G2});\ \frac{1}{3}\beta_3 = 2(W_{G2} - W_{G1})\ , \tag{28}$$

and, consequently,

$$W_{G1} = \tfrac{1}{2}\alpha_2 + \frac{1}{3}\,\alpha_3\ , \tag{29a}$$

and

$$W_{G2} = \tfrac{1}{2}\beta_2 + \frac{1}{3}\,\beta_3 \tag{29b}$$

These same results were obtained implicitly by Carlson and Colburn (1942).

The derivatives of $\bar{G}_T$ with respect to composition (X_2) for binary solutions conforming to the two-coefficient Margules model can be determined from the equation of state (24):

$$\left(\frac{d\bar{G}_T}{dX_2}\right)_{P,T} = (\mu_2 - \mu_1) = (\mu_2^\circ - \mu_1^\circ) + RT\ \ln\frac{X_2}{X_1} +$$

$$+ W_{G1}\ X_2\ (3X_1 - 1) - W_{G2}\ X_1\ (3X_2 - 1)\ ;$$

$$\left(\frac{d^2\bar{G}_T}{dX_2^2}\right)_{P,T} = \frac{RT}{X_1X_2} - 2\ [W_{G1}\ (3X_2 - 1) + W_{G2}\ (3X_1 - 1)]$$

(compare equations 9 and 10); and

$$\left(\frac{d^3\bar{G}_T}{dX_2^3}\right)_{P,T} = \frac{d^3\bar{G}_{ID}}{dX_2^3} + \left[\frac{d^3\bar{G}_{EX}}{dX_2^3}\right]$$

$$= \frac{RT(X_2 - X_1)}{(X_1X_2)^2} + [6(W_{G2} - W_{G1})]$$

(The reader will recall our discussion of the conditions for the binary spinode (equation 10) and the significance of the inflection points, where $(d^2\bar{G}_T/dX_2^2)_{T,P} = 0$, in bracketing regions of compositional stability (or metastability) and instability with respect to unmixing). It is well-known (Prigogine and Defay, 1954, pp.239-243) that at the critical (or consolute) point

$$\left(\frac{d^2\bar{G}_T}{dX_2^2}\right)_{T,P} = 0 \text{ and } \left(\frac{d^3\bar{G}_T}{dX_2^3}\right)_{T,P} = 0 .$$

The expressions obtained for the second and third derivatives of $\bar{G}_T$ can thus be solved simultaneously to give

$$\frac{W_{G1}^{(c)}}{RT^{(c)}} = \frac{9\,X_1^cX_2^c - (1 + X_1^c)}{6(X_1^cX_2^c)^2} \tag{30a}$$

and

$$\frac{W_{G2}^{(c)}}{RT^{(c)}} = \frac{9X_1^cX_2^c - (1 + X_2^c)}{6(X_1^cX_2^c)^2} \tag{30b}$$

where superscript c denotes a parameter measured or calculated at the critical temperature and composition (see Thompson, 1967, equations 88). The extraction of explicit values for W_{G1} and W_{G2} is considerably more involved for non-critical conditions (see below).

Figure 1 shows several free energy - composition curves and their derivatives, generated by means of eqn. (24) and the equations given above for the first and second derivatives of $\bar{G}_T$ with respect to X_2.

The behaviour of the activity coefficients for the two-parameter Margules model at infinite dilution becomes clear from equations (17) and (29) or (26):

As $X_1 \rightarrow 1$, $X_2 \rightarrow 0$ and $RT \ln \gamma_2 \rightarrow \frac{1}{2}\beta_2 + \frac{1}{3}\beta_3 = W_{G2}$.

As $X_2 \rightarrow 1$, $X_1 \rightarrow 0$ and $RT \ln \gamma_1 \rightarrow \frac{1}{2}\alpha_2 + \frac{1}{3}\alpha_3 = W_{G1}$.

Thus, as Hardy (1952, p.209) has noted, the Margules parameters W_{G1} and W_{G2} can be determined by extrapolating tangents to the activity-composition or excess free energy-composition curves at the compositional extremes. A geometrical interpretation of this convenient property is given in Figure 2.

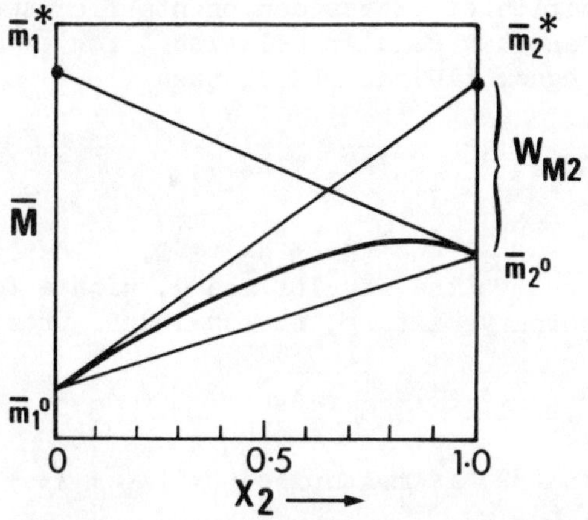

Fig. 2. Plot of a general molar variable, $\bar{M}_T$, as a function of composition in the binary mixture 1-2. The lines $\bar{m}_1^o$ - $\bar{m}_2^*$ and $\bar{m}_2^o$ - $\bar{m}_1^*$ are tangents to the curve $\bar{M}_T'(X_2)$ at the compositional extremes. The function M excludes any chemical mixing term.

This approach to understanding the significance of the W-parameters is by no means limited to binary mixtures. Although the notation becomes progressively more complicated as components are added, the technique of projecting the tangent of $\bar{M}_T'$ at infinite dilution to the opposite end-member in a bounding binary system remains unchanged.

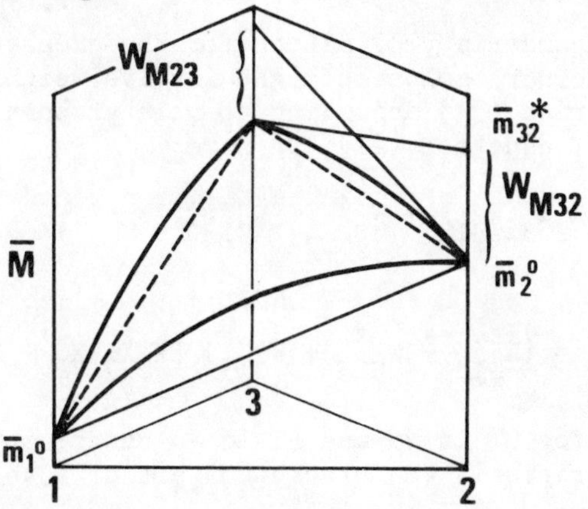

Fig. 3. Plot of a general molar variable, $\bar{M}_T$, as a function of composition in the ternary mixture 1-2-3.

It is possible to formulate expressions analogous to (18) for a ternary system by following the method suggested by equations (19) - (20). Before attempting this, however, it is instructive to develop a somewhat simpler model. In a binary system, the unweighted addition of excess components from two regular solutions gives rise to another regular solution. For example, if $\bar{G}^{(R)}_{EX}$ = $\bar{G}^{(1)}_{EX}$ + $\bar{G}^{(2)}_{EX}$, eqns. (19) and (20), then

$$\bar{G}^{(R)}_{EX} = AX_1X_2 = X_1X_2(AX_2 + AX_1), \quad (31)$$

where $A = (W_{G1} + W_{G2})$ and $(X_1 + X_2) = 1$. Clearly if $W_{G1} = W_{G2}$, eqn. (18) reduces to the regular model, with a form identical to (31). For a ternary mixture, the analogous procedure gives

$$\bar{G}^{(R)}_{EX} = \alpha_{12} X_1X_2 + \alpha_{23} X_2X_3 + \alpha_{13} X_1X_3 . \quad (32)$$

If each term in (32) is multiplied by $(X_1 + X_2 + X_3)$ [= 1] and expanded, we find

$$\bar{G}^{(R)}_{EX} = \alpha_{12}(X_1X_2^2 + X_2X_1^2) + \alpha_{23}(X_2X_3^2 + X_3X_2^2) + \alpha_{13}(X_3X_1^2 + X_1X_3^2) + X_1X_2X_3(\alpha_{12} + \alpha_{23} + \alpha_{13}), \quad (33)$$

which is identical to the strictly regular ternary solution model of Prigogine and Defay (1954; see p.257, eqn. 16.90, *et seq.*, cf. eqn. 5c). The coefficients α_{ij} are equivalent to the W_{Gij}, used above, and $\alpha_{ij} = \alpha_{ji}$.

A second-order approximation for the excess free energy of mixing of a ternary solution might involve weighing the individual terms in equation (32) according to mole proportions. Following the example of equation (21), we have

$$\bar{G}_{EX} = X_1X_2X_3(W_{G12} + W_{G23} + W_{G13}) \quad (34a)$$

and

$$RT \ln \gamma_1 = (W_{G12} + W_{G23} + W_{G13}) [X_2X_3(X_2 + X_3 - X_1)]. \quad (34b)$$

(Expressions for RT ln γ_2 and RT ln γ_3 can be obtained by permuting the subscripts in the bracketed part of (34b)).

A third expression for the excess free energy of mixing in a ternary solution may be obtained by summing terms, without compositional weighing factors, that represent "subregular" mixing in the constituent binary systems (e.g. equation (18). Thus,

$$\bar{G}_{EX} = \sum_{i=1}^{3} \sum_{j=1}^{3} W_{Gij}(X_i^2X_j) \delta_{ij} , \quad (35)$$

where δ_{ij} is a discontinuous factor that is equal to zero when i=j and to one when i≠j (Kronecker δ). (Compare Thompson, 1967, equation 109).

A fourth, and perhaps more realistic approximation for $\bar{G}_{EX}$ in a ternary solution, may be derived by combining three "sub-regular" binary mixing terms, weighted in proportion to the molar composition:

$$\bar{G}_{EX} = X_1 [X_2X_3(W_{G32}X_3 + W_{G23}X_2)] + \\ + X_2 [X_1X_3(W_{G31}X_3 + W_{G13}X_1)] + \\ + X_3 [X_1X_2(W_{G21}X_2 + W_{G12}X_1)] \quad , \qquad (36)$$

or

$$\bar{G}_{EX} = X_1X_2X_3 [(W_{G12} + W_{G13})X_1 + (W_{G21} + W_{G23})X_2 + \\ + (W_{G31} + W_{G32})X_3] \qquad (37)$$

(cf. equation (18)). Expressions for the chemical potentials or activity coefficients of individual components may also be determined for each of the models, (32, 33), (34a), (35) and (36, 37). Additional Margules expressions for the excess free energy and activity coefficients of ternary or higher-order mixtures are given by Wohl (1946, 1953), King (1969), Brown and Skinner (1974), Currie and Curtis (1976) and Grover (in preparation).

The successful determination of values for W_{Gi} (or, in general, W_{Mi}) depends very much on the number of components in the system of interest and the number and quality of data available. Thompson (1967, pp.353-356, especially equation 87), Luth and Fenn (1973), Saxena (1973) and Blencoe (in press) have provided algorithms for calculating W_{G1} and W_{G2} for a binary system from the compositions of the two coexisting phases that define the equilibrium position of the bimodal curve at fixed temperature and pressure. These formulations assume, necessarily, that the coexisting phases share a common equation of state; with this assumption, the standard-state chemical potentials for the end-members can be cancelled when the chemical potentials $\mu_i(T,P,X_i)$, [or activities, $a_i(T,P,X_i)$] of the coexisting phases are equated. However, this simplification is by no means appropriate for all systems in which chemical unmixing occurs (see Lerman, 1965; Warner and Luth, 1974; Saxena and Nehru, 1975; Grover, Lindsley and Schweitzer, 1976, and Blencoe, in press).

Values for W_{G1} and W_{G2} can be determined from the activity coefficients in a binary system at $X_1 = X_2 = 0.5$ by means of equation (26). For the composition $X_2 = 0.5$, (26a) reduces to

$$\ln \gamma_1 = W_{G2}/4RT,$$

and (26b) to

$$\ln \gamma_2 = W_{G1}/4RT.$$

This result is independent of the values of W_{G1} and W_{G2}, and can be used to show that W_{G1} and W_{G2} may also be determined from the excess free energy curve at $X_2 = 0.5$:

$$W_{G2} = 4(\bar{G}_{EX})_{T,P,X_2 = 0.5} - 2(d\bar{G}_{EX}/dX_2)_{T,P,X_2 = 0.5}.$$

A similar expression can be devised for W_{G1}, although calculation of either W_{G1} or W_{G2} from a $\bar{G}_{EX}$ curve at $X_2 = 0.5$ is difficult owing to the need for a measurement of slope.

Equations requiring an accurate knowledge of the surface of some molar property $\bar{M}_T$ are rarely useful for calculating values of $W_{Mi}(T,P)$ because, in order to use them for this purpose, the surface $\bar{M}'_T(T,P,X_1,X_2,\ldots,X_c)$ must be well-known, especially for a range of compositions at and near the compositional end-members and in the binary joins between them. Even when such data are available, the procedure for determining the W_{Mi} graphically is inexact. A method that is far more precise involves fitting all available experimental data to an equation of state in $\bar{M}_T$ or $\bar{M}'_T$ by means of least-squares or other statistical methods. The W_{Mi} terms in the equation of state can either be expressed explicitly (equation 24) or included implicitly in the coefficients of a polynomial expansion in the (c-1) independent composition parameters. Waldbaum and Thompson (1968, pp.2001-2004) show for a binary system how such a polynomial in $\bar{V}_T$ may be related to the Margules parameters W_{V1} and W_{V2}. Thompson (1967, p.359) points out in this connection that it is necessary to include at least third-order terms in composition in the polynomial expression in order to represent asymmetric ternary behaviour (cf. Grover, in preparation).

An additional technique for determining Margules coefficients should also be examined. When two or more phases coexist in equilibrium at fixed temperature and pressure, the chemical potentials for each of the several components are equal in all of the phases. If a and b denote two phases coexisting in equilibrium in the binary system 1-2, the equivalence of chemical potentials, $\mu_i^a = \mu_i^b$, leads to equilibrium expressions having a form

$$\mu_1^a + \mu_2^b = \mu_2^a + \mu_1^b \, . \tag{38}$$

When this condition is combined with a model-dependent equation for chemical potential, such as (25), we can derive a general expression for the distribution coefficient K_D:

$$\begin{aligned} -RT \ln K_D = {} & (\mu_2^{ao} - \mu_1^{ao}) - (\mu_2^{bo} - \mu_1^{bo}) + \\ & + [W_{G2}^a (X_1^a)^2 - W_{G1}^a (X_2^a)^2 + 2(W_{G1}^a - W_{G2}^a) X_1^a X_2^a] - \\ & + [W_{G2}^b (X_1^b)^2 - W_{G1}^b (X_2^b)^2 + 2(W_{G1}^b - W_{G2}^b) X_1^b X_2^b] \end{aligned} \tag{39}$$

where $K_D = (X_2^a X_1^b)/(X_1^a X_2^b)$ and the two-coefficient Margules model is assumed to apply for both phases a and b. Such an equation is required when the structures of the two phases at equilibrium are different (that is, when a and b do not obey the same equation of state). In this circumstance, it is convenient to define the standard state for component i as the pure, end-member i-bearing phase stable at the temperature and pressure of interest. Thus, $\mu_1^{ao} = \mu_2^{bo} = 0$. However, six unknowns still remain and it is generally not possible to extract values for them from reversed two-phase data at P and T (but see Grover, Lindsley and Schweitzer, 1976). If both phases obey the same equation of state, the expression (and the extraction of values for the W coefficients) is simplified considerably: $W_{G1}^a = W_{G1}^b$; $W_{G2}^a = W_{G2}^b$; and $\mu_i^{ao} = \mu_i^{bo}$.

The most important application for equation (39) is in connection with a <u>reciprocal</u> (or metathetical) ternary system such as 1a - 2a - 1b - 2b. In a system of this kind, the reciprocal end-members are related by an exchange equation,

$$1^a + 2^b \rightleftarrows 2^a + 1^b .$$

Several examples of geologically important reciprocal systems are:

(1) alkali feldspar - alkali halide:

$$KAlSi_3O_8 + NaCl \rightleftarrows NaAlSi_3O_8 + KCl \; ;$$

(2) orthopyroxene - $(Mg,Fe^{2+})Cl_2$:

$$Mg_2Si_2O_6 + Fe^{2+}Cl_2 \rightleftarrows Fe_2Si_2O_6 + MgCl_2; \text{ and}$$

(3) orthopyroxene - high-calcium pyroxene:

$$Mg_2Si_2O_6 + 2CaFeSi_2O_6 \rightleftarrows Fe_2Si_2O_6 + 2CaMgSi_2O_6 \, .$$

If one of the constituent mixtures, a or b, in a reciprocal system is ideal ($W_{G1} = W_{G2} = 0$) or if its solution properties are well-known, equation (39) can be of considerable use in determining the solution properties of the other (coexisting) phases. Specifically, by equilibrating a and b experimentally at fixed temperature and pressure for a variety of bulk compositions, it is possible to establish numerous chemical tie lines between two coexisting members of each solution series, a and b. (At equilibrium, the compositions and proportions of the coexisting phases must balance through the bulk composition in accordance with the lever rule). If we know W^a_{G1} and W^a_{G2} or can assume a particular kind of mixing behaviour for solution a, it is convenient to write equation (39) in an alternate form:

$$[-RT \ln K_D - f(X^a_1, X^a_2)] = A_o + A_1 X^b_2 + A_2 X^{b2}_2 , \qquad (40)$$

where $A_o = [(\mu^{ao}_2 - \mu^{ao}_1) - (\mu^{bo}_2 - \mu^{bo}_1) + W^b_{G2}]$, a constant for T,P;

$A_1 = 2(W^b_{G1} - 2W^b_{G2})$; and

$A_2 = 3(W^b_{G2} - W^b_{G1})$.

The expressions for A_1 and A_2 may also be solved simultaneously to give:

$$W^b_{G1} = -\tfrac{1}{2} A_1 - \frac{2}{3} A_2 \; ; \qquad (41a)$$

$$W^b_{G2} = -\tfrac{1}{2} A_1 - \frac{1}{3} A_2 \qquad (41b)$$

(compare equations 26, 29a and 29b; A_o, A_1 and A_2, like W^b_{G1} and W^b_{G2}, are independent of composition). The experimentally-determined compositions for coexisting phases constitute independent sets of values for X^a_1, X^a_2, X^b_1 and X^b_2, and these can be used to plot $[-RT \ln K_D - f(X^a_1, X^a_2)]$ versus X_2 (cf. equation 40). If the resulting curve is linear or parabolic, the coefficients A_o, A_1 and A_2 can be determined from equation (40) by means of a least-squares polynomial fitting technique. W^b_{G1} and W^b_{G2} can then be calculated using equations (41). (Notice that if the function is linear in X_2, $A_2 = 0$ and $W^b_{G1} = W^b_{G2}$). If the function is neither parabolic nor linear, a two- (or one-) coefficient Margules expression for the excess free energy does not apply and another solution model should be sought. Thompson and Waldbaum (1968) provide a thorough discussion of both the theoretical and practical aspects of this approach, and give an example of its use involving the statistical treatment of ion-exchange data for high-temperature alkali feldspars and alkali chlorides.

Use of the Margules method for treating non-ideal behaviour

in systems of geologic importance has increased dramatically since Lerman (1965), Thompson (1967) and Thompson and Waldbaum (1967) published their pioneering work. The following list of geologic publications utilizing (or explaining) the Margules equations or their equivalents is ordered chronologically:

AN ANNOTATED LIST OF GEOLOGICAL PUBLICATIONS UTILIZING THE MARGULES EQUATIONS OR THEIR EQUIVALENTS

Lerman (1965). A two-coefficient Margules expression analogous to equation (24) is used to fit experimental data on the compositions of coexisting calcite and dolomite for a variety of temperatures. (For lack of appropriate data it is assumed that there is a single equation of state applicable both to calcite and dolomite, although Lerman recognizes that this cannot be true). The stability of natural biogenic dolomites and magnesium-calcites is considered in terms of the $\bar{G}_T$-X relation derived.

Thompson (1967). This paper is a principal source for the equations, justification and development of binary Margules theory. Thompson presents algorithms to solve explicitly for W_{G1} and W_{G2}, given the compositions of two coexisting phases at temperature and pressure.

Thompson and Waldbaum (1967). Orville's hydrothermal ion-exchange data on the compositions of high-temperature alkali feldspars coexisting with aqueous NaCl-KCl solutions are used to derive two-coefficient Margules expressions for excess free energy and molar volume. These expressions are used to calculate the critical mixing curve and the univariant curve for two feldspars coexisting with jadeite plus quartz (to 30 kbar).

Waldbaum (1968). Solution calorimetric data are given for the standard enthalpy and Gibbs free energy of formation of albite glass and selected end-member feldspars having different structural states. Waldbaum presents a Margules formulation for $\Delta H^o_{f(T)}$ of microcline-low albite solutions at intermediate compositions.

Thompson and Waldbaum (1968). A two-coefficient expression for $\bar{G}_{EX}$ is determined from a statistical analysis of ion-exchange data for disordered alkali feldspars. Theory concerning the distribution coefficient for exchange between aqueous salt solutions and feldspar (K_D) is developed and a detailed analysis of the geometric properties of RT ln K_D - ion exchange curves is given.

Waldbaum and Thompson (1968). A second-degree Margules expression for the excess molar volume is derived from a statistical analysis of molar volume data for disordered (high-temperature) alkali feldspars. The two-coefficient equation for binary mixtures is derived from a third-order polynomial expansion in X_2.

Thompson and Waldbaum (1969a). Two-feldspar data from several sources are analyzed critically in terms of Margules expressions, and used to obtain the binodal curve bounding the two-phase region in the system $NaAlSi_3O_8$ (albite) - $KAlSi_3O_8$ (sanidine).

Waldbaum and Thompson (1969). The mixing properties for sanidine - high albite solutions derived by Thompson and Waldbaum (1969a) are used to construct feldspar phase diagrams. The equations of state for alkali feldspar are used to predict equilibria involving reactions among feldspar and other phases (including feldspar liquids) at various temperatures and pressures.

Thompson and Waldbaum (1969b). Mixing parameters - and the binodal curve - for the subsolidus system NaCl-KCl are derived using two independent analytic methods. The first method involves plots of functions in composition <u>versus</u> $1/T$, and may be used to calculate Margules parameters directly. The other method, developed in detail, is independent of any equation of state and gives mixing parameters by means of an examination of the critical properties of the solution.

Waldbaum (1969). The thermodynamic mixing properties of NaCl-KCl liquids are treated by combining observed liquidus data with the calculated results of Thompson and Waldbaum (1969b) for subsolidus mixing in the alkali halide system. The analysis follows the equations used by Waldbaum and Thompson (1969), but Waldbaum casts the Margules equation in terms of the composition of the liquidus minimum in order to utilize the best data available.

Thompson (1969). Margules' theory is applied in principle to <u>homogeneous</u> crystalline solutions, with the alkali feldspars, Fe(II)-Mg orthopyroxenes, Na-K-Ca-Fe-Mg amphiboles and Ca-Mg carbonates used as examples. For an expansion of this theory and an application to Al-Si ordering as a function of temperature in monoclinic and triclinic feldspars, see Hovis (1974) and Thompson, Waldbaum and Hovis (1974).

Thompson (1970). This paper expands the theory presented by Thompson (1969) and gives equations that can be used to derive the model-dependent bulk properties of a homogeneous solution, G_{EX} and γ_i (compare Grover, 1974).

Green (1970b). Mixing properties for the subsolidus system NaCl-KCl are obtained by means of an asymmetric quasi-chemical model (Guggenheim, 1935; 1952) and compared with those obtained using the two-coefficient Margules model (Thompson and Waldbaum, 1969b). Green contends that the reduced mixing enthalpy and reduced excess entropy predicted by the quasi-chemical model are in better agreement with experimental values than are the predictions of the Margules model.

Broecker and Oversby (1971). [Chapter 11] . Several good problems requiring familiarity with Margules equations are given.

Bachinski and Müller (1971). High-temperature, metastable portions of the two-phase region in the system microcline - low albite were determined at 600° to 900°C and 1 atm by means of alkali ion-exchange and exsolution - dissolution experiments. Compositional data are fitted to two-coefficient Margules expressions using the method of Thompson (1967) and the "r-s" method of Thompson and Waldbaum (1969b).

Waldbaum and Robie (1971). Enthalpies of fusion, Al-Si ordering and Na-K mixing for various alkali feldspars, measured calorimetrically, are treated according to a Margules formulation and compared to similar equations of state derived from equilibrium data on the compositions of coexisting high-temperature feldspars.

Matsumoto (1971). Binary and ternary regular solution models (without ternary coefficients) are used to treat Fe-Mg-Co exchange in coexisting olivine and pyroxene. On the basis of his theoretical formulation and natural composition data, the author attempts to estimate the temperature and pressure of equilibration for selected peridotite inclusions.

Grover, Lindsley and Turnock (1972). A two-coefficient Margules equation is fitted to reversed experimental data on the compositions of coexisting augite and pigeonite at 15 kbar in the pseudobinary system $Ca_y(Fe_{0.75}Mg_{0.25})_{2-y}Si_2O_6$ $(0 \leq y < 1)$. The variation in the temperature of the consolute point with composition and pressure is calculated using data on conjugate pairs and calculated molar volumes.

Eugster, Albee, Bence, Thompson and Waldbaum (1972). Electron microprobe data on the compositions of synthetic muscovite - paragonite pairs produced in hydrothermal experiments are fitted to a third-order Margules expression for the excess Gibbs free energy using the technique of Thompson and Waldbaum (1969b). The composition data were obtained for a temperature interval of 300° to 600°C at a pressure of approximately 2 kbar; W_{Gi} is given in the form: $W_{Gi} = A_i + B_iT + C_iP$, so that the critical temperature (833°) and composition (39 mole percent Mu) may be calculated from the Margules formulation.

Blencoe and Luth (1973). A two-coefficient (asymmetrical) Margules expression, including the temperature and pressure dependence of W_{Gi}, is used to fit largely unreversed two-phase data obtained at 400°-580°C and 2, 4 and 8 kbar for the binary system muscovite - paragonite.

Saxena (1973). A general account of the mixing properties of crystalline solutions, with detailed discussions of the van Laar, regular (one coefficient, symmetric), "sub-regular" (two-coefficient, asymmetric) and quasi-chemical models. Margules and other equations for $\bar{G}_{EX}$, μ_i and γ_i are presented for binary and ternary mixtures (without ternary coefficients), and applied in one form or another to a number of mineral systems.

Ganguly (1973). An expression for the activity coefficient of a component in a quaternary regular solution (without ternary or quaternary terms; cf. Prigogine and Defay, 1954, p.257) is simplified on the basis of observed behaviour in the system diopside (plus hedenbergite) - jadeite - acmite - CATS molecule. The final form gives the activity coefficient of the jadeite component as a function of one adjustable Margules parameter, W_{12}, and the mole fractions X_{Di}, X_{Hd} and X_{Jd}. An approximate value for W_{12} is calculated from available experimental and natural data.

Warner and Luth (1973). A symmetric (one-coefficient) Margules equation is used to represent subsolidus data on the composition of coexisting forsterite and monticellite, determined experimentally at 2, 5 and 10 kbar and 800° to 1300°C. W_G is given as an explicit function of T, T^2, P and PT.

Luth and Fenn (1973). A method for the direct calculation of binar[y] solvi, based on the equality of activities for a given chemical component in two coexisting phases, is used to evaluate existing experimental isobaric data for the compositions of coexisting sanidine and high albite. The algorithm given expresses the activities in terms of Margules coefficients.

Luth (1974). This paper presents a thorough critical analysis of the experimental data on phase separation in the alkali feldspars and the theoretical treatments of those data published to date. Luth concludes that the solvi determined in a peralkaline experimental environment are the most nearly internally consistent.

Luth, Martin and Fenn (1974). The compositions of two-phase

assemblages in the system $KAlSi_3O_8$ - $NaAlSi_3O_8$ are determined in experiments using peralkaline starting materials at a variety of temperatures and 1.25, 2.5, 5 and 10 kbar pressure. The results are analyzed according to a two-coefficient Margules model using the method of Luth and Fenn (1973), and compared to previous experimental and theoretical determinations.

Brown and Skinner (1974). Symmetric Margules functions are used to represent the excess Gibbs free energy of non-ideal mixtures in this iterative approach to the rapid calculation of the equilibrium compositions of coexisting phases in multicomponent systems ($c > 15$).

Powell (1974). This paper presents a brief review of the properties of the van Laar, asymmetric quasi-chemical and Margules one-coefficient (regular) and two-coefficient ("sub-regular") models for chemical mixing in a binary system, and compares the predictions of these models for the systems: NaCl-KCl, Zn-Co olivine, and $KAlSi_3O_8$ (sanidine) - $NaAlSi_3O_8$ (high albite). (cf. Carlson and Colburn, 1942, and Green, 1970a, b).

Ganguly and Kennedy (1974). Margules equations for a "simple" ternary mixture (without specific ternary terms; see Prigogine and Defay, 1954) are applied to the quasi-ternary system: pyrope - grossular - (almandine + spessartine) and approximate values for the mixing coefficients (W terms) are derived from existing natural and experimental data.

Powell and Powell (1974). In order to calibrate a clinopyroxene - olivine geothermometer, the authors use ternary regular solution theory (Wohl, 1946; Prigogine and Defay, 1954) to represent the thermodynamic mixing properties of $Mg-Fe^{2+}-Al$ ($=Al+Ti+Cr+Fe^{3+}$) clinopyroxene. Olivine is assumed to be ideal. The crystallization temperatures of several intrusions, xenoliths and lavas are estimated on the basis of the model and known pressures.

Warner and Luth (1974). Experimental data for the compositions of coexisting orthoenstate and diopside in the system $Mg_2Si_2O_6$ - $CaMgSi_2O_6$ at 900° to 1300°C and 2, 5 and 10 kbar are treated according to a two-constant Margules equation. (Both ortho-enstatite and diopside are assumed to obey a single equation of state; cf. Saxena and Nehru, 1975; and Grover, Lindsley and Schweitzer, 1976). The temperature- and pressure-dependence of the W_G terms is given explicitly. An extrapolated two-phase region, calculated for 30 kbar pressure, is in good agreement with data obtained independently at that pressure.

Thompson (1974). A.B. Thompson uses the thermodynamic equations of state determined by Eugster, et al. (1972) for muscovite-paragonite mixtures and by Waldbaum and J.B. Thompson (1969) for sanidine - high albite mixtures, in conjunction with dehydration data for end-member compositions, to construct binary phase diagrams (in projection) for the dehydration of white mica.

Chatterjee and Froese (1975). Reactions involving the devolitization of muscovite and paragonite in the presence of quartz are used to derive projected binary phase diagrams for the dehydration of white mica. The methods are different but the results are similar to those of Thompson, 1974; the activities of alkali feldspar, muscovite and paragonite are represented by means of the Margules expressions given by Waldbaum and Thompson (1969) and Eugster, et al., 1972.

Saxena and Nehru (1975). Independent equations of state for diopside and orthoenstatite solutions are assumed to obey regular solution models (one-constant Margules equations where $W_{G1} = W_{G2}$). The W parameter for each solution is calculated from assumed differences in the standard free energies of formation for end-member phases and published experimental data on the compositions of coexisting phases. Temperatures for natural assemblages of enstatite and diopside are estimated from Margules expressions for <u>homogeneous</u> mixtures in which activities are presumed to be related to individual site occupancies.

Stormer (1975). An expression for temperature as a function of pressure and the compositions of the phases (as X_{albite}) in an equilibrium assemblage containing alkali and plagioclase feldspar is derived from the expressions given by Thompson and Waldbaum (1969a) for the W_G coefficients of the alkali feldspars as a function of temperature and pressure. The two-feldspar geothermometer is applied to several natural examples.

Currie and Curtis (1976). Extended Margules equations applicable to quaternary mixtures are derived through the use of the Gibbs-Duhem equation in conjunction with series expansions representing the activity coefficients of individual components to the third power in composition. These equations are applied to reactions involving jadeitic pyroxene, nepheline (both treated as "subregular" solutions) and albite (treated as an ideal solution), where the adjustable parameters are obtained from analyses of published experimental data. The limits on the composition of omphacitic pyroxene predicted by the model agree with observations of natural assemblages. In an appendix, the authors extend their asymmetric, "subregular" model to a system having c components. It is possible, using their formulation, to predict the activity coefficient for any component, i=1,2, ... , c, from a knowledge of the mixing properties of the constituent binary systems. This model can contain ternary or higher-order coefficients and simultaneously satisfy Raoult's Law at end-member compositions and obey the Gibbs-Duhem relation.

Ganguly (1976). Approximate thermodynamic mixing properties of the (Ca_3Al_2)-$(Ca_3Fe_2^{3+})$-$(Ca_3Cr_2^{3+})$ garnets are given in terms of a ternary regular solution model (Prigogine and Defay, 1954, pp.248-249). Crystal field theory is used to calculate heats of mixing.

Grover, Lindsley and Schweitzer (1976). Independent equations of state for the structurally distinct crystalline solutions diopside$_{(ss)}$ - pigeonite$_{(ss)}$ and orthoenstatite$_{(ss)}$ are calculated from experimental data on phase compositions at the invariant temperature and pressure where the three phases coexist stably in the system $CaMgSi_2O_6$ - $Mg_2Si_2O_6$.

Blencoe (in press). This paper reviews the characteristics of the two-parameter Margules, van Laar and quasi-chemical solution models for binary mixtures, and presents explicit algorithms in the form of FØRTRAN IV subroutines for calculating the adjustable coefficients for the several models from (experimental) data on the compositions and equilibration temperatures of two coexisting phases. Blencoe discusses the role of crystal structure, and associated problems, in the calculation of solution parameters.

Grover (in preparation). A third-order Margules model for the solution properties of a ternary reciprocal system is based on a ten-term polynomial expansion for G_T and G_{EX} in terms of the compositional parameters $X_2/(X_1+X_2)$ and X_3. This formulation is consistent with classic (additive) ternary solution theory but allows asymmetric mixing behaviour solubility of the four reciprocal components in any appropriate number of coexisting phases.

REFERENCES

Bachinski, S.W. and Müller, G. J. Petrol., 12, 329 (1971).
Blander, M. and Braunstein, J. Ann. N.Y. Acad. Sci., 79, 838 (1960).
Blander, M. and Topol, L.E. Inorg. Chem., 5, 1641 (1966).
Blander, M. and Yosim, S.J. J. Chem. Phys., 39, 2610 (1963).
Blencoe, J.G. Computation of thermodynamic mixing parameters for isomorphous, binary crystalline solutions using solvus

experimental data. Computers and Geosci. (In press).
Blencoe, J.G. and Luth, W.C. Geol. Soc. Amer. Abstr., 5, 553, (1973).
Broecker, W.S. and Oversby, V.M. Chemical Equilibria in the Earth. McGraw-Hill, New York, 318 p. (1971).
Brown, T.H. and Skinner, B.J. Amer. J. Sci., 274, 961 (1974).
Carlson, H.C. and Colburn, A.P. Indus. Eng. Chem., 34, 581 (1942).
Chatterjee, N.D. and Froese, E. Amer. Mineral., 60, 985 (1975).
Currie, K.L. and Curtis, L.W. J. Geol., 84, 179 (1976).
Darken, L.S. J. Amer. Chem. Soc., 72, 2909 (1950).
Denbigh, K. The Principles of Chemical Equilibrium (Second Ed.), Cambridge Univ. Press, 494 p.
Eugster, H.P., Albee, A.L., Bence, A.E., Thompson, J.B. Jr., and Waldbaum, D.R. J. Petrol., 13, 147 (1972).
Ganguly, J. Earth Planet. Sci. Lett., 19, 145 (1973).
Ganguly, J. Contrib. Mineral. Petrol., 55, 81 (1976).
Ganguly, J. and Kennedy, G.C. Contrib. Mineral. Petrol., 48, 137 (1974).
Gibbs, J.W. Trans. Conn. Acad., 3, 108 (1876); 343 (1878) (1961).
Green, E.J. Geochim. Cosmochim. Acta, 34, 1029 (1970a).
Green, E.J. Amer. Mineral., 55, 1692 (1970b).
Grover, J. Geochim. Cosmochim. Acta, 38, 1527 (1974).
Grover, J., Lindsley, D.H. and Schweitzer, E.L. Geol. Soc. Amer. Abstr. (1976, in press).
Grover, J., Lindsley, D.H. and Turnock, A.C. Geol. Soc. Amer. Abstr., 4, 521 (1972).
Guggenheim, E.A. Proc. Royal Soc. London, Series A, 148, 304 (1935)
Guggenheim, E.A. Trans. Faraday Soc., 33, 151 (1937).
Guggenheim, E.A. Mixtures, Clarendon Press, Oxford, 270 p. (1952).
Guggenheim, E.A. Thermodynamics (Fourth Ed.), North-Holland Pub. Co., Amsterdam, 476 p. (1959).
Hardy, H.K. Acta Metall., 1, 202 (1953).
Hildebrand, F.B. Advanced Calculus for Applications. Prentice-Hall, Englewood Cliffs, New Jersey, 646 p. (1962).
Hildebrand, J.H. J. Amer. Chem. Soc., 51, 66 (1929).
Hildebrand, J.H. and Scott, R.H. Regular Solutions. Prentice-Hall, Englewood Cliffs, New Jersey, 180 p. (1962).
Hill, T.L. An Introduction to Statistical Thermodynamics. Addison-Wesley, Reading, Mass., 508 p. (1960).
Hovis, G.L. A solution calorimetric and X-ray investigation of Al-Si distribution in monoclinic potassium feldspars, 114-144 in The Feldspars (W.S. MacKenzie and J. Zussman, eds.), Crane Russak and Co., New York, 717 p. (1974).
King, M.B. Phase Equilibrium in Mixtures. Pergamon Press, Oxford, 584 p. (1969).
Lerman, A. Geochim. Cosmochim. Acta, 29, 977 (1965).
Lewis, G.N. and Randall, M. Thermodynamics (revised by K.S. Pitzer and L. Brewer) (Second Ed.), McGraw-Hill, New York, 723 p. (1961).
Luth, W.C. Analysis of experimental data on alkali feldspars:

unit cell parameters and solvi, 249-296 in The Feldspars (W.S. MacKenzie and J. Zussman, eds.). Crane Russak and Co., New York, 717 p. (1974).

Luth, W.C. and Fenn, P.M. Amer. Mineral., 58, 1009 (1973).

Luth, W.C., Martin, R.F. and Fenn, P.M. Peralkaline alkali feldspar solvi, 297-312 in The Feldspars (W.S. MacKenzie and J. Zussman, eds.), Crane Russak and Co., New York, 717 p. (1974).

Margules, M. Akad. Wiss. Wien, Sitzungs., Math.-Naturwiss. Classe, 104, Abt. IIa, 1243 (1895).

Matsumoto, T. Geochem. J., 4, 111 (1971).

Mellor, J.W. Higher Mathematics for Students of Chemistry and Physics (Fourth Ed.). Dover, 641 p. (1955).

Powell, M. and Powell, R. Contrib. Mineral. Petrol., 48, 249 (1974).

Powell, R. Contrib. Mineral. Petrol., 46, 265 (1974).

Prigogine, I. and Defay, R. Chemical Thermodynamics (translated by D.H. Everett), Longmans Green, New York, 543 p. (1954).

Redlich, O. and Kister, A.T. Indus. Eng. Chem., 40, 341 (1948a).

Redlich, O. and Kister, A.T. Indus. Eng. Chem., 40, 345 (1948b).

Rushbrooke, G.S. Proc. Royal Soc. London, Series A, 166, 296 (1938).

Rushbrooke, G.S. Introduction to Statistical Mechanics. Clarendon Press, Oxford, 334 p. (1949).

Saxena, S.K. Lithos, 4, 345 (1971).

Saxena, S.K. Thermodynamics of Rock-forming Crystalline Solutions. Springer-Verlag, New York, 188 p.

Saxena, S.K. Amer. Mineral., 61, 643 (1976).

Saxena, S.K. and Ghose, S. Amer. Mineral., 56, 532 (1971).

Saxena, S.K. and Nehru, C.E. Contrib. Mineral. Petrol., 49, 259 (1975).

Saxena, S.K. and Ribbe, P. Contrib. Mineral. Petrol., 37, 131 (1972).

Sharkey, R.L., Pool, M.J. and Hoch, M. Metall. Trans., 2, 3039 (1971).

Stormer, J.C. Jr. Amer. Mineral., 60, 667 (1975).

Thompson, A.B. Contrib. Mineral. Petrol., 44, 173 (1974).

Thompson, J.B. Jr. Thermodynamic properties of simple solutions, 340-361 in Researches in Geochemistry, Vol. 2 (P. Abelson, ed.), Wiley, New York, 663 p. (1967).

Thompson, J.B. Jr. Amer. Mineral., 54, 341 (1969).

Thompson, J.B. Jr. Amer. Mineral., 55, 528 (1970).

Thompson, J.B. Jr. and Waldbaum, D.R. Trans. Amer. Geophys. Union, 48, 230 (1967).

Thompson, J.B. Jr. and Waldbaum, D.R. Amer. Mineral., 53, 1965 (1968).

Thompson, J.B. Jr. and Waldbaum, D.R. Amer. Mineral., 54, 811 (1969a).

Thompson, J.B. Jr. and Waldbaum, D.R. Geochim. Cosmochim. Acta, 33, 671 (1969b).

Thompson, J.B. Jr., Waldbaum, D.R. and Hovis, G.L. Thermodynamic properties related to ordering in end-member alkali feldspars, 218-248 in The Feldspars (W.S. MacKenzie and J. Zussman, eds.), Crane Russak and Co., New York, 717 p. (1974).
Waldbaum, D.R. Contrib. Mineral. Petrol., 17, 71 (1968).
Waldbaum, D.R. Geochim. Cosmochim. Acta, 33, 1415 (1969).
Waldbaum, D.R. and Robie, R.A. Zeit. Krist., 134, 381 (1971).
Waldbaum, D.R. and Thompson, J.B. Jr. Amer. Mineral., 53, 2000 (1968).
Waldbaum, D.R. and Thompson, J.B. Jr. Amer. Mineral., 54, 1274 (1969).
Warner, R.D. and Luth, W.C. Amer. Mineral., 58, 998 (1973).
Warner, R.D. and Luth, W.C. Amer. Mineral., 59, 98 (1974).
Wilson, G.M. J. Amer. Chem. Soc., 86, 127 (1964).
Wohl, K. Trans. Amer. Inst. Chem. Eng. (Chem. Eng. Progress), 42, 215 (1946).
Wohl, K. Chem. Eng. Progress, 49, 218 (1953).

STUDY PROBLEMS

Useful Equations

The two coefficient Margules excess function for any molar property of state, $\bar{M}$, in the binary system 1 - 2:

$$\bar{M}_{EX} = X_1 X_2 (W_{M1} X_2 + W_{M2} X_1). \quad \text{(P-1)}$$

Relation among Margules coefficients:

$$W_{Gi} = W_{Ei} + P\, W_{Vi} - T\, W_{Si} ; \quad \text{(P-2a)}$$

$$W_{Hi} = W_{Ei} + P\, W_{Vi} \quad (i = 1, 2). \quad \text{(P-2b)}$$

The two-coefficient Margules equation of state for a phase in the binary system 1-2:

$$\bar{G}_T = \mu_1^o + (\mu_2^o - \mu_1^o)\, X_2 + RT \quad X_1 \ln X_1 + X_2 \ln X_2 \quad +$$

$$X_1 X_2 (W_{G1} X_2 + W_{G2} X_1) \quad \text{(P-3)}$$

Combined equation of state for two phases, a and b, coexisting in equilibrium at constant temperature and pressure in the binary system 1 - 2 or the reciprocal system a_{ss} (1-2) - b_{ss} (1-2):

$$- RT \ln K_D = (\mu_2^{ao} - \mu_1^{ao}) - (\mu_2^{bo} - \mu_1^{bo}) +$$
$$W_{G2}^a (X_1^a)^2 - W_{G1}^a (X_2^a)^2 + 2(W_{G1}^a - W_{G2}^a) X_1^a X_2^a -$$
$$W_{G2}^b (X_i^b)^2 - W_{G1}^b (X_2^b)^2 + 2(W_{G1}^b - W_{G2}^b) X_1^b X_2^b ,$$
$$\text{(P-4)}$$

where $K_D = (X_2^a X_1^b)/(X_1^a X_2^b)$.

1(a) Give the dimensions of the following thermodynamic parameters in S.I. units:

X_i, μ_i^o, γ_i, W_{Gi}, W_{Si}, W_{Vi}, $P\,W_{Vi}$, W_{Hi}.

Note: i denotes the ith component in an n-component system.

(b) Show by means of dimensional analysis that equations (P-2a), (P-3) and (P-4) are internally consistent.

2(a) Show that equation (P-4) can be written in an alternate (linear) form such that each of the four Margules parameters, W_{G1}^a, W_{G2}^a, W_{G1}^b, and W_{G2}^b, appears as the single coefficient of a term in composition.

(b) Assume $(\ _2^{ao} - \ _1^{ao}) - (\ _2^{bo} - \ _1^{bo}) = C(T,P)$, a constant.

(i) If phases a and b have different equations of state, what is the minimum number of pairs of coexisting phases of known composition necessary to determine the several Margules parameters uniquely?

(ii) Assume that the minimum number of compositions data needed (part 2b(i)) are available and solve equation (P-4) analytically for $C(T,P)$, W_{G2}^a and W_{G1}^b.

Suggestion: Independent sets of data for the compositions of coexisting phases may be indexed by j = 1, 2, ... n. Use Cramer's Rule for the solution of systems of linear equations and leave your answer in determinant form.

(c) Equation (P-4) is a combined equation of state for two phases in equilibrium at constant temperature and pressure. In light of this, discuss your answer for part (b) critically, showing how theory and experiment are incompatible if the system in question is a binary solution.

(d) Explain how the difficulty you have discussed in part (c) is eliminated for a reciprocal system consisting of two coexisting binary solutions, a(1-2) and b(1-2).

(e) Give the expression analogous to equation (P-4) that applies when phases a and b can be represented by a single equation of state. Use Cramer's Rule to solve the applicable system of equations analytically.

3. Two coexisting, non-colinear binary solutions, a(1-2) and b(1-2), form a ternary reciprocal system at 650°C and 2000 bars. At this temperature and pressure there is not ternary solution away from the binary joins. The following values have been determined for the Margules parameters of solutions a and b:

$$W^b_{G1} = W^b_{G2} = 0$$

$$W^a_{G1} = 20{,}100 \text{ (S.I. units/mole)}$$

$$W^a_{G2} = 10{,}900 \text{ (S.I. units/mole)}$$

In addition, R = 8.31470 J/°K mole and $C = (\mu_2^{ao} - \mu_1^{ao}) - (\mu_2^{bo} - \mu_1^{bo}) = 6300$ (appropriate S.I. units).

(a) Solve for (X_1^b/X_2^b) as a function of W^a_{G1}, W^a_{G2}, X_1^a and X_2^a.

(b) Using the values given for W^a_{G1}, W^a_{G2} and C, determine and plot tie lines for a variety of compositions, X_1^a, in this system at 650°C and 2000 bars. Use a rectangular (reciprocal) compositions base and plot a sufficient number of tie lines that trend in the compositions of coexisting phases in the system (including the boundaries of possible three-phase fields) are clear.

(c) Use the isobaric-isothermal plot of tie lines in the reciprocal system, part (b), to construct a distribution diagram, X_1^b versus X_1^a, for this system.

4. The binary system 1 - 2 has a two-phase region, A + B, bounded by solvi that are linear in T-X space from 200° to approximately 650°C. The consolute point is at $X_2 = 0.37$ and 850°C for P = 1000 bars. The low-temperature solubility limits were determined reversibly in this system to be:

limit of component 2 in phase A, $(X_2^A)_s = (1.25 \times 10^{-4})T^{o}\underline{C} + 0.11875$;

limit of component 1 in phase B, $(X_1^B)_s = (-3.333 \times 10^{-4})$ $T^o\underline{C} + 0.81667$.

(a) Determine values for W_{G1} and W_{G2} at T = 650°C, 500°C, 350°C and 200°C.

<u>Suggestion</u>: Use the following expressions to derive two tractable equations in the unknown W_{G1} and W_{G2} <u>or</u> see Thompson, 1967, equations 87a and 87b:

$$\mu_i^A = \mu_i^{Ao} + RT \ln X_i^A + RT \ln \gamma_i^A ;$$

$$RT \ln (\gamma_2^A/\gamma_1^A) = W_{G2}^A(X_1^A)^2 - W_{G1}^A(X_2^A)^2 + 2(W_{G1}^A - W_{G2}^A)X_1^A X_2^A;$$

and $\mu_i^A = \mu_i^B$ (i = 1,2) at equilibrium.

(b) Determine W_{S1} and W_{S2} for this system at P = 1000 bars. Using your best values for W_{S1} and W_{S2} at the appropriate temperatures, calculate W_{H1} and W_{H2} for T = 650°, 500°, 350° and 200°C.

(c) The excess molar volume in this system has been measured at room temperature for two compositions:

$$X_2 = 0.85;\ \bar{V}_{EX} = 0.9\ cm^3/mole$$

$$X_2 = 0.10;\ \bar{V}_{EX} = 0.8\ cm^3/mole.$$

Calculate W_{V1} and W_{V2} (STP).

(d) Determine W_{G1} and W_{G2} for T = 400°C and P = 2000 bars. Assume that the excess molar volume remains essentially constant to 500°C and 3000 bars. State explicitly any other assumptions that you must make in order to complete this calculation.

(e) Plot $\bar{G}_T$ as a function of composition in the system 1 - 2 at

(i) 500°C, 1000 bars and
(ii) 400°C, 2000 bars

Take, as the standard state, the pure end member phases stable at the temperature and pressure of interest.

DETERMINATION OF ATOMIC OCCUPANCIES

E.J.W.Whittaker

Department of Geology & Mineralogy, Oxford.

The most important and most reliable method of determining the ordering of different atomic species over different structural sites is undoubtedly that based on crystal structure refinement using X-ray or neutron diffraction. The reason for its pre-eminence is essentially due to the adequacy of the models that we have available for relating the observational data to structural features, and also to the fact that we are able to make a very large number of independent measurements, much larger in number than the number of unknowns in the model. The disadvantage of the method is that for accurate work it requires elaborate equipment and a very large expenditure of time and effort, and even if it is desired to determine only atomic ordering parameters it is impossible to derive these on their own - a large proportion of the effort goes into determining other parameters which, although of interest in other ways, may not be of much interest to the thermodynamicist.

A few years ago it seemed that many of these disadvantages might be eliminated by the use of Mössbauer spectroscopy. For this the equipment, time and effort required are all less, the number of independent observations is still fairly large, and the number of extraneous parameters of the model is much smaller. Unfortunately some of the hopes raised by this method are not being fulfilled, and the reasons (as we shall see) are that the model is less well defined and the relationships between the data and the parameters sought are less well assorted.

D. G. Fraser (ed.), Thermodynamics in Geology, 99-113.

Somewhat similar attempts have been made to use the infrared OH frequencies of hydroxyl bearing materials to determine the ordering of the cations to which the OH is coordinated. Since each OH is coordinated to more than one cation the method can in principle give information about short range ordering between neighbouring equivalent sites, but as we shall see this information may be subject to very serious errors.

THE DIFFRACTION METHOD

A single uniform crystal of optimum size measures about 0.1 to 0.2 mm, because in a larger crystal the X-rays would be seriously reduced in their passage through it. It is set up in an exactly known orientation and completely bathed in an X-ray beam. It is successively set in a variety of orientations to satisfy the Bragg equation

$$\sin\theta = \lambda/2d_{hkl}$$

where θ is the angle between the incident beam and the crystallographic plane (*hkl*) for each of several thousand different planes. At each of these orientations the crystal is slowly turned in a defined manner through the exact reflecting position, and the X-ray photons reflected by the crystal are counted. Counts of background radiation are also made at each side of the reflecting position. The reflected counts, minus the background, constitute the primary data, I(*hkl*). These have to be treated to eliminate the effect of various well-understood geometrical factors depending on angular relationships between the incident and reflected rays and the axes of rotation of the crystal. They must also be corrected for absorption in the crystal, though this can usually only be approximately computed for each reflection, unless the crystal has been ground to a simple shape like a sphere.

The corrected intensities vary over a very wide range - a few may approach 10^6 counts, the weakest will be a few tens of counts and statistically indistinguishable from zero. The latter are rejected according to pre-selected criteria (e.g. if they are $<3\sigma$ on counting statistics). The errors in the data for the strongest reflections will be predominantly from errors in absorption correction and from a phenomenon called extinction; for the weakest ones they will arise from counting statistics. They will on average be of the order of 3%. The square root of each intensity is taken, and denoted F_{obs} (*hkl*), the observed structure factor.

Interpretation is based on the fact that a crystal built up of unit cells containing on average atoms of type i at sites (x_i, y_i, z_i) in the cell and vibrating about these positions with a mean square amplitude u^2_{hkl} in a direction perpendicular to the plane (hkl) will give structure factors proportional to

$$F_{calc}(hkl) = \sum_i f_i(hkl) \exp 2\pi i(hx_i + ky_i + lz_i)\exp(-2\pi^2 u^2_{hkl}/d^2_{hkl})$$

where $f_i(hkl)$ is the scattering power of the average atom occupying the i th site, in the direction of the hkl reflection, relative to the scattering power of a classical electron. If there are m symmetry-independent sites in the unit cell the least squares refinement process then refines

(1) an arbitrary scale factor relating F_{obs} to F_{calc}

(2) the $3m$ coordinates x_i, y_i, z_i

(3) the $6m$ vibration parameters, describing the amplitudes and orientation of the vibration ellipsoids of the atoms, and

(4) occupancy parameters for those sites where the atomic species is in doubt (i.e. usually the cation positions only).

In order to do this it is necessary to render F_{calc} into pseudo-linear dependence on small adjustments from assumed values of these parameters in the form of the approximation

$$F_{calc} = F_{ass} + \sum_j \xi_j \frac{\partial F_{ass}}{\partial \xi_j}$$

where the ξ_j represent all the adjustable parameters. Thus it is necessary to calculate for each and every one of the thousand or so reflections the values of F_{ass} and $\frac{\partial F_{ass}}{\partial \xi_j}$ for every parameter. Least squares solutions are then obtained for the adjustments to the parameters ξ_j; these adjustments then give new assumed parameters, and the process is repeated. Estimates of the standard deviations of the parameters are obtained at the same time. Various constraints may have to be applied to ensure that the process converges; it may be necessary for example to adjust only the occupancy parameters <u>or</u> the vibration parameters at a given site in any given cycle of refinement and to alternate between them in alternate cycles. However these are tactical details, and essentially the strategy is to continue the refinement until the calculated shifts are small relative to the estimated standard deviations, and some residual such as

$$\frac{\sum_{hkl} \left| |F_{obs}| - |F_{calc}| \right|}{\sum_{hkl} |F_{obs}|}$$

cannot be further reduced. The final value of this depends both on the accuracy of the observations and the validity of the model, and is likely to be ∿0.05 for a well refined structure of a reasonably complicated mineral. The number of independent observations may be 10 times the number of adjusted parameters.

What we are directly concerned with are the occupancy parameters. These come into the scattering factors f_i which give the scattering power at the ith atomic site. If a site i is partially occupied by various different cations, j, with mole fractions n_j then

$$f_i = \sum_j n_{ij} f_j$$

The way atomic scattering factors depend on the nature of the atom, and their angular dependence for two mineralogically important elements (Mg^{2+} and Fe^{2+}) are shown in Fig. 1.

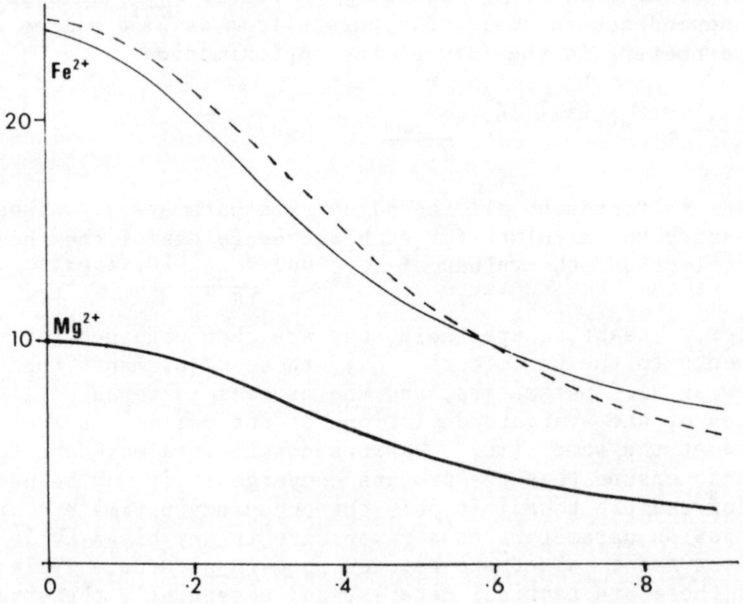

Fig.1. Scattering factors in electron units (——) for Mg^{2+}, Fe^{2+} and (---) for Mg^{2+} scaled by the mean ratio f_{Fe}/f_{Mg} to show the relative dependence on sin θ/λ.

The first thing to notice is that the scattering factors of the two elements are very different at all angles, so that we can discriminate well between their effects. In fact if we have simply a mixture of Mg^{2+} and Fe^{2+} on a site, so that

$$f_i = nf_{Fe} + (1-n)f_{Mg}$$

then the precision with which we can determine n has a standard deviation of the order of 0.01 in normal circumstances. It is convenient to take this as a standard and relate the precision of other determinations to it. Such a precision is usually about as good as is justified by the chemical analysis.

The second point to note is that the fall off with angle even for atoms as different as Mg and Fe is remarkably similar, as is shown by scaling up f_{Mg} by the average of f_{Fe}/f_{Mg}. This means that we are primarily dependent for discrimination on the general level of an f curve and not on its shape. The general level is approximately proportional to the atomic number of the element, and therefore the discrimination between pairs of neighbouring elements such as Mg-Al or Al-Si, in which we are often interested, is very poor. Precision in the determination of n in such cases is likely to be of the order of ± 0.25, which is scarcely worth while. The similarity of shape also means that any combination of three atoms that give the same mean atomic number will be indistinguishable from one another by their scattering factor. Thus $Mg_{0.58}Fe_{0.42}$, $Al_{0.62}Fe_{0.38}$ and all intermediates like $Mg_{0.30}Al_{0.30}Fe_{0.40}$ have virtually identical scattering factors. Thus in a typical ferro-magnesian silicate we cannot refine the occupancies of a site directly in terms of more than two species, usually "equivalent Mg" and "equivalent Fe", and interpretation in terms of the actual occupancy by other atoms (e.g. Al, Mg etc) has to be dependent on the chemistry. If there are only 2 sites to be considered then the known total composition enables us to allocate 3 different atomic species between the two sites, but for more than 3 species the solution becomes indeterminate unless some other constraints are available. If there are more than 2 sites the solution is indeterminate even for 3 species (Hawthorne 1973).

By way of parenthesis it may be added that although a change of oxidation state by 1 unit might be expected to change the scattering factor of an atom by as much as a change of 1 unit of atomic number, this is only true at very low angles. The tighter packing of the electrons in a more highly charged ion gives rise to a less rapid fall of scattering power with angle, and this offsets the smaller number of electrons. Thus X-ray diffraction is very insensitive to oxidation state.

From the foregoing considerations it would therefore appear that we can only determine by diffraction with reasonable precision distributions of 2 atomic species, over any number of sites, if they differ in atomic number by at least 5 or so; and of 3 atomic species over 2 sites only provided that at least two of them differ to that extent. Fortunately however there are some further possibilities open to us.

(i) Use of bond lengths

The fact that we have to refine the atomic coordinates at the same time as the occupancies can be turned to good account, since the average bond length from a site M to its coordinating anions depends on the mean ionic radius of the cations at the site. In a well refined structure the mean bond length in a coordination polyhedron will be determined with a precision of $\sim$ 0.005 Å. This amounts to a few percent of the difference in radius between Mg and Al, or between Al and Si, and it can therefore be used to determine ordering of these ions with a precision of perhaps 0.03. It can also be used as an additional control to permit the assignment of 3 octahedral cations over 3 sites.

(ii) Use of neutron diffraction

Diffraction of neutrons is essentially similar to that of X-rays but is dependent only on nuclear properties and not on electron density. There can be dramatic differences between adjacent elements (e.g. Cr, Mn, Fe as shown in Fig.2) and both +ve and -ve values are possible. Unfortunately there is little improvement in discrimination between Al and Si, though there is an appreciable improvement for Mg and Al.

However the scattering powers for neutrons are generally an order of magnitude lower than for X-rays, and available neutron beams are much weaker than X-ray beams. Thus specimens have to be very much larger (up to 1 cm across), and suitable crystals of minerals are often not available. Some refinement can be done from powder diffractometer data, but even then the provision of a specimen of sufficient size may be a problem. By combining the results of X-ray and neutron refinements we can increase by one the number of species that can be assigned unambigously to any given number of sites, because the species have different effects on the structure factors for the different radiations.

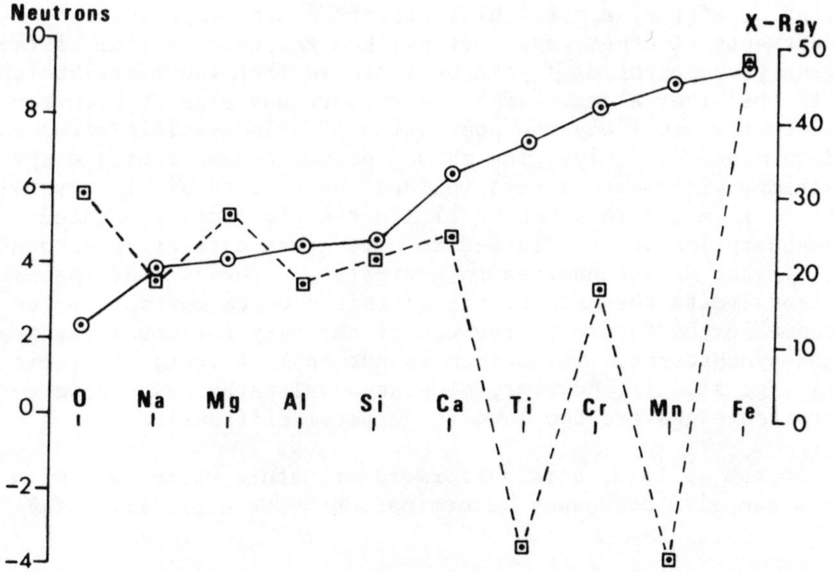

Fig.2. Comparison of element-to-element variation of scattering factor for neutrons (⊡) with that for X-rays at sin θ/λ = 0.3 (⊙). The scale for the neutron data is expended by a factor of 5 relative to the X-ray data. Scattering lengths in units of 10^{-15}m.

(iii) Use of more than one X-ray wavelength

The f values we have discussed so far have been for X-rays of much shorter wavelength than any absorption edge of the atom. If the wavelength is close to an absorption edge the scattering factor is modified, and use of such a wavelength can enable us to increase the discrimination between adjacent elements. Also if we combine the data obtained with suitable different radiations we can increase the number of species that are assignable in principle. For example suitable choice of radiation can change $|f_{Fe}| - |f_{Mn}|$ (f becomes complex in these circumstances) at sin θ/λ = 0.3 from 1.0 to about 5.0. By use of such a radiation therefore, we could perhaps discriminate between Mn and Fe with a precision of about ± 0.02in n. The accuracy would, however, be likely to be relatively poorer because high absorption affects are inevitably introduced at such wavelengths.

Better accuracy than this is claimed by Johnston & Duncan (1975) for a method which makes use of the fact that f

is complex. If reflections measured at a wavelength close to an absorption edge of a particular element A are compared with measurements at other wavelengths, then the contribution of the imaginary component of f_A can be isolated from the contributions of all the other atoms. If the structure has already been refined in the usual way the population of A in specific sites can be determined directly. The method proposed uses energy dispersive detection, with white X-rays incident on a large single crystal plate ($\sim$ 1 cm^2). This method eliminates problems associated with absorption and is claimed to have given a relative accuracy of 3% in the Mn occupancies in tourmaline. The size of specimen required limits the method, though this problem may perhaps be overcome in the future by the use of the very intense X-ray beam from a synchrotron. The method is not applicable to light elements (Mg, Al, Si), however, because wavelengths near their absorption edges are too long for crystal diffraction.

To sum up then, straightforward structure refinement with X-rays can give occupancy determinations with a precision (σ) of:

0.01 directly for 2 species as different in atomic number as Mg and Fe in one site;

0.03 via bond lengths for 2 species as different in ionic radius as Mg and Al

If a second radiation is used (neutrons or a special X-ray wavelength related to a particular species) then additional transition elements can be determined independently. Provided the chemistry is known to the same accuracy then the number of elements whose distribution can be determined is given below

No. of sites	No. of methods used 1 (f)	2 (f+r or f_1+f_2)	3 (f_1+f_2+r)
2 sites	3	4	5
3 or more sites	2	3	4

where f_1 and f_2 denote scattering factors for different radiations and r denotes the ionic radius criterion.

MÖSSBAUER SPECTROSCOPY

The resonant frequency of the nucleus of ^{57}Fe (and of

specific isotopes of some other less mineralogically important elements) is modified by the properties of the electric field in which it lies. It is therefore different for Fe atoms in different sites and in different oxidation states. Thus occupancies by Fe(II) and Fe(III) can be determined in principle in any number of sites whose environments differ enough to separate the frequencies sufficiently. The specimen is in powder form and usually amounts to a few hundred milligrams. The incident radiation consists of γ-rays from the excited state of ^{57}Fe derived by radioactive decay of ^{57}Co, and its frequency is scanned through the range of interest by use of the Doppler effect associated with a vibrating source. The γ-rays in several hundred narrow frequency ranges are counted and accumulated in a multi-channel analyser. The extent of the absorption has to be restricted to a few percent, so counts in each channel have to be of the order of 10^6 to give statistically acceptable spectra.

In such minerals as silicates each Fe species in each site gives rise to a pair of absorption lines which should ideally be of equal intensity and breadth and have a Lorentzian profile. Unfortunately the relationship between the breadths of the lines and the relative displacement of lines due to Fe at different sites, is such that they are at best poorly resolved (Fig.3). They therefore have to be interpreted by a least squares refinement (similar to the X-ray case) of the parameters

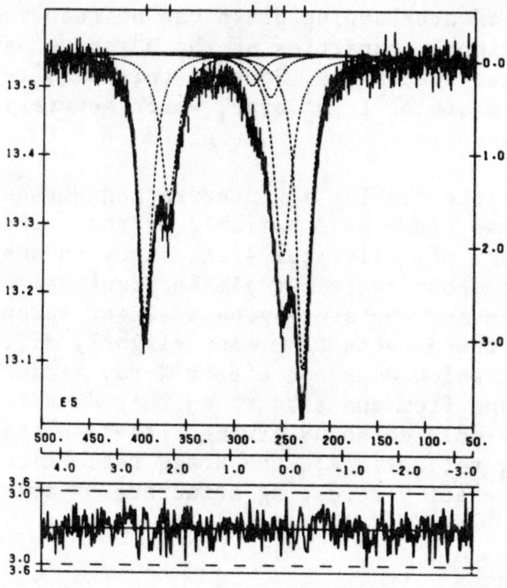

Fig.3. Mössbauer spectrum of holmquistite fitted with six peaks.

(position, breadth and magnitude) of a postulated set of $2n$ Lorentzians where

n = no. of sites containing Fe(II) + no of sites containing Fe(III)

Assignment of particular pairs of absorptions to particular sites usually depends on experience with known structures.

In reasonably simple cases (like pyroxenes) the statistical precision of the occupancy may be quite good ($\sigma \sim 0.02$ or less), and Bancroft (1973) gives comparative X-ray and Mössbauer results for pyroxenes and clino-amphiboles which, surprisingly, agree more closely than this. In more complex cases such as ortho-amphiboles serious difficulties can arise because the parameters of the constituent Lorentzians become highly correlated and the standard deviation of an occupancy parameter can easily rise to about $\sigma = 0.10$ (Burns 1970). In a number of cases Hawthorne (1973) has shown that although a highly satisfactory resolution of a Mössbauer spectrum can be obtained if the areas of the peaks are constrained to fit the X-ray results, a free refinement leads to substantially different results. The high correlations between parameters means that such results are not trustworthy. The problems involved have been analysed in some detail by Dollase (1975).

It has recently been shown by Duncan & Johnston (1974) that the problem of overlapping peaks can be resolved by making use of the directional properties of the Mössbauer absorption in a single crystal specimen. However this requires an oriented crystal plate of 1 cm^2 size, which severely limits its applicability.

One cause of the problem of interpreting Mössbauer spectra is that the peaks have variable widths. This may be because crystallographically equivalent sites in substitutionally disordered compounds are not all in identical environments, and this variation may therefore give a slight spread to the peaks, which sum the effects from many slightly different sites. This is a problem which does not affect X-ray refinements. Although deviations from one site to another do affect X-ray diffraction, they only do so by giving rise to background between the Bragg reflections; the Bragg reflections themselves correspond to an exactly repeating structure of averaged sites.

INFRA-RED METHODS

The fact that crystallographically equivalent positions

are locally different in substitutionally disordered compounds was turned to good account by Burns & Strens (1966), who showed that the stretching frequencies of OH groups coordinated to different cations could be resolved in the infra-red. Each OH in amphibole is coordinated by 3 ions in two M1 sites and an M3 site, and its stretching frequency decreases as the mean charge and electronegativity on these 3 ions increases. The relative intensities of the corresponding absorption bands are then proportional to the statistical probabilities of all possible combinations of three ions from the population of the various ions present in the M1 + M3 sites. The pattern of relative intensities can therefore be related to the occupancies of the M1 + M3 sites taken together if it is assumed that all combinations are equally probably, and that there is no ordering preference between these two sites. Since a system of two ionic species (e.g. Mg and Fe(II)) has one parameter (the ratio of Mg to Fe) and gives four bands from the combinations Mg_3, Mg_2Fe, $MgFe_2$ and Fe_3, the system is overdetermined. It is therefore possible in principle to use departures from the expected ratios of band intensities to give a measure of relative occupancy between the two sites, and to give measures of clustering or anti-clustering (Strens, 1966). For example Fe (II) ions might conceivably tend to cluster together, while high charge ions like Al^{3+} would be likely to avoid one anothers' neighbourhood. This latter type of information is unobtainable from diffraction and Mössbauer methods, and would obviously be extremely valuable. Unfortunately it has been shown by Law (1976) by model calculations that such deductions are extremely unreliable, the variation in intensity ratios due both to partition of the species between M1 and M3 and to clustering being very insensitive to the relevant parameters. It has also been shown by Rowbotham & Farmer (1973) and Law (private communication) that the model is not valid for, and does not even qualitatively explain, the spectra of minerals only a little more complex than those originally studied.

THE MEANING OF OCCUPANCY

With the exception of some infra-red results that must, as we have seen, be considered doubtful, all occupancy measurements give statistical occupancies averaged over a set of crystallographically equivalent sites. They give no information as to the presence or absence of clustering or anti-clustering of like ions in adjacent sites, or of local ordering of one ion into a sub-set of the nominally equivalent sites. If such ordering were more than local it would manifest itself by a lowering of the space-group symmetry or the formation of a super-lattice with a larger cell, and would be detectable by the diffraction methods that have been considered above. However if it occurred on a basis of small domains, or very short range order it could easily go undetected by X-ray diffraction while still causing an appreciable reduction of entropy.

In any situation where there is substitutional disorder of any kind in a crystal the diffraction pattern consists of two parts: the diffraction pattern of the averaged structure, which is confined to the Bragg reflections and whose analysis in terms of occupancy I have discussed; and superimposed on this the diffraction pattern of the difference structure i.e. a distribution of electron density everywhere equal to the difference between the average and the actual structure. Even in the case of complete disorder this will be non-zero, but its diffraction pattern will be very weak everywhere because it will be diffused over all diffraction directions. Only if there is some kind of ordering on subsets of the crystallographic sites, which subsets define a superlattice of the crystal lattice, will the intensity of the difference pattern concentrate into definite directions and be easily detectable and interpretable. However if such a superlattice regularity is confined to 1 or 2 dimensions or to very small 3-dimensional domains then the difference diffraction will be smeared out. It may then be difficult to detect, and even when detected it can be difficult to interpret unambiguously.

Both problems may be eased by using electron diffraction. Electrons are much more strongly scattered than X-rays and give strong diffraction patterns in the electron microscope even from crystals of the order of 1000Å x 1000Å x 100Å. Non-Bragg diffuse reflections may therefore be enhanced in visibility. The problem of interpretation may also be overcome. If the resolution of the microscope is good enough and the diffuse scattering is included in the radiation forming the electron microscope image the actual structure can be imaged instead of the average structure. The best resolution available for this sort of work is only $\sim$ 3Å, so we still cannot resolve atoms and see which ones lie where; but it might well be possible to distinguish unit cells (or parts of cells) containing clustered Fe from others containing clustered Mg for example.

So far as is known such a result has not yet been seen in a silicate. However sub-unit cell structure has been resolved in amphiboles (Hutchison, Irusteta and Whittaker, 1975) which can be associated with the form of the amphibole double chain. This has permitted the direct observation of an intercalated layer of triple chains in some thin crystals. From a compositional point of view such an intercalation corresponds to a small amount of solid solution of talc in amphibole, so that we have in this case a direct observation of the mechanism of a solid solution which occurs by a strictly localised intergrowth that does not affect the average structure.

REFERENCES

Bancroft, G.M., Mössbauer Spectroscopy (book) p. 206. McGraw-Hill. London (1973).

Burns, R.G., Geol. Soc. America, Abstracts with Programs. 2, No. 7, 509-511 (1970).

Burns, R.G., and Strens, R.G.J., Science 153, 890-892 (1966).

Dollase, W.A., Amer. Mineral. 60, 257-264 (1975).

Duncan, J.F., and Johnston, J.H., Australian J. Chem. 27, 249-258 (1974).

Hawthorne, F.C., The Crystal Chemistry of the Clino-Amphiboles Ph.D. Thesis. McMaster University (1973).

Hutchison, J.L., Irusteta, M.C. and Whittaker, E.J.W., Acta Cryst. A31, 794-801 (1975).

Johnston, J.H., and Duncan, J.F., J. App. Cryst 8, 469-472 (1975).

Law, A.D., The Physics & Chemistry of Minerals and Rocks (ed. R.G.J. Strens), Wiley, 677-686 (1976).

Rowbotham, G., and Farmer, V.C., Contr. Min. Pet. 38, 583-592 (1973).

Strens, R.J.G., Chem. Communications 519-520 (1966).

STUDY PROBLEM

The intensity of a reflection from the plane hkl of a centrosymmetric crystal is given by

$$I\,(hkl) \propto F(hkl)^2$$

where $$F(hkl) = \sum_j f_j \cos 2\pi\,(hx_j + hy_j + lz_j).$$

The summation is over all the atoms in the unit cell, whose positions are x_j, y_j, z_j, and scattering power f_j. The scattering power varies with angle of scattering, but at any given angle is to a first approximation proportional to the atomic number of the atom.

Find an approximation for maximum and average changes in intensity of reflections from a pyroxene $FeMgSi_2O_6$, showing no

FeMg ordering for an exchange of 0.01 Mg and Fe atoms between M1 + M2. Express the result as a %age of the intensity of an average reflection at the same angle.

Taking account of the variation of errors in intensity determination due to counting statistics, and the falling of intensity with angle, which of the following classes of reflections would you expect to contain the most reliable information on atomic ordering:

The strongest reflections

Weak medium angle reflections

High angle reflections.

SOLUTION TO PROBLEM

$$I_{hkl} = \Big(\sum_j f_j \cos 2\pi(hx_j + ky_j + lz_j)\Big)^2$$

Put $A_j = 2\pi(hx_j + hy_j + lz_j)$

$$\text{Hence } I = \sum_j f_j^2 \cos^2 A_j + 2 \sum_{j=1}^{n} \sum_{J=j+1}^{n} f_j f_J \cos A_j \cos A_J$$

All the terms are identical before and after the exchange of atoms except those involving j=1 (M1) or j=2 (M2) or both. f_{Mg}=12 and f_{Fe}=26 (in each case multiplied by an angularly dependent factor that is assumed to be the same). Thus before the change f_1=f_2=19, and f_1 increases by 0.14 and f_2 decreases by 0.14. Hence the change in intensity can be shown to be

$$\Delta I = 0.14 \times 38.14 \cos^2 A_1 - 0.14 \times 37.86 \cos^2 A_2 - 2 \times 0.14^2 \cos A_1 \cos A_2$$
$$+ 0.28 (\cos A_1 - \cos A_2) \sum f_J \cos A_J$$

Order of magnitude calculations show that the fourth term is the dominant one, so the others are neglected.

If we approximate by putting $f_J = \overline{f}$, then

$$\Delta I = 0.28 \overline{f} (\cos A_1 - \cos A_2) \sum_{j=3}^{10} \cos A_j$$

Hence

$$|\Delta I| = 0.28\,\bar{f}\,|\cos A_1 - \cos A_2| . |\sum_{j=3}^{10} \cos A_j|$$

and

$$|\Delta I| = 0.28\,\bar{f}\,|\cos A_1 - \cos A_2| . |\sum_{j=3}^{10} \cos A_j|$$

Now if the A terms can be regarded as statistically independent

$$\overline{|\cos A_1 - \cos A_2|} = 2\overline{|\sin \tfrac{1}{2}(A_1 + A_2)|} . \overline{|\sin \tfrac{1}{2}(A_1 - A_2)|}$$

$$= \frac{8}{\pi^2}$$

and

$$\overline{|\sum_3^{10} \cos A_j|} = 8\overline{|\cos A_j|}$$

$$= \frac{16}{\pi}$$

Thus $\overline{|\Delta I|} \sim 11$

The maximum value of I is $(\Sigma f_j)^2 \sim 13000$

and the mean value of I is $\Sigma f_j^2 \sim 1500$

If the largest intensities are assumed to give about 10^6 counts it then follows that the mean intensities will give about 6.10^4 counts, and hence that $\overline{|\Delta I|}$ will be of the order of several times the statistical uncertainty in $\bar{I}$. This indicates that a transfer of 0.01 of an atom between M1 and M2 should be detectable in its effect on the average reflections.

The strongest reflections will have the least %age effect from the interchange, and are often subject to the worst systematic errors. High angle reflections are the most influenced by errors in atomic positions. Thus the weaker medium angle reflections contain the most reliable information on ordering.

THE ACCURACY AND PRECISION OF CALCULATED MINERAL DEHYDRATION EQUILIBRIA

G.M. Anderson

Department of Geology, University of Toronto

INTRODUCTION

The calculation of dehydration equilibria at pressures greater than 2 Kb has only become possible fairly recently with the publication of reliable free energy data for water by Burnham, Holloway and Davis (1969). The converse operation, the extraction of free energy data for minerals from P-T curves, has also received considerable attention. It is now well known that small errors in thermochemical data lead to large errors in calculated mineral equilibria, and that mineral equilibria at high P and T can be used to derive thermochemical data having uncertainties quite comparable to those obtained through calorimetry. In this paper I intend to examine the factors affecting the accuracy and precision of calculated mineral equilibria, and the more general relationship between calorimetry, phase equilibria determinations and thermochemical data. The discussion centres on dehydration equilibria, but the same principles are involved for decarbonation and solid-solid equilibria. Uncertainties introduced by order-disorder, non-stoichiometry, solid solution and so on are not considered.

DEFINITIONS

Accuracy will be used here to mean the degree of agreement between a measured quantity and the "true" value, without regard to the precision attached to the measurement.

Precision will mean the "reproduceability" of that measure-

D. G. Fraser (ed.), Thermodynamics in Geology, 115-136.

ment, that is, some measure of the dispersion of measurements, the mean of which is reported as the measured quantity. There are several ways of reporting the precision.

While these definitions are, as far as I can tell, the ones always used by geoscientists, it may be worth pointing out here that this definition of accuracy is not universally agreed on. Many authors, including some most knowledgeable in the subject (e.g. Eisenhart, 1968, 1969; Murphy, 1969) use bias or systematic error for the deviation of a measurement from the true value, and reserve accuracy to mean an over-all concept including both closeness to the true value and reproduceability or precision. We would not normally call a marksman "accurate" if he splayed his shots all over the target even if after 100 or 1000 shots he showed no tendency to favor any particular area, that is, even if his "average shot" was in the bull's-eye. Similarly, a duck hunter who puts his first barrel behind the duck and his second ahead of the duck has a dubious satisfaction in knowing that "on the average" he would be having duck for supper (Eisenhart, 1969). Yet this is the implication of the commonly accepted definition I have given. I use it because it is simple and widely understood.

Uncertainty is a term I use in a general way to mean some index of precision or precision plus systematic error, in contexts where the exact meaning is either clear or not important. It is an ambiguous term, but can be useful in general discussion in spite of this.

ACCURACY OF CALCULATED EQUILIBRIA

Only two factors are involved in the accuracy of calculated equilibria, i.e. the closeness of the calculated equilibrium temperature at a given pressure to the (accurately) measured temperature or temperature bracket. They are the accuracy of the thermochemical data employed, and the adequacy of the calculation procedure. (We will assume that both the thermochemical and experimental data refer to the same mineral compositions, state of ordering, and so on.)

Effect of Inaccuracies in Thermochemical Data

Calculated equilibria are in general extremely sensitive to inaccuracies in free energy or enthalpy data, and to a much lesser extent to those in entropy data. Molar volumes of minerals are generally very well known and errors in them have a negligible effect. At 2 Kb for example the temperature of the muscovite-sanidine-corundum-water equilibrium is shifted by 5.2°C and the brucite-periclase-water equilibrium by 7.1°C per 100 calories in the free energy of reaction. These values gain perspective when you realize that free energy determinations are often considered good or successful if they are within 500 or even 1000 calories of the true value. (This often represents 0.5% or less of a free energy of formation.)

Effect of Calculation Method

Calculation methods are discussed in this volume by Chatterjee. As he points out, apart from combinations or extrapolations of known equilibria, the calculation of dehydration equilibrium temperature involves primarily a choice as to whether to use heat capacities of minerals or whether to assume that these cancel, i.e. that ΔCp of the solid phases is zero and that the solid phase enthalpy and entropy changes are thus constant. In those cases where the heat capacities of all the minerals involved are known, the choice is clear - their use gives the closest thing to an error-free calculation of a high-temperature free energy of reaction that can be achieved. In cases where some or all of the heat capacities are not known, they can either be estimated or $\Delta Cp = 0$ for the solids can be assumed. Chatterjee (this volume) shows that the latter assumption works very well in some cases. Ulbrich and Merino (1974) show that estimated heat capacities result in errors of from zero to over a kilocalorie in calculating the free energies of various minerals at 700°C, but they do not compare these errors with those one would obtain using $\Delta Cp = 0$. It seems likely that in some cases estimated heat capacities work very well and in other cases an assumption of $\Delta Cp = 0$ would give a smaller error.

In two of the three equilibria used as examples here, heat capacities are available for all phases so the curves were calculated using both approaches, with the results shown in figure 1 and table 1. The thermochemical data and equations used are listed in Appendix 1. We see that at 2 Kb the effect of using heat capacities is to shift the curve position about 23° lower for brucite and about 10° higher for muscovite. Translated into free energies or enthalpies of individual phases which might be extracted from these curves this makes a difference of 300 and 200 calories in the two cases respectively, using the 7.1 and 5.2°C/100 calorie figures mentioned earlier. These effects are well within the normal "limits of error" but they should not be neglected. There is no reason not to use heat capacities if they are available.

PRECISION OF CALCULATED EQUILIBRIA

If we can accurately calculate an equilibrium at P and T, why should we care about its precision? The reason of course is

Table 1

Calculated equilibrium temperatures

P(Kb)	Const. ΔS method	Heat capacity method
	brucite = periclase + water	
1	570	551
2	604	581
4		625
6	688	660
8	716	
10	746	
	muscovite = sanidine + corundum + water	
1	609	616
2	646	656
4		718
6	748	771
8	788	

that even with greatly improved thermodynamic data we will rarely be able to calculate equilibria with 100% accuracy, and as petrology becomes increasingly quantitative and predictive, it will be important to know the envelope of uncertainty around calculated curves. It will be important for example to be able to decide whether two curves fairly close together could with reasonable probability actually be reversed, eliminating the stability field of a mineral. Predictions and deductions about rocks based on experimental data will only be convincing if the uncertainties are known.

If we are calculating the temperature of an equilibrium assemblage given the pressure on the system, there are four main factors which govern the precision attached to that temperature.

1. The precision of the free energy or enthalpy of the mineral phases (free energies are used with the constant ΔS method, enthalpies with the heat capacity method; see equations in Appendix 1). The precision of the entropies is often a significant but smaller factor. Since these appear in a $T\Delta S$ term, they are more significant the higher the temperature.

2. The number of mineral phases in the reaction.

3. The absolute value (not the imprecision) of the entropy of reaction.

4. The degree of correlation or covariance between the free energies or enthalpies of the minerals.

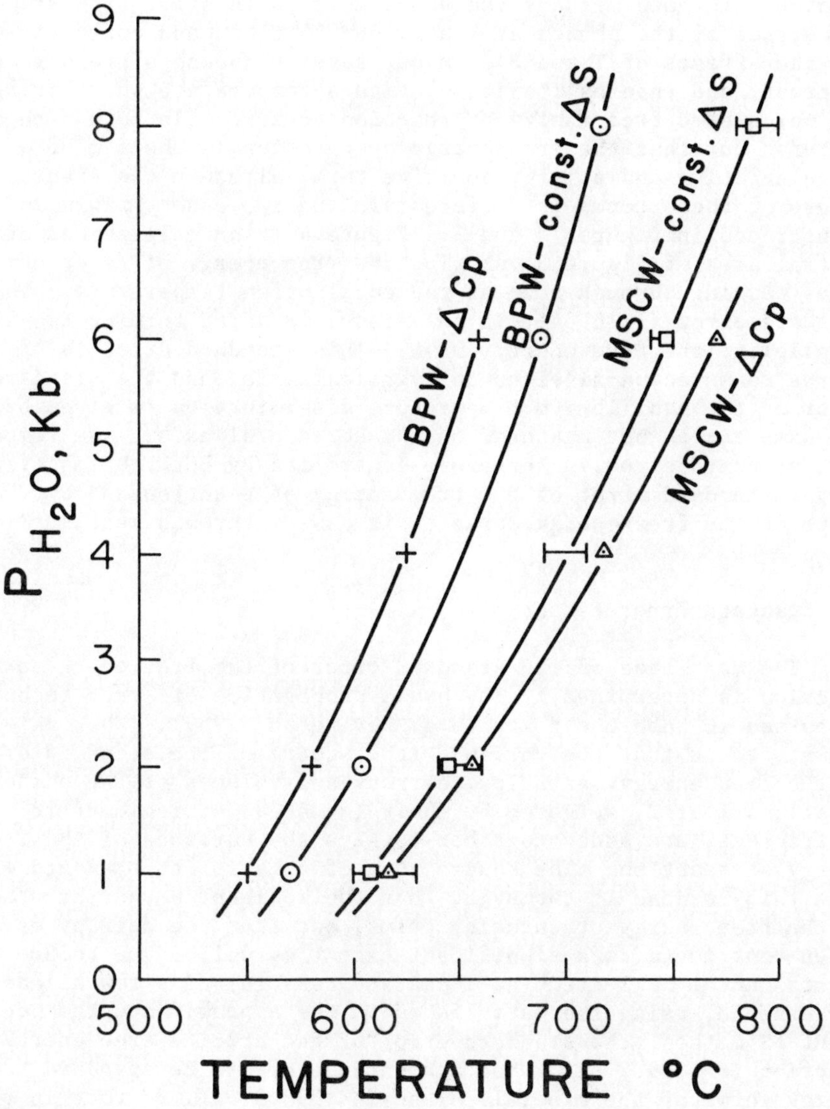

Figure 1. Calculated equilibria using constant ΔS and heat capacity methods. BPW - brucite + periclase + water; MSCW - muscovite + sanidine + corundum + water. Temperature brackets for MSCW equilibrium from Chatterjee and Johannes (1975).

Each of these factors is highly significant, the net result being controlled by their interaction. Before considering them, to let us examine exactly how the equilibrium temperature is calculated. In both methods the basic idea is to start with standard properties of the phases at 1 atm. and 25°C and add corrections for the effects of T and P. In our case we choose a pressure of interest, and then by iteration, find a temperature which brings the calculated free energy of reaction to zero. In both methods it turns out that the free energy of reaction is the sum of a large negative and a large positive term, although the exact nature of these terms is different in the two cases. This is illustrated in figures 2 and 3. Figure 4 is an enlargement of the central area of figure 3, showing the free energy of reaction curve passing through zero at the equilibrium temperature. The standard error (1116 cal. in this case) is added forming two lines parallel to the free energy curve. This standard error is of course measured parallel to the vertical axis, and the standard error of the equilibrium temperature determination is given by the same curves but measured on the horizontal axis. The temperature standard error is therefore controlled by both the magnitude of the standard error of the free energy of reaction and by the slope of the free energy curve as it passes through zero.

The Standard Error Factor

The magnitude of the standard error of the free energy of reaction is determined by the usual propagation-of-error techniques discussed in numerous texts (e.g. Bevington, 1969). What this amounts to is that the variance (the square of the standard error) of the free energy, enthalpy, entropy and volumes of the phases in equations 1 or 2, weighted by their (squared) stoichiometric coefficients are added together to give the variance of the free energy of reaction. The square root of this is its standard error. When this is done it turns out that the dominant errors are those of the free energy or enthalpy terms, but that the entropy errors often contribute to a significant extent as well. The volume errors can be neglected, at least at pressures of a few kilobars. For example, using the data listed in the appendix and the constant ΔS method, the standard error of the brucite free energy of reaction is ±399.2 cal. The standard error of the standard free energy terms of the minerals alone is ±398.8 cal. so that in this case the contribution of the entropy terms is negligible. In the muscovite equilibrium (and using the heat capacity method) the standard error of the enthalpy terms is 960.6 cal. and of the entropy terms 0.611 cal./deg. The net standard error, $(960.6^2 + (656 + 273)^2 \times 0.611^2)^{\frac{1}{2}}$, is 1116 cal. so that in this case the entropy uncertainties are significant. I have not included error terms for the heat capacity integrals or the free energy of water terms. These are relatively small contributors.

Whether the entropy uncertainties are significant or not,

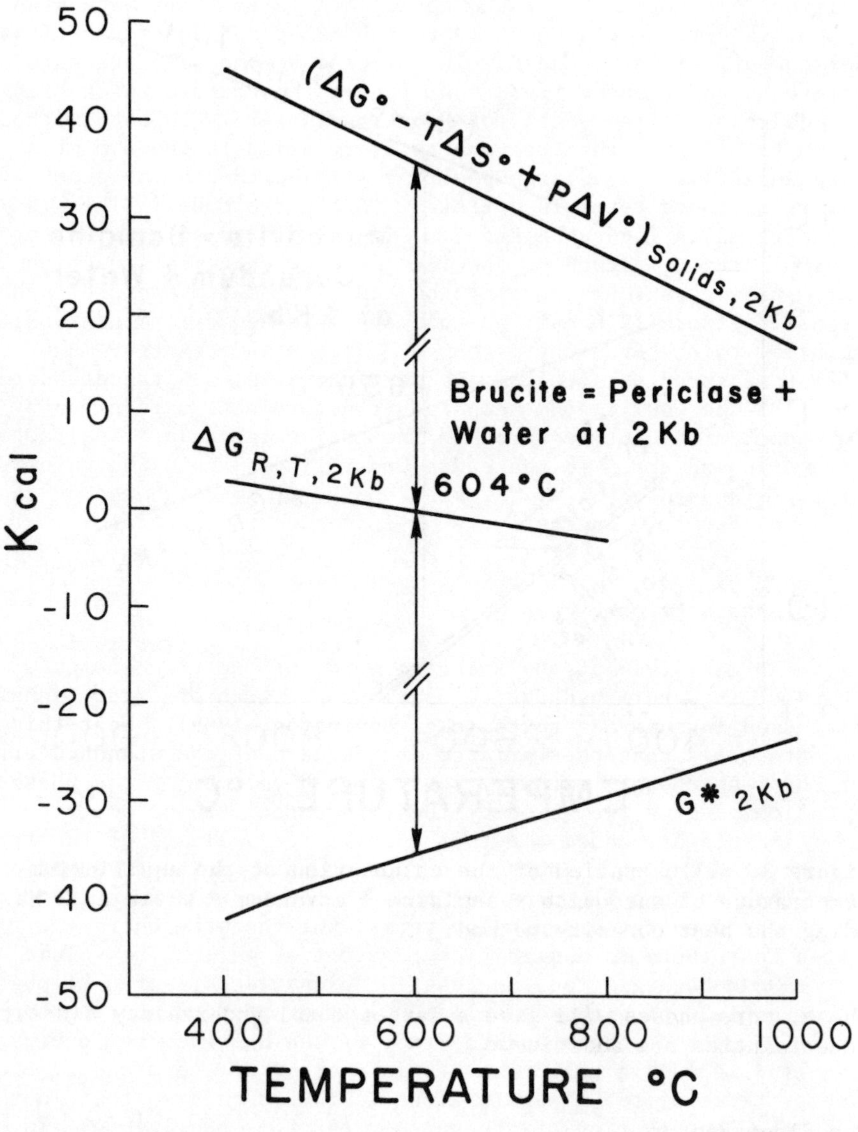

Figure 2. Illustration of the calculation of the equilibrium temperature of brucite + periclase + water at 2 Kb. by the constant ΔS method.

obviously the larger the uncertainties in the individual mineral terms, the larger the uncertainty of the total free energy of reaction. A contributing factor is however the number of phases involved in the reaction. For a given level of uncertainty per

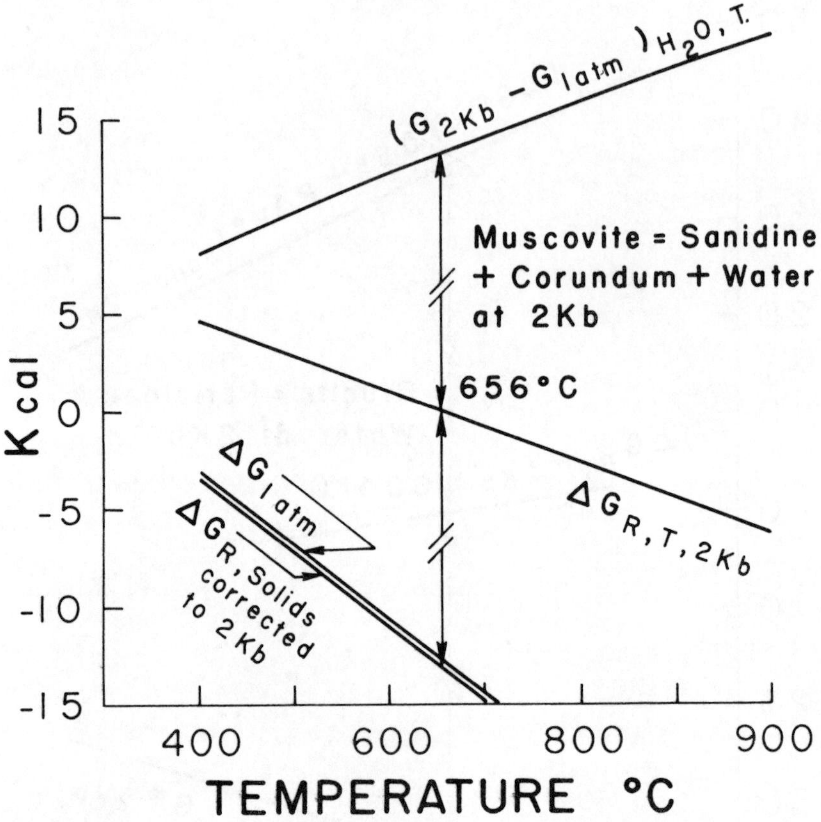

Figure 3. Illustration of the calculation of the equilibrium temperature of muscovite + sanidine + corundum + water at 2 Kb using the heat capacity method.

phase, more phases will give a larger total uncertainty since the uncertainties are additive.

The Slope Factor

The third factor previously mentioned is the slope of the free energy curve as it passes through zero. Figure 4 illustrates the fact that for a given value of the standard error of the free energy term, the standard error of the temperature will be small if the free energy curve has a steep slope and vice versa. The numerical value of this slope is the entropy of reaction. Note that this is the entropy of reaction at P and T, and so cannot be evaluated from one atmosphere data such as those in Robie and

Waldbaum (1968). It is, however, quite easily calculated, for example by first calculating the entropy of reaction at 1 atm. and T from tables and adding to this the slope of $(G_P - G_{1\ atm})$ at T from Burnham et al. (1969) or Helgeson and Kirkham (1974, table 29), where T here refers to the reaction equilibrium temperature at the chosen pressure. Alternatively the complete free energy expression can be evaluated at two temperatures near the equilibrium temperature, preferably above and below it, and the slope calculated from these numbers. The standard error of the

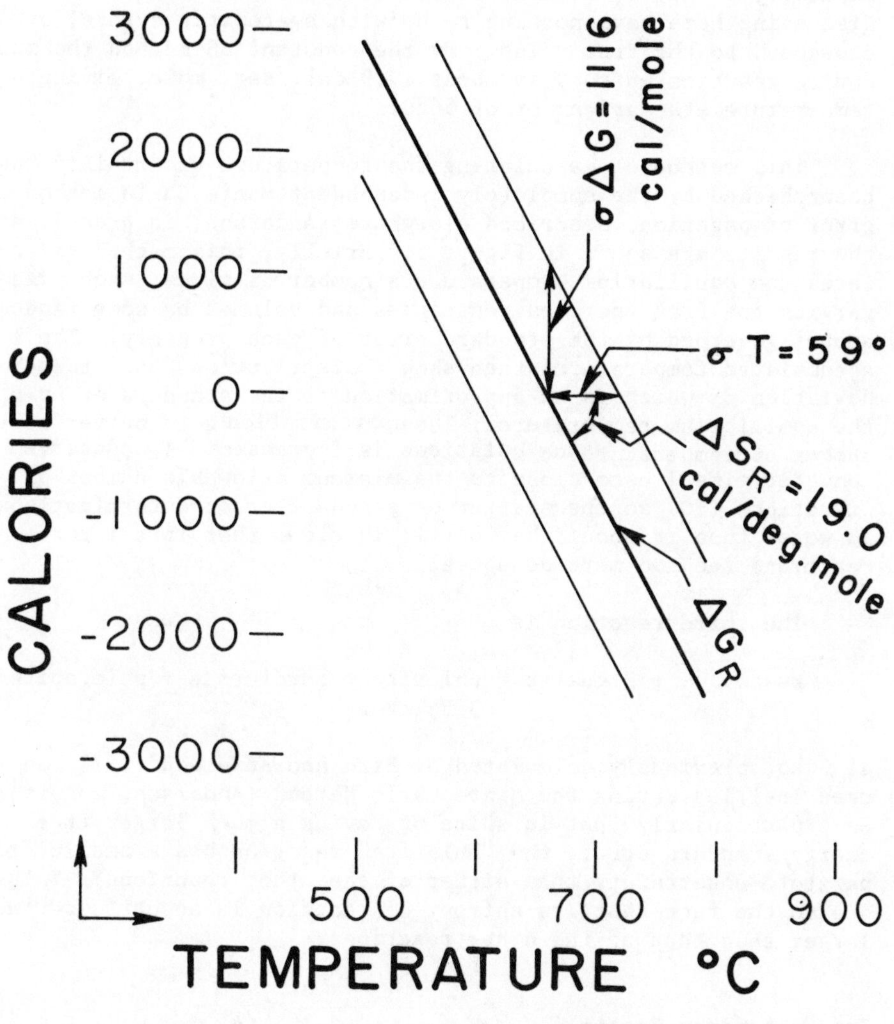

Figure 4. Enlargement of the central portion of figure 3, showing the relationship between free energy standard error, entropy of reaction, and temperature standard error.

equilibrium temperature is then the free energy standard error divided by the reaction entropy or slope. For example, at 2 Kb and 656°C the entropy of the muscovite reaction is about 19.0 cal./deg. mole. With a free energy standard error of 1116 cal., the temperature standard error is then 1116/19.0 or 59°C.

Note that the constant ΔS and the heat capacity methods of calculating equilibrium temperatures will give somewhat different estimates of the entropy of reaction and will therefore have slightly different temperature uncertainties. This simply adds emphasis, if any is needed, to the fact that the errors we are discussing here have nothing to do with systematic errors, or closeness to the true value. By the constant ΔS method the muscovite reaction entropy is about 19.9 cal./deg. mole, giving a temperature standard error of 56°C.

This method of calculating the temperature uncertainty has been checked by the completely independent Monte Carlo method of error propagation, described elsewhere (Anderson, in press), and the results are shown in figure 5. Briefly, this method calculates the equilibrium temperature a number of times, each time varying the free energies, entropies and volumes by some random amount governed by the standard error of each property. The accumulated temperatures then show a distribution, the standard deviation of which is an approximation to the standard error of the equilibrium temperature. The approximation gets better as the number of temperature calculations is increased. To conserve computer time I used close to the minimum allowable number of iterations (30) so the scatter of standard error determinations is worse than it should be, but it is clear that in all three reactions the two methods agree.

The third reaction is

Muscovite + 2 quartz = chlorite + cordierite + phlogopite + 3.5 water

at 5 kb, previously calculated by Bird and Anderson (1973) and used in illustrating the Monte Carlo Method (Anderson, in press). Note particularly that in spite of having a much larger free energy standard error, the "chlorite" reaction has a smaller temperature uncertainty than either of the other reactions. This is due to the fact that its entropy of reaction is about four times larger than that of the other reactions.

The Covariance Factor

The three precision-controlling factors discussed above are sufficient to give the final temperature uncertainty as long as

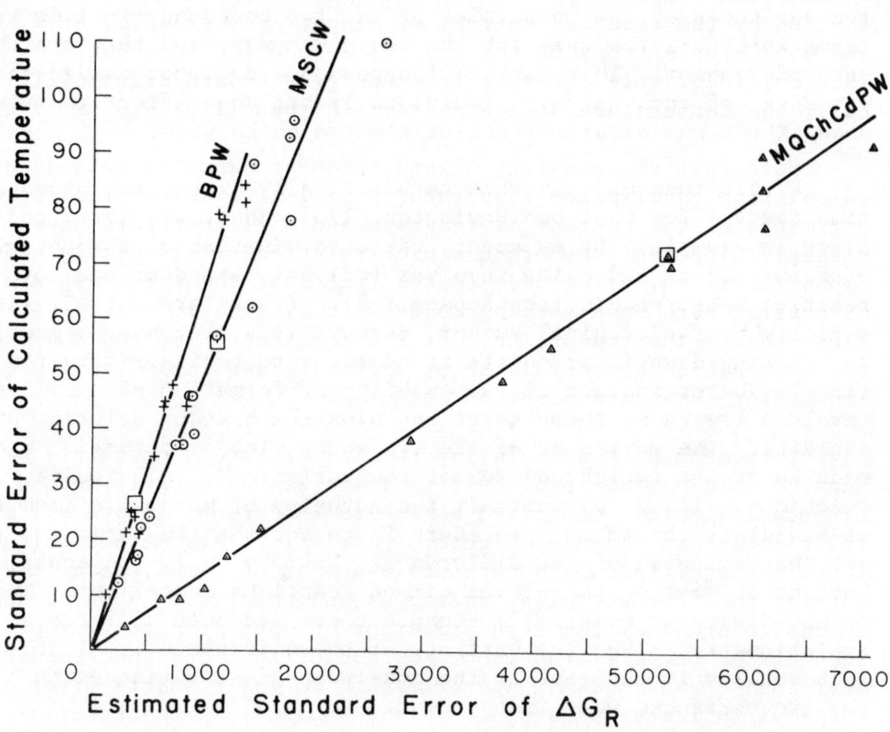

Figure 5. Effect of free energy standard error (in calories) on temperature standard error for the brucite (+), muscovite (o) and chlorite (Δ) equilibria. In each case the large square indicates the temperature standard error calculated from the entropy of reaction and the actual standard error of the free energy of reaction. Other points on each curve were calculated by the Monte Carlo Method using arbitrarily assigned errors of the individual minerals.

the uncertainties of all the thermochemical input data are independent of one another. This is rarely the case.

The determination of the enthalpy of formation of a compound by solution calorimetry involves the measurement of a number of heats of solution or heats of reaction, the number getting larger as the number of elements in the compound increases. The final enthalpy of formation is the algebraic sum of all the individual heats of reaction, and its variance is the sum of the variances of the individual heats.

Two different compounds which contain some of the same elements will require many of the same individual heats of reaction in the determination of their enthalpies of formation. Therefore the variances of the enthalpies of the two compounds include many terms which are the same for the two compounds, and they are thus not independent. This lack of independence does not change the variance of each enthalpy individually but does affect the variance of the sum or difference of the two enthalpies.

Rather than explain this mathematically with error propagation theory (for that see Bevington, 1969, chap. 4), let's consider an example. By my count, the determination of the enthalpy of formation of muscovite involves twelve independent heat of reaction measurements (see Appendix 2). (These are not all listed directly by the original author, Barany, 1964, but must be derived by breaking down intermediate standards into their primary reactions). Determination of the enthalpy of formation of sanidine involves eleven of these reactions plus the heat of solution of sanidine. The variances of the two enthalpies of formation are made up of the (weighted) sum of the variances of all twelve reactions. If we now subtract the enthalpy of muscovite from that of sanidine, the normal procedure is to add the two variances to get the variance of the difference. But obviously the contributions of most of the eleven common reactions cancel out. They do not disappear completely because there are some differences in stoichiometry. Thus the variance of the difference is clearly much smaller in the case of the non-independent variances than if the two variances were independent.

However, if we add the enthalpies of muscovite and sanidine (as we would if they were on the same side of a reaction), the individual heats of reaction do not cancel, and their variances, being correlated, actually contribute to making the variance of the sum of enthalpies greater than it would be if the variances were independent. To sum up with a simplified example:

Let $\Delta H_A = x(\Delta H_1 \pm \sigma_1) + y(\Delta H_2 \pm \sigma_2) + z(\Delta H_3 \pm \sigma_3)$

$\Delta H_B = m(\Delta H_4 \pm \sigma_4) + n(\Delta H_2 \pm \sigma_2) + p(\Delta H_3 \pm \sigma_3)$

then $\sigma^2_{\Delta H_A} \; (=\sigma^2_A) = x^2\sigma^2_1 + y^2\sigma^2_2 + z^2\sigma^2_3$

$\sigma^2_{\Delta H_B} \; (=\sigma^2_B) = m^2\sigma^2_4 + n^2\sigma^2_2 + p^2\sigma^2_3$

and $\sigma^2(\Delta H_A - \Delta H_B) = \sigma^2_A + \sigma^2_B - 2yn\sigma^2_2 - 2zp\sigma^2_3$

$= x^2\sigma^2_1 + m^2\sigma^2_4 + (y^2+n^2-2yn)\sigma^2_2 + (z^2+p^2-2zp)\sigma^2_3$

$= x^2\sigma^2_1 + m^2\sigma^2_4 \quad \text{if } y=n;\ z=p$

$$\sigma^2(\Delta H_A+\Delta H_B) = \sigma_A^2 + \sigma_B^2 + 2yn\sigma_2^2 + 2zp\sigma_3^2$$

$$= x^2\sigma_1^2 + m^2\sigma_4^2 + (y^2+n^2+2yn)\sigma_2^2 + (z^2+p^2+2zp)\sigma_3^2$$

$$= x^2\sigma_1^2 + m^2\sigma_4^2 + 4y^2\sigma_2^2 + 4z^2\sigma_3^2 \qquad \text{if } y=n;\ z=p$$

$$= \sigma_A^2 + \sigma_B^2 + 2y^2\sigma_2^2 + 2z^2\sigma_3^2$$

As shown in Appendix 2, the covariance of muscovite and sanidine enthalpies reduces the overall standard error of the enthalpy of reaction from 960.6 cal. to 278.6 cal., but the standard error of the free energy of reaction is only reduced from 1116 cal. to $(278.6^2 + 929^2 \times 0.611^2)^{1/2}$ or 632.3 cal. and the temperature standard error is reduced from 59°C to 632.3/19.0 or 33°C. These numbers check very well with Monte Carlo simulations. The determination of the enthalpy of formation of corundum was by combustion calorimetry and not solution calorimetry, so its variance is independent of those of the other phases.

In more complex reactions, such as the chlorite reaction, with several phases having common elements on each side of the reaction, determining the covariance will be a tedious business but theoretically very straightforward, as long as sufficient data on the individual calorimetric experiments is available in the literature. Whether the covariance will lead to an increase or a decrease in the temperature standard error is hard to say until the calculation is performed.

THERMOCHEMICAL DATA FROM P-T BRACKETS

In recent years many free energies of formation have been deduced from phase equilibria determinations at elevated pressures and temperatures. This is an attractive procedure because a reaction which has been "reversed" over a temperature range of say 20°C provides a very tight control on the free energy of reaction. Consider again the free energy vs. temperature curve where it passes through zero (figure 6). If we have a temperature bracket A-B where $\Delta T(=T_B - T_A)$ is 20°C, and if the entropy of reaction ΔS_R is 15 cal./deg.mole, then the maximum uncertainty introduced in free energy by the temperature range itself is $\Delta T \Delta S_R$ or 300 calories.

Thus if the corrections from P and T to 1 atm. and 25°C could be carried out without error, and if the free energies of all but one of the reacting phases were perfectly known, then the free

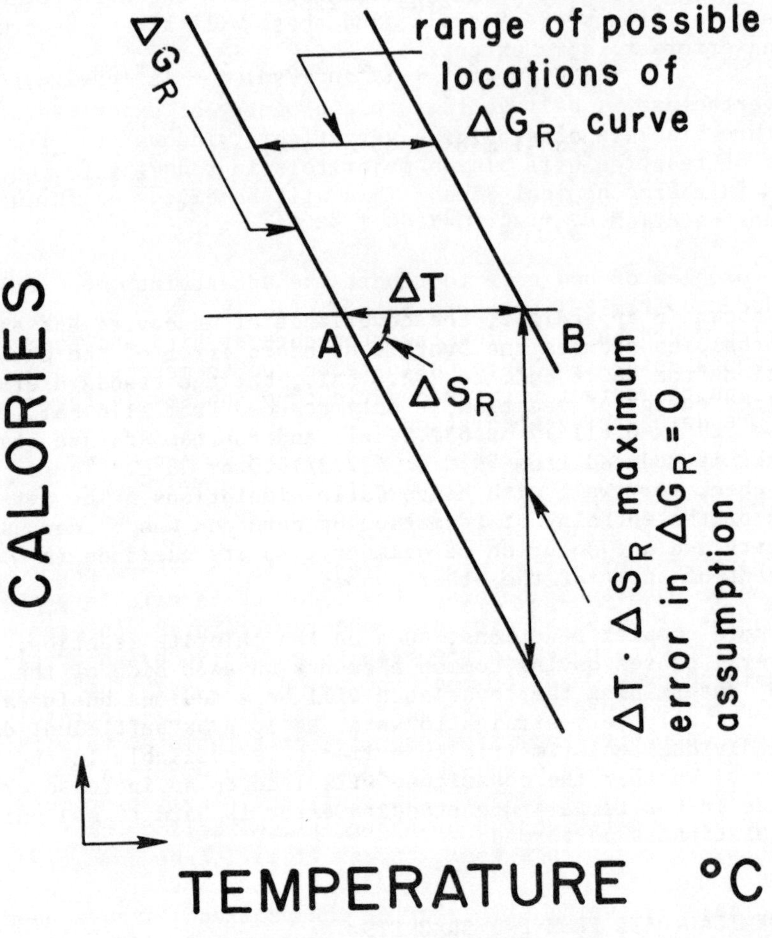

Figure 6. Schematic illustration of the effect of the size of an experimental temperature bracket on the possible error introduced in the calculation of a mineral free energy.

energy of the "unknown" phase could be deduced within a range of 300 calories.

Naturally the reduction to standard conditions cannot be made without error, although if heat capacities are known there is no reason why this source of error should not be negligible. Similarly, the free energies of all the other phases in the reaction cannot be assumed to be perfectly known. Nevertheless this procedure has proved very useful so far in tracking down major inconsistencies in the thermochemical data of groups of minerals. The major limitation of this method as a source of data for individual minerals is of course that it can deal only with the difference in free energy of groups of minerals, and there will always be compensating errors to some extent.

Nevertheless, I believe that this property of experimental temperature brackets of providing very tight "windows" for free energies of reaction will play a major role in reducing the uncertainties in thermochemical data. This will be discussed further in the next section.

The problem of how best to report the uncertainty of a free energy deduced from P-T brackets is a difficult one. My present inclination would be to follow the procedure of Bird and Anderson (1973) though using the Monte Carlo technique to make the procedure simpler, and to correct for the covariance of phases in the reaction before doing this calculation.

DISCUSSION AND CONCLUSIONS

As earth scientists we are interested in the thermochemical properties of substances because they allow us to calculate phase relationships at elevated pressures and temperatures. An examination of the uncertainties in the results generated by uncertainties in the data quickly leads to the conclusion that even the best data are none too good. To calculate phase diagrams accurately it is not good enough to know free energies or enthalpies to within a kilocalorie - we must eventually know them to within a few hundred calories. This is a tall order, given typical calorimetric uncertainties of a few tenths of a percent. I believe that we will eventually reach this goal, though it will take some time.

When major (several kilocalorie) inconsistencies have been ironed out, we will be left with a long process of "fine-tuning" the data so that reasonably complex equilibria can be calculated in a variety of systems to accuracies of 10 or 20 degrees. This process will consist of a continual dialogue between calorimetric determinations and phase equilibrium determinations. Eventually a growing group of "bedrock" thermochemical determinations will be decided upon, meaning values known (or perhaps believed) to 200 calories or better. The reason these values will be believed to that level will probably be that when combined with one another they invariably produce calculated curves which pass through tightly reversed experimental brackets. To calculate a curve which passes through an experimental bracket which limits $\Delta G_{R,T}$ to within 300 calories, $\Delta G^{o}_{R,298}$ must be known to better than 300 calories, to allow for errors introduced by the calculation.

This is undoubtedly the case for the muscovite reaction, as shown in Figure 1, but because of probable compensating errors and the fact that the data for muscovite used in the calculation were derived by Chatterjee and Johannes (1974) from the same P-T brackets, it does not follow that we know the free energies of each of the phases to uncertainties which would add up to a total uncertainty of less than 300 calories.

If, however, each of these minerals could be combined in a variety of reactions with other minerals, and the fit with tightly reversed experimental brackets was always good, then we could conclude that all those minerals were very close to exactly known, no matter what the nominal standard errors attached to them were.

In other words, phase equilibrium determinations will eventually be used to place much more realistic (smaller) standard error estimates on the thermochemical values of minerals. Even at the present time, many thermochemical values are probably very close to their true values, even though they may carry rather large standard errors. There is no necessary connection between the size of a standard error attached to a value and the accuracy of the value. The process of continuous improvement of thermochemical values and more realistic evaluation of uncertainties will keep us all busy for a long time to come.

Finally, since I have not used up all the space allotted, I'd like to make a plea for more attention to be paid to the reporting of uncertainties by earth scientists (in spite of the fact that I see examples of every sin conceivable in this regard in my own work). It's not difficult to see reasons for the generally poor treatment accorded uncertainties in the literature. Apart from the tedious nature of error propagation calculations and covariance problems, the simple fact is that uncertainty estimates have rarely been used by anyone in any meaningful way, so there is no incentive to report uncertainties in an understandable way. Journal editors and referees don't seem to care, so why bother?

It is my contention that as the art and science of calculating geologically interesting equilibria matures, the need for realistic uncertainty estimates will become very clear. The two most essential elements of a report on uncertainty are: 1) a clear separation between an estimate of possible systematic error or bias and an estimate of precision; and 2) a statement as to what index of precision is being used (i.e. range, standard deviation, confidence intervals, etc.).

The most useful publications for anyone interested in this area are two publications of the National Bureau of Standards, Natrella (1963) and Ku (1969). I should also like to mention the excellent article by Frinck and Waggoner (1968), which although not dealing with the present topic, contains the only explicit treatment of covariance in experimental data I have yet come across.

It deals with determining equilibrium constants in aqueous solutions.

ACKNOWLEDGMENT

This research is supported by a grant from the National Research Council of Canada.

APPENDIX 1

Formulae and Data Used in Calculations

The two expressions for evaluating the free energy of reaction at P and T are:

1. The constant-ΔS method:

$$\Delta G_R = \Delta G_S^o - \Delta S_S^o(T-25) + P\Delta V_S^o + G^*$$

where ΔG_R is the free energy of reaction at P and T and is put equal to zero to solve for the equilibrium T at a chosen P.

ΔG_S^o, ΔS_S^o, ΔV_S^o are the free energy, entropy and volume differences between the product and reactant solid phases only, at 25°C, 1 atm. (Note that entropies of formation must be used to be compatible with G^*.)

T is the temperature in degrees celsius

P is the pressure in bars

G^* is from Fisher and Zen (1971) and is evaluated by a polynomial in T in the calculations reported here.

This equation is equivalent to equation (20) in Chatterjee (this volume).

2. The heat capacity method:

$$\Delta G_R = \Delta H_R^o - T\Delta S_R^o + 10519 + \int_{298}^{T} \Delta Cp\,dt - T\left(28.39 + \int_{298}^{T} \frac{\Delta Cp}{T}dt\right)$$

$$+ P\Delta V_S^o + (G_P - G_{1\,atm})_{H_2O,T}$$

where ΔG_R, ΔV_S^o, P are described above

ΔH_R^o, ΔS_R^o are the enthalpy and entropy of reaction (including water) at 25°C, 1 atm. (Third law entropies are preferred, having slightly smaller standard errors than entropies of formation)

T is the temperature in degrees kelvin

ΔCp is the difference in heat capacity between products and reactants

$(G_P - G_{1\ atm})_{H_2O,T}$ is the difference in free energy between water at P,T and water at 1 atm.,T, evaluated by a polynomial in T based on the data in Helgeson and Kirkham (1974, table 29)

10519 is the difference in calories between $(H_T - H_{298})$ of steam and water at T, 1 atm.

28.39 is the difference between the entropies of steam and water at 25°C, 1 atm.

These two numbers are used because the heat capacity coefficients for H_2O refer to steam, whereas water is the reference substance.

This equation is equivalent to equation (1) in Ulbrich and Merino (1974). The heat capacity integrals are evaluated from the integrated form of the Maier-Kelley equation:

$$\int_{298}^{T} \Delta Cp\,dt = a(T-298) + \frac{b}{2}(T^2-298^2) - c(1/T - 1/298)$$

$$\int_{298}^{T} \frac{\Delta Cp}{T}dt = a\ \ln(T/298) + b(T-298) - \frac{c}{2}(1/T^2 - 1/298^2)$$

(298 is an abbreviation of 298.15)

The coefficients a, b and c are from Kelley (1960) and other publications of the U.S. Bureau of Mines, and are listed below:

	a	$bx10^3$	$cx10^{-5}$
microcline S	63.83	12.90	-17.05
corundum C	27.49	2.82	-8.38
muscovite M	97.56	26.38	-25.44
steam W	7.30	2.46	0
brucite B	13.04	15.80	0
periclase P	11.40	0.98	-2.46
(S+C+W-M)	1.06	-8.20	0.01
(P+W-B)	5.66	-12.36	-2.46

Thermochemical data used in calculating curves

* Values from Chatterjee and Johannes (1975); all others from Robie and Waldbaum (1968). Uncertainties are one standard error.

	S^{o}_{298} cal/deg.m.	$\Delta S^{o}_{f,298}$ cal/deg.m.	$\Delta H^{o}_{f,298}$ Kcal/m.	$\Delta G^{o}_{f,298}$ Kcal/m.	V cal/bar
sanidine	53.26* ±0.5	-178.474* ±0.502	-945.686* ±0.465	-892.474* ±0.490	2.6029 ±0.00025
corundum	12.18 ±0.015	-74.854 ±0.0261	-400.400 ±0.15	-378.082 ±0.155	0.61126 ±0.000085
water	16.71 ±0.015	-38.996 ±0.016	-68.315 ±0.005	-56.688 ±0.010	-
muscovite	69.0 ±0.35	-305.474 ±0.354	-1428.488* ±0.827	-1337.411* ±0.650	3.3654 ±0.00035
periclase	6.44 ±0.02	-25.868 ±0.0251	-143.800 ±0.05	-136.087 ±0.055	0.26883 ±0.00007
brucite	15.09 ±0.015	-72.924 ±0.0166	-221.200 ±0.39	-199.458 ±0.395	0.58867 ±0.0008

Data for minerals in the chlorite reaction are given by Bird and Anderson (1973), except that for the error calculations here, one standard error for the entropy of chlorite was taken as 0.5 rather than 5 cal/deg.m.

APPENDIX 2

Calculation of Sanidine-Muscovite Covariance

Independent Calorimetric Measurements:

		$H \pm 2\sigma$ cal.
1.	$2AlCl_3 \cdot 6H_2O = 2Al^{3+} + 6Cl^- + 12H_2O$	36,040± 80
2.	$70.386\ H_2O = 70.386\ H_2O$	59,000±130
3.	$6HCl \cdot 12.731H_2O = 6H^+ + 6Cl^- + 76.386H_2O$	63,300± 80
4.	$2Al + 6HCl \cdot 12.731H_2O = 2AlCl_3 \cdot 6H_2O + 64.386H_2O + 3H_2$	-234,000±260
5.	$H_2 + \frac{1}{2}O_2 = H_2O$	-68,317± 10
6.	$3SiO_2 + 18HF = 3H_2SiF_6 + 6H_2O$	99,870±240
7.	$KCl = K^+ + Cl^-$	1,620± 70
8.	$KAl_3Si_3O_{10}(OH)_2 + 18HF + 10H^+ = K^+ + 3Al^{3+} + 3H_2SiF_6 + 12H_2O$	-219,640±260
9.	$HCl \cdot 12.731H_2O = H^+ + Cl^- + 12.731H_2O$	10,830± 90
10.	$K + \frac{1}{2}Cl_2 = KCl$	-104,180±100
11.	$Si + O_2 = SiO_2$	-217,750±340
12.	$\frac{1}{2}H_2 + \frac{1}{2}Cl_2 + 12.731H_2O = HCl \cdot 12.731H_2O$	38,900± 50
13.	$KAlSi_3O_8 + 18HF + 4H^+ = K^+ + Al^{3+} + 3H_2SiF_6 + 5H_2O$	-146,804±232

Note: Reactions 1-12 are from Barany (1964) or from references there. Reaction 13 is from Waldbaum (1966, table 3-3 and Appendix E).

Subscripts referring to temperature and concentration have been omitted to simplify the reactions. This renders some of them incorrect as written (e.g. no. 2, which involves warming and dilution), but they serve the purpose here.

$$\Delta H^{o}_{f,musc} = 1.5H_1 + 1.63825H_2 - 1.5H_3 + 1.5H_4 + 6H_5 + H_6 + H_7 - H_8 - H_9 + H_{10} + 3H_{11} - H_{12}$$
$$= -1421.225 \pm 1.192 \text{ Kcal}$$

$$\Delta H^{o}_{f,san} = 0.5H_1 + 0.63825H_2 - 0.5H_3 + 0.5H_4 + 2H_5 + H_6 + H_7 - H_9 + H_{10} + 3H_{11} - H_{12} - H_{13}$$
$$= -946.453 \pm 1.098 \text{ Kcal}$$

This heat of formation is slightly different from that in Waldbaum (1968) and Robie and Waldbaum (1968) (-944.378 Kcal) presumably because small corrections for calorimeter temperature and acid strength must be made which are not included here. Now,

$$\Delta H^{o}_{f,san} - \Delta H^{o}_{f,musc} = -H_1 - H_2 + H_3 - H_4 - 4H_5 + H_8 - H_{13}$$
$$= 474772 \pm 470 \text{ cal. } (2\sigma)$$

The result obtained treating these two enthalpies as independent is

$$\Delta H^{o}_{f,san} - \Delta H^{o}_{f,musc} = (-946453 \pm 1098) - (1421225 \pm 1192)$$
$$= 474772 \pm 1620 \text{ cal. } (2\sigma)$$

The difference between $(1620/2)^2$ and $(470/2)^2$ cal. is twice the covariance of the enthalpies of muscovite and sanidine.

Since 470 cal. is the "real" 2σ of the difference in enthalpy between sanidine and muscovite, $(235^2/2)^{\frac{1}{2}}$ or 166 cal. is given to each mineral as its "one σ" value for calculating the total variance of ΔG_T and of the equilibrium temperature. Thus the "real" standard error of ΔH^{o}_{R} is

$$\sigma_{\Delta H^{o}_{R}} = (166^2 + 150^2 + 5^2 + 166^2)^{\frac{1}{2}}$$
$$= 278.6 \text{ cal.}$$

as compared with 960.6 cal. if 827 and 465 are used instead of

166 and 166 for the standard errors of muscovite and sanidine respectively.

REFERENCES

Anderson, G.M. Geochim. Cosmochim. Acta (in press).
Barany, R. U.S. Bur. Mines Rep't Invest. 6356 (1964).
Bevington, P.R. Data Reduction and Error Analysis for the Physical Sciences, McGraw Hill (1969).
Bird, G.W. and Anderson, G.M. Amer. Jour. Sci., 273, 84-91 (1973).
Burnham, C.W., Holloway, J.R. and Davis, N.F. Geol. Soc. Amer. Spec. Paper 132 (1969).
Chatterjee, N. and Johannes, W. Contrib. Mineral. Petrol., 48, 89-114 (1974).
Eisenhart, C. Science, 160, 1201-1204 (1968).
Eisenhart, C. Nat. Bur. Standards Spec. Pub. 300, vol. 1, 21-48 (1969).
Frinck, C.E. and Waggoner, P.E. Bull. Connecticut Agricultural Experiment Station, New Haven, No. 696 (1968).
Helgeson, H.C. and Kirkham, D.H. Amer. Jour. Sci., 274, 1089-1198 (1974).
Kelley, K.K. U.S. Bur. Mines Bull. 584 (1960).
Ku, H.H. (ed.) Nat. Bur. Standards Spec. Pub. 300, vol. 1 (1969).
Murphy, R.B. Nat. Bur. Standards Spec. Pub. 300, vol. 1, 357-360 (1969).
Natrella, M.G. Nat. Bur. Standards Handbook 91 (1963).
Robie, R.A. and Waldbaum, D.R. U.S. Geol. Survey Bull. 1259 (1968).
Ulbrich, H.H. and Merino, E. Amer. Jour. Sci., 274, 510-542 (1974).
Waldbaum, D.R. Unpub. Ph.D. thesis, Harvard Univ. (1966).
Waldbaum, D.R. Contr. Mineral. Petrol., 17, 71-77 (1968).

STUDY PROBLEMS

1. Show that the standard error given for the free energy of formation of muscovite is not consistent with the errors given for the entropy and enthalpy of formation of the same mineral. (This was my error, by the way, not that of the original authors, and does not affect anything in the article). What should the error for the free energy term be?

2. Calculate the heat of formation of muscovite and its standard error from the data in Appendix 2.

3. If $(G_p - G_{1\ atm})_{H_2O,\ 2Kb} = a + bT + cT^2 + dT^3$

where $a = -1874.3$
$b = 28.537$
$c = -0.0107$
$d = 3.1321 \times 10^{-6}$

calculate the equilibrium temperature of the muscovite sanidine corundum water equilibrium at 2Kb.

THERMODYNAMICS OF DEHYDRATION EQUILIBRIA

Niranjan D. Chatterjee

Institute of Mineralogy
Ruhr University
D-4630 Bochum, Germany

INTRODUCTION

Experimental studies of phase equilibria at elevated temperatures and pressures have had a tremendous impact on deciphering petrogenetic processes during the last three decades. It was soon realized, however, that complete understanding of rock-forming processes can hardly be achieved by empirical experiments alone. As a result, thermodynamic analysis of phase equilibria emerged as a promising field of research.

The present paper is tailored to serve the dual purpose of (1) putting together the general thermodynamic theory of heterogeneous equilibria as applied to *dehydration reactions at elevated temperatures and pressures*, and (2) to discuss aspects of *extracting thermodynamic data* of minerals from phase equilibria experiments and conversely, to *calculate phase diagrams* from available thermodynamic data. The state of the art will be discussed for the benefit of those who might seek a lead to get involved in this active field of research. Needless to say, the discussions will reflect the author's experiences, preferences and prejudices. The immediately related problem of consistency and errors of retrieved thermodynamic data or calculated phase relations will not be discussed; this will be the topic of another paper in this volume by G.M.Anderson.

THEORETICAL BACKGROUND

Let us consider a chemical equilibrium among several solids and a fluid, whereby *both* the solids and the fluid are regarded as

D. G. Fraser (ed.), Thermodynamics in Geology, 137-159.

solutions of several chemical species. For such solutions, solid or fluid, the Gibbs energy G of the phase α is given by the relation

$$G^{\alpha} = \sum_i n_i^{\alpha} \bar{G}_i^{\alpha} \qquad (1),$$

where n_i^{α} refers to the mole number of the i th species and $\bar{G}_i^{\alpha}$ indicates the partial molar Gibbs energy (or chemical potential) of the i th species, both in the phase α. $\bar{G}_i^{\alpha}$, in turn, is defined by the relation

$$\bar{G}_i^{\alpha} = (\partial G^{\alpha}/\partial n_i^{\alpha})_{T,P,n_j} \qquad (2),$$

with temperature indicated by T, pressure by P and finally n_j indicating the mole numbers of all other species except n_i.

If a chemical reaction takes place between the various species of the solid phases and the fluid, then a *mass balance* equation may be set up, such that

$$\nu_1 M_1 + \nu_2 M_2 = \nu_3 M_3 + \nu H_2O \qquad (3),$$

where M_1, M_2 etc. indicate the species of the solids, and H_2O that species of the fluid actively participating in the reaction, while ν_1, ν_2 and so on refer to the corresponding stoichiometric coefficients. Under the condition of chemical equilibrium, $\Delta G_T^P = 0$. For the chemical reaction of eq.(3), at the temperature T and pressure P of equilibrium

$$\Delta G_T^P = \sum_i \nu_i \bar{G}_i = 0 \qquad (4).$$

Evaluation of the *energy balance* of such chemical reactions requires specification of standard and reference states. For our purpose, the standard states of the condensed phases will be chosen as the pure solid at 1 bar pressure and a given temperature T (^{o}K), that of the fluid phase as the pure gas species* at a fugacity of 1 bar and stated temperature T (^{o}K). The reference

---

* Note that under isothermal conditions, a pure non-ideal gas at 1 bar fugacity has the same molar Gibbs energy as it would have if it were in an ideal state at 1 bar pressure.

states for the standard molar entropy of formation $\tilde{S}^o_{f,T}$, standard molar enthalpy of formation $\tilde{H}^o_{f,T}$, and standard molar Gibbs energy of formation $\tilde{G}^o_{f,T}$ are the constituent elements in their respective stable forms and standard state at 1 bar and specified temperature T (^{o}K). In these notations, the tilde "~" refers to a molar property, the superscript "o" to the 1 bar standard state, while the subscript "T" indicates the temperature in ^{o}K, and finally, the subscript "f" denotes formation from the stable elements, i.e. the chosen reference state.

Given these specifications, at the temperature T and pressure P of equilibrium, the partial molar Gibbs energy of the i th species of a solid is given by the relation

$$\bar{G}_{M_i} = \tilde{G}^o_{f,T,M_i} + \int_1^P \tilde{V}_{T,M_i}(P)dP + RT\ln a_{M_i} \qquad (5),$$

where $\tilde{V}_{T,M_i}$ indicates the molar volume of the pure substance M_i at a temperature T, and a_{M_i} the activity of the species M_i in the crystalline solution at P and T of interest. Similarly, for the gas species H_2O

$$\bar{G}_{H_2O} = \tilde{G}^o_{f,T,H_2O} + RT\ln f^*_{H_2O}/f^o_{H_2O} + RT\ln a_{H_2O} \qquad (6),$$

with $f^*_{H_2O}$ and $f^o_{H_2O}$ referring respectively to the fugacity of *pure* H_2O at a given P and T and at the standard state, and a_{H_2O} indicating the activity of the species H_2O in the fluid at P and T. Recalling that at standard state the pure gas has a fugacity of 1 bar, the eq.(6) reduces to

$$\bar{G}_{H_2O} = \tilde{G}^o_{f,T,H_2O} + RT\ln(f^*_{H_2O}.a_{H_2O}) \qquad (7).$$

At P and T of interest, $a_{H_2O} \equiv f_{H_2O}/f^*_{H_2O}$, i.e. the activity is the ratio of fugacity of the species H_2O in the *gas mixture*, f_{H_2O}, to its fugacity in the *pure* state, $f^*_{H_2O}$, so that

$$f_{H_2O} = f^*_{H_2O}.a_{H_2O} \qquad (8).$$

Substituting eq.(8) in (7), we recover the general relation

$$\bar{G}_{H_2O} = \tilde{G}^o_{f,T,H_2O} + RT\ln f_{H_2O} \qquad (9).$$

For the chemical reaction under consideration (eq.3), therefore,

$$\Delta G_T^P = \Delta G_T^o + \int_1^P \Delta V_{T,s}(P)dP + RT\ln\Pi_i a_i^{\nu_i} + RT\ln f_{H_2O}^{\nu} \quad (10),$$

with $\Delta V_{T,s}$ implying the volume difference due to the *solids only*, at a temperature T. The last two terms in eq.(10) are identical to RTln*K*, where *K* stands for equilibrium constant involving activities and fugacities at P and T; thus eq.(10) reduces to

$$\Delta G_T^P = \Delta G_T^o + \int_1^P \Delta V_{T,s}(P)dP + RT\ln K \quad (11).$$

At equilibrium $\Delta G_T^P = 0$, so that

$$\Delta G_T^o = - \int_1^P \Delta V_{T,s}(P)dP - RT\ln K \quad (12).^*$$

Eq.(12) is a basic relation of general validity, and has been commonly used in geological research. Some of its uses will be illustrated in the following.

If not too large a range of pressure and temperature is considered, a safe assumption is that $(\partial\Delta V_s/\partial T)_P \simeq 0 \simeq (\partial\Delta V_s/\partial P)_T$. Therefore, the $\int_1^P \Delta V_{T,s}(P)dP$ term in eq.(12) may be approximated by $\Delta V^o_{298,s}(P-1)$.** Introducing this approximation, eq.(12) may be rearranged to

$$RT\ln K + \Delta V^o_{298,s}(P-1) = - \Delta G_T^o = - \Delta H_T^o + T\Delta S_T^o \quad (13).$$

Dividing eq.(13) by RT -

$$\ln K + \Delta V^o_{298,s}(P-1)/RT = - \Delta H_T^o/RT + \Delta S_T^o/R \quad (14).$$

---

* Note that only when the standard state of the solids is chosen at a pressure P, does the eq.(12) reduce to its more familiar form $\Delta G_T^o = - RT\ln K$.

** The subscript 298 is an abbreviation for 298.15°K, and will be used throughout this paper.

Therefore, a plot of $\ln K + \Delta V^{o}_{298,s}(P-1)/RT$ vs $T^{-1}(^{o}K)$ yields a curve with a slope equivalent to $-\Delta H^{o}_{T}/R$, so that ΔH^{o}_{T} of a solid-gas reaction may be evaluated as a function of T. For most equilibria involving *pure* solids and *pure* H_2O, such plots turn out to be straight lines, implying temperature-independence of ΔH^{o} of reaction. Now, $(\partial \Delta H^{o}_{T}/\partial T)_P = \Delta C^{o}_{p} = T(\partial \Delta S^{o}_{T}/\partial T)_P$, and as such, temperature-independent ΔH^{o} means $\Delta C^{o}_{p} = 0$, and consequently, ΔS^{o} is also independent of temperature. The $\ln K + \Delta V^{o}_{298,s}(P-1)/RT$ vs $T^{-1}(^{o}K)$ plot thus provides a means of obtaining ΔH^{o} and ΔS^{o} of reactions following eq.(14); some geological examples are given by Eugster and Wones (1962), Orville and Greenwood (1965) and Wones (1967).

An alternative way of obtaining ΔH^{o} and ΔS^{o} of solid-gas equilibria is by applying the eq.(13). In this case, $RT\ln K + \Delta V^{o}_{298,s}(P-1)$ is plotted as a function of T (^{o}K). Within the range of experimental control, again a straight line results, the slope of which gives the temperature-independent value of ΔS^{o}. Substituting ΔS^{o} in eq.(13), $-\Delta H^{o}$ of the reaction is obtained. This method of extracting ΔH^{o} and ΔS^{o} of solid-gas reactions has been advocated by Weisbrod (1968) and Froese (1976).

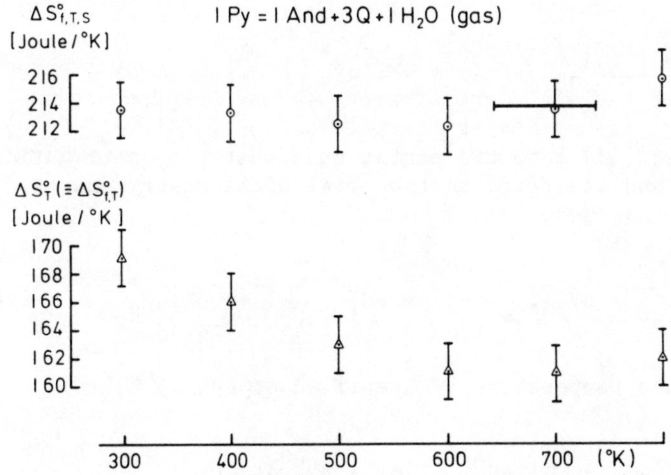

Figure 1. A plot of ΔS^{o}_{T} and $\Delta S^{o}_{f,T,s}$ of the reaction 1 pyrophyllite = 1 andalusite + 3 quartz + 1 H_2O gas as a function of temperature. Source of data indicated in the text. The uncertainties (thin vertical bars) are only those due to $\tilde{S}^{o}_{298}$ of pyrophyllite. The thick horizontal bar indicates the range of temperature experimentally explored by Haas and Holdaway (1973) and the retrieved value of $\Delta S^{o}_{f,298,s}$ obtained by the curve-fitting method (see later).

The methods of calculation detailed above take advantage of two simplification: (i) $\Delta V_s^o \neq f(P,T)$ and (ii) $\Delta C_p^o = 0$. In absence of data on the equations of state of geologically interesting solids, i.e $V = V(P,T)$, and owing to the fact that within a restricted range of P and T $(\partial \Delta V_s/\partial T)_P \simeq 0 \simeq (\partial \Delta V_s/\partial P)_T$, the former approximation will be tolerated. But how about $\Delta C_p^o = 0$? Figure 1 illustrates a plot of ΔS_T^o of a simple reaction among pure solids and pure H_2O as a function of T. It is seen that ΔS_T^o does vary significantly with T, so that $\Delta C_p^o = 0$ would appear to be a rather poor approximation. By contrast, $\Delta S_{f,s}^o$, the difference of the entropies of *formation* of the solids, turns out to be a remarkably stable quantity. The same holds true in many other cases, so that if anything, $\Delta S_{f,s}^o \neq f(T)$ would be a more reasonable approximation (cf. Fisher and Žen,1971; Chatterjee,1975). Therefore, it makes sense to recast the partinent relations in term of $\Delta S_{f,s}^o$, rather than using ΔS^o, the entropy difference due to the whole reaction. In the following, we shall go yet another step forward to see if some approximation of this kind is at all necessary. In doing so, for the sake of simplicity we shall restrict ourselves to equilibria among pure solids and pure H_2O.

For equilibria involving *pure* solids and *pure* H_2O, we may then rewrite the eq.(11) as follows -

$$\Delta G_T^P = \Delta G_T^o + \Delta V_{298,s}^o(P-1) + \nu RT\ln f_{H_2O}^* \qquad (15).$$

Splitting up eq.(15) into two parts, such that the contributions of the *solids* and the *fluid* to the total Gibbs energy change can be separately assessed,

$$\Delta G_T^P = \Delta G_{T,s}^o + \Delta V_{298,s}^o(P-1) + \nu\tilde{G}_{f,T,H_2O}^o + \nu RT\ln f_{H_2O}^* \qquad (16).$$

If the reference temperature is specified at 298.15^oK,

$$\Delta G_T^P = \Delta G_{298,s}^o - \int_{298}^{T} \Delta S_{f,s}^o(T)dT + \Delta V_{298,s}^o(P-1) + \nu\tilde{G}_{f,T,H_2O}^o + \nu RT\ln f_{H_2O}^* \qquad (17).$$

As has been suggested above, if $\Delta S_{f,s}^o$ were indeed temperature-independent, the eq.(17) could be easily evaluated. However, pending a demonstration of this, $\Delta S_{f,s}^o$ has to be regarded as a function of temperature, rather than as a constant. To evaluate the variation of $\Delta S_{f,s}^o$ as a function of temperature, we introduce the relation

$$\Delta S^{o}_{f,T,s} = \Delta S^{o}_{f,298,s} + \int_{298}^{T} \frac{\Delta C^{o}_{pf,s}}{T}\, dT \qquad (18),$$

where $\Delta C^{o}_{pf,s}$ indicates the difference of standard heat capacities of *formation* of the solids; a quantity obtainable from $\tilde{C}^{o}_{p}$ data of the solids and the constituent elements. Substituting eq.(18) in eq.(17) -

$$\Delta G^{P}_{T} = \Delta G^{o}_{298,s} - \int_{298}^{T} (\Delta S^{o}_{f,298,s} + \int_{298}^{T} \frac{\Delta C^{o}_{pf,s}}{T}\, dT)\, dT$$

$$+ \Delta V^{o}_{298,s}(P-1) + \nu \tilde{G}^{o}_{f,T,H_2O} + \nu RT\ln f^{*}_{H_2O} \qquad (19).$$

The relation (19) can be evaluated explicitly only when $\tilde{C}^{o}_{p}$ data of all the solids and the constituent elements are available. This is not always the case. In absence of such data, an attempt may be made to *estimate* $\tilde{C}^{o}_{p}$ of the relevant phases and to obtain $\Delta C^{o}_{pf,s}$ from these estimates. Clearly, this approach will be justified only if very accurate estimates are possible.

Fortunately, as demonstrated in Figure 1, the variation of $\Delta S^{o}_{f,s}$ with temperature often does not exceed the *range of uncertainties* of $\Delta S^{o}_{f,s}$ (cf.Chatterjee,1975,Fig.2). In such cases, $\Delta S^{o}_{f,s} \neq f(T)$ or $\Delta C^{o}_{pf,s} = 0$ is an adequate approximation. Where this is a valid assumption, eq.(19) would reduce to

$$\Delta G^{P}_{T} = \Delta G^{o}_{298,s} - \Delta S^{o}_{f,298,s}(T-298.15) + \Delta V^{o}_{298,s}(P-1)$$

$$+ \nu \tilde{G}^{o}_{f,T,H_2O} + \nu RT\ln f^{*}_{H_2O} \qquad (20).$$

This relation is identical to that recommended by Fisher and Zen (1971) and Zen (1969,1972) for extracting thermodynamic data from phase equilibrium studies (Note that $\tilde{G}^{o}_{f,T,H_2O} + RT\ln f^{*}_{H_2O}$ of eq. (20) is identical to $G^{*}_{H_2O}(T,P)$ as defined by Fisher and Zen(1971). And finally, substituting the relation $\Delta H^{o}_{298,s} - 298.15\, \Delta S^{o}_{f,298,s}$ for $\Delta G^{o}_{298,s}$ in eq.(20), we recover

$$\Delta G^{P}_{T} = \Delta H^{o}_{298,s} - T\Delta S^{o}_{f,298,s} + \Delta V^{o}_{298,s}(P-1)$$

$$+ \nu \tilde{G}^{o}_{f,T,H_2O} + \nu RT\ln f^{*}_{H_2O} \qquad (21).$$

The eq.(21) has been used by the author in several cases (e.g. Chatterjee (1970,1975).

RETRIEVAL OF THERMODYNAMIC DATA

Calorimetric, spectroscopic and electrochemical measurements are the classical sources of thermodynamic data. In the present section it will be demonstrated that thermodynamic data of *comparable quality* can be extracted from the results of phase equilibria experiments at elevated pressures and temperatures. The *prerequisite* for data extraction are that

(i) phase equilibria experiments be undertaken with well characterized phases of known compositions, preferably synthetic end members;

(ii) the equilibrium be demonstrated by reaction reversal;

(iii) narrow P-T brackets be obtained over as large a range of temperature as possible. For retrieval of thermodynamic data it is immaterial whether or not the reaction investigated is a stable one.

Suppose P-T data are available for an equilibrium involving *pure* solids and *pure* H_2O of the type

$$\nu_1 M_1 + \nu_2 M_2 = \nu_3 M_3 + \nu H_2O \qquad (3).$$

If the thermodynamic data of H_2O and all but one condensed phases are available, then those of the *"unknown"* phase may be calculated. Depending on whether or not $\Delta C^o_{pf,s}$ data are available as a function of temperature, the eq.(19) on the one hand and eq.(20) or (21) on the other would apply.

Retrieval of thermodynamic data has been commonly achieved by the so-called *point-by-point method* (e.g. Robie and Stout,1963; Fisher and Zen,1971; Zen,1969,1972). In this method, $\tilde{S}^o_{f,298}$ of *all* the condensed phases have to be known, enabling one to solve for $\tilde{G}^o_{f,298}$ of the unknown phase, setting the relation (20) equals zero. If on the other hand $\Delta C^o_{pf,s}$ data are also available, the eq.(19) is used instead. For the purpose of computations, $f^*_{H_2O}$ data are obtained from the tabulation by Burnham et al.(1969), while $\tilde{G}^o_{f,T,H_2O}$ are taken from Robie and Waldbaum (1968) *adjusted* to the 1 bar standard state following Fisher and Zen (1971,p.301). *Typically*, the two extreme P-T points of each bracket are used to obtain two values of $\tilde{G}^o_{f,298}$, yielding in all twice as many different $\tilde{G}^o_{f,298}$ as there are brackets; the average of these values is the preferred $\tilde{G}^o_{f,298}$ of the unknown phase.

An alternative method of data extraction - dubbed the *curve-fitting method* - is based on eq.(21). In this case the term

$\Delta V^o_{298,s}(P-1) + \nu G^o_{f,T,H_2O} + \nu RT\ln f^*_{H_2O}$ is similarly computed for the two extreme points of each of the brackets. These are then plotted as a function of T (°K) and a straight line is fitted through these points.* The slope of this straight line is equivalent to $\Delta S^o_{f,298,s}$, and its intercept at 0°K gives $-\Delta H^o_{298,s}$; whence $\tilde{S}^o_{f,298}$ and $\tilde{H}^o_{f,298}$ of the unknown phase is *simultaneously* obtained. Therefore, contrary to the point-by-point method, *no prior knowledge* of $\tilde{S}^o_{f,298}$ of the unknown phase is necessary for the application of the curve-fitting method. The merits and limitations of these methods will be discussed below with the help of a few examples culled from the literature. The thermodynamic data used as input in the subsequent calculations are reproduced in Table 1.

Let us start by considering the P_{H_2O} - T equilibrium data of the reaction

$$\underset{\text{pyrophyllite}}{1\ Al_2Si_4O_{10}(OH)_2} = \underset{\text{andalusite}}{1\ Al_2SiO_5} + \underset{\text{quartz}}{3\ SiO_2} + \underset{\text{steam}}{1\ H_2O} \qquad (22),$$

as given by Haas and Holdaway (1973). Pyrophyllite will be treated as the unknown phase here, for which $\tilde{H}^o_{f,298}$ and hence, $\tilde{G}^o_{f,298}$ are sought. Fortunately, the standard molar entropy $\tilde{S}^o_{298}$(King and Weller) and the $\tilde{C}^o_p$ data to 773°K (Kuskov,1973) of pyrophyllite are available. Furthermore, similar data are also available for quartz and andalusite (Robie and Waldbaum,1968). From these $\Delta S^o_{f,s}$ and ΔS^o of the reaction (22) have been computed to 800°K and plotted as a function of T (°K) in Figure 1. It is seen that the change of $\Delta S^o_{f,s}$ with T remains *within the range of uncertainty* (although ΔS^o does change significantly); therefore, the eqs.(20) or (21) are applicable. Figure 2 shows the results of evaluation of this equilibrium, both by the curve-fitting (Fig.2a) and the point-by-point (Fig.2b) methods, assuming in both the cases $\Delta S^o_{f,s}$ is not a function of T. By contrast, Figure 2c displays the same data, using $\Delta S^o_{f,s}$ as a function of T(cf.Fig.1). Three things emerge out of these results. First, the retrieved thermodynamic data of pyrophyllite are very similar, no matter which method of calculation

---
*

The curve-fitting method has traditionally made use of linear regression technique. For reasons discussed in Chayes (1968,p.360), least squares technique would appear to be *formally* inappropriate; therefore, linear regression might be replaced by the method of linear discriminant functions for curve-fitting. However, our own calculations demonstrate that the *results* of curve-fitting do not differ materially whichever method is used, if *sufficiently narrow* reversal brackets are available.

Table 1. Standard thermodynamic data of the condensed phases

Phase	Composition	$H^o_{f,298.15}$ (KJ/mol)	$S^o_{f,298.15}$ (J/°K.mol)	$G^o_{f,298.15}$ (KJ/mol)	$V^o_{298.15}$ (cm^3)	$S^o_{298.15}$ (J/°K.mol)
Andalusite	$Al_2O(SiO_4)$	- 2591.528[a)] (± 2.971)	- 494.758	- 2444.017[a)] (± 3.012)	51.530[a)] (± 0.04)	93.22[a)] (± 0.4)
Muscovite (Al/Si *disordered*)	$KAl_2(AlSi_3O_{10})(OH)_2$	- 5976.800[b)] (± 5.200)	- 1260.329*	- 5601.033	140.808[c)] (± 0.029)	306.47[b)] (± 0.6)
Muscovite (Al/Si *ordered*)			- 1279.029†			287.77[g)] (± 0.6)
Pyrophyllite	$Al_2Si_4O_{10}(OH)_2$		- 1255.719		127.610[d)] (± 0.06)	236.81[e)] (± 2.09)
Quartz	SiO_2	- 910.648[a)] (± 1.674)	- 182.489	- 856.239[a)] (± 1.715)	22.688[a)] (± 0.001)	41.34[a)] (± 0.08)
Sanidine	$K(AlSi_3O_8)$	- 3959.590[b)] (± 4.000)	- 736.675	- 3739.950	108.929[f)] (± 0.038)	232.90[f)] (± 0.4)

References: a) Robie and Waldbaum (1968); b) Robie (personal communication, July, 1976); c) Chatterjee and Johannes (1974); d) Taylor and Bell (1970); e) King and Weller (1970); f) Openshaw et al. (1976); g) see text.

* Includes an ideal molar configurational entropy term amounting to 18.70 J/°K.mol (see text)

† Configurational entropy due to Al/Si disordering in muscovite structure *ignored* ("Al/Si ordered")

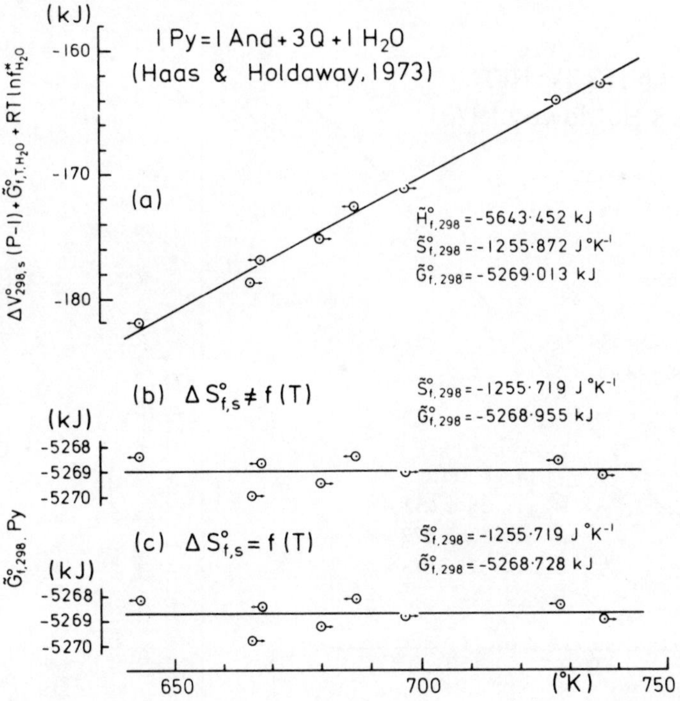

Figure 2. Retrieval of thermodynamic data of pyrophyllite from experimental phase equilibria study of the reaction (22) by Haas and Holdaway (1973). The data points correspond to the extremes of the reversal brackets, the arrows indicating the direction of the reaction. (a) curve-fitting technique(eq.21); (b) point-by-point method, assuming $\Delta S^o_{f,s} \neq$ (T) (eq.20); and (c) point-by-point calculation taking into account $\Delta S^o_{f,s}$ as a function of temperature (eq.19).

is used. Second, the retrieved value of $\tilde{S}^o_{f,298}$ for pyrophyllite obtained by the curve-fitting method agrees excellently with calorimetric data (cf. Table 1). Third, whether or not $\Delta S^o_{f,s}$ is considered as a function of T, the results of data extraction do not differ much.

Figure 3 shows the calculated equilibrium curve for the reaction (22), based on the retrieved thermodynamic data of pyrophyllite. As long as the proper combination of input data and the method of calculation is retained, identical curves result. Only when these are interchanged, do the resulting curves differ somewhat. For example, if the retrieved data obtained in Figure 2c is used as input, but $\Delta S^o_{f,s}$ is assumed independent of T, the calcu-

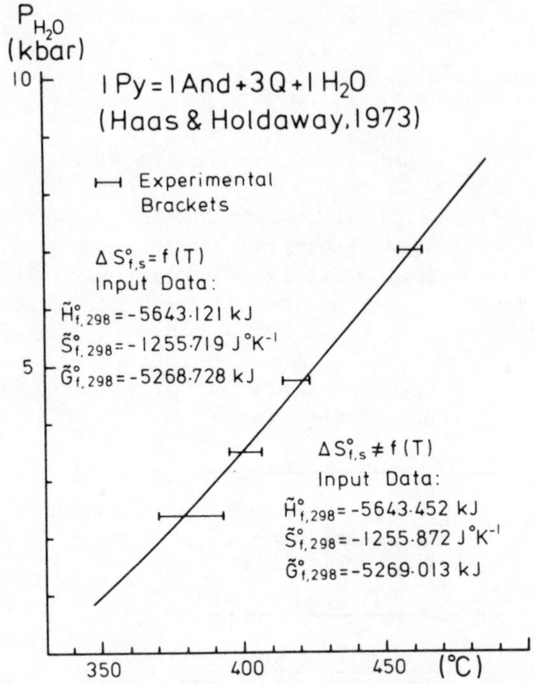

Figure 3. Equilibrium curve of the reaction 1 pyrophyllite = 1 andalusite + 3 quartz + 1 H_2O, *recalculated* on the basis of the displayed input data. Note, that both sets of data coupled with the appropriate method of calculation lead to the identical curve.

lated curve shifts by three degrees to lower temperature. This observation reconfirms our earlier conclusion that if $\Delta S^o_{f,s}$ remains within the range of uncertainty, it may be safely handled as a constant. Furthermore, if $\Delta S^o_{f,s}$ happens to be independent of temperature, evaluations by both curve-fitting and point-by-point methods yield identical results, as they should. The basic difference between the two methods is that the point-by-point method is dependent on a prior knowledge of $\tilde{S}^o_{f,298}$ of the unknown phase, while the curve-fitting method is not. If no calorimetric data on $\tilde{S}^o_{f,298}$ of the unknown phase is available, point-by-point calculations might be attempted on the basis of *estimated* $\tilde{S}^o_{f,298}$. As pointed out earlier, use of such estimated values would be meaningful only if very accurate estimates are possible. It is noteworthy that previous estimates of $\tilde{S}^o_{298}$ and $\tilde{C}^o_p(T)$ for pyrophyllite were off by as much as 10-15% (against later measured values for these quantities), so that Kuskov (1973,p.406) cautions against uncritical use of estimated data.

For another example of data retrieval, let us consider the reaction

$$KAl_2(AlSi_3O_{10})(OH)_2 + SiO_2 = KAlSi_3O_8 + Al_2SiO_5 + H_2O \quad (23)$$

muscovite quartz sanidine andalusite steam

based on experimental phase equilibria work by Chatterjee and Johannes (1974). Thermodynamic data of all the phases involved in eq.(23) are known (Table 1), including their high temperature $\tilde{C}_p^o$ data. The only exception to this is sanidine, whose high temperature $\tilde{C}_p^o$ measurement is currently under way (R.A.Robie, personal comm., July,1976). Therefore, the following calculations had to be done under the assumption $\Delta S_{f,s}^o$ is independent of temperature.

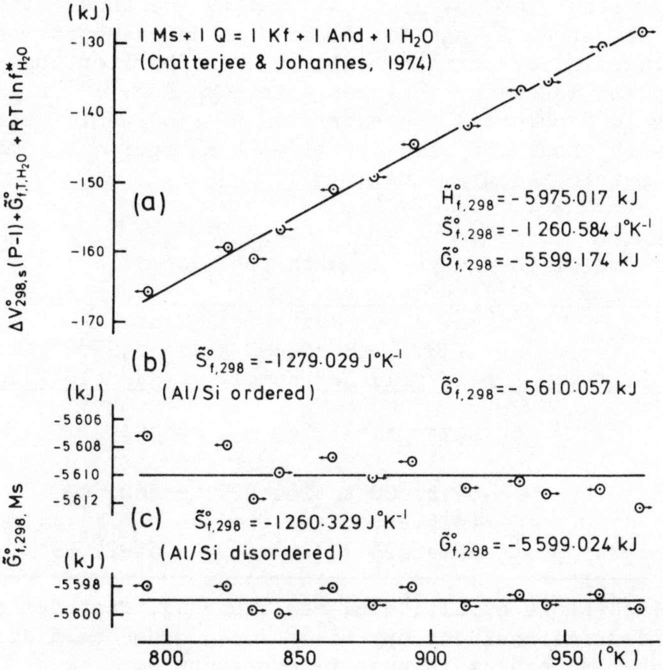

Figure 4. Retrieval of thermodynamic data of muscovite from experimental phase equilibria study of the reaction (23) by Chatterjee and Johannes (1974). The significance of the data points are analogous to those in Figure 2. (a) curve-fitting technique; (b) point-by-point technique, disregarding configurational entropy due to Al/Si disordering in muscovite; and (c) point-by-point calculation, taking into account the effect of Al/Si disordering on the $\tilde{S}_{298}^o$ of muscovite.

In going through this exercise, muscovite will be treated as the unknown phase and the results of data extraction will be compared to those obtained by calorimetry. Figure 4 displays the results. The curve-fitting method yields $\tilde{S}^o_{f,298}$ for muscovite equalling -1260.584 J/°K (Fig.4a), disagreeing sharply with the calorimetric value of -1279.029 J/°K (uncorrected for Al/Si disordering in the muscovite structure: labeled "Al/Si ordered" muscovite in Table 1). This disagreement also shows up in the trend of $\tilde{G}^o_{f,298}$ values for muscovite obtained via point-by-point method, using an $\tilde{S}^o_{f,298}$ values of -1279.029 J/°K as input (Fig.4b). It is well known, however, that residual configurational entropy due to Al/Si disordering escapes detection by cryogenic calorimetry. To account for this, an ideal molar configurational entropy of 18.70 J/°K ($\tilde{S}^o_{cfg} = -4R(0.75\ln 0.75 + 0.25\ln 0.25) = 18.70$ J/°K) may be added to the calorimetric value, giving a *preferred* $\tilde{S}^o_{f,298}$ of -1260.329 J/°K for muscovite (labeled "Al/Si disordered" in Table 1), agreeing excellently with -1260.584 J/°K obtained by the curve-fitting method. If the preferred $\tilde{S}^o_{f,298}$ for Al/Si disordered muscovite is used as input in point-by-point calculation, the earlier observed trend in $\tilde{G}^o_{f,298}$ for muscovite vanishes entirely (Fig.4c, "Al/Si disordered). Table 2 compares the retrieved thermodynamic data for muscovite with those obtained by Robie et al (person. comm., 1976) by calorimetric measurements.

Table 2. Comparison of thermodynamic data for muscovite

Method	$H^o_{f,298}$ (KJ/mol)	$S^o_{f,298}$ (J/°K.mol)	$G^o_{f,298}$ (KJ/mol)
Curve-fitting	- 5975.017	- 1260.584	- 5599.174
Calorimetry	- 5976.800 (± 5.200)	- 1260.329* (± 0.6)	- 5601.033
Point-by-point	- 5974.791	- 1260.329†	- 5599.029

* Preferred value (Robie et al.,1976, see Table 1), includes an ideal molar configurational entropy of 18.70 J/°K due to disordering of tetrahedral Al and Si in muscovite structure.

† Used as input, to calculate $\tilde{G}^o_{f,298}$ by the point-by-point method.

Out of the above discussions, two points appear worthy of note. First, the extracted thermodynamic data (Table 2) for muscovite are comparable to those obtained by calorimetric methods. This fact lends credence to suitably extracted thermodynamic data of other phases. Second, the curve-fitting method - where applicable - yields reliable results without prior knowledge of $\tilde{S}^o_{f,298}$

of the unknown phase. Furthermore, this method also provides a correct information on the magnitude of the configurational entropy, if any. By contrast, based on a previous knowledge (or estimation) of $\tilde{S}^o_{f,298}$, the point-by-point method calculates the relatively insensitive function $\tilde{G}^o_{f,298}$ only.

In summary, then, the following points should be taken into account for retrieval of thermodynamic data:

1) The equilibrium curve used to extract thermodynamic data should be established by unequivocal reaction reversal with well characterized starting materials.

2) Care should be taken to verify how far the *mass balance* holds all along the equilibrium curve; this may not be true if phases with variable H_2O contents (e.g. zeolites, cordierite etc) or else solid solutions are involved. A case in point is the anomalous behaviour of the phase wairakite (cf. Chatterjee,1976).

3) Where high-temperature $\tilde{S}^o_{f,T}$ of all the phases have been *measured,* and to the extent that $\Delta S^o_{f,T,s}$ varies with T *beyond* the range of their uncertainties, $\Delta S^o_{f,s}$ has to be considered explicitly as a function of temperature and the calculation executed by the *point-by-point* method. Lack of $\tilde{S}^o_{f,T}$ data has been occasionally remedied by estimated ones. In absence of criteria to judge just how reliable such estimates are, it is recommended to use the approximation $\Delta S^o_{f,s} \neq f(T)$ in such cases.

4) Whenever $\Delta S^o_{f,s} \neq f(T)$ is an adequate approximation, the *curve-fitting* technique of data retrieval should be preferred, because it is not dependent on a previous knowledge of $\tilde{S}^o_{f,298}$ of the unknown phase. Much rather, it provides an independent check on $\tilde{S}^o_{f,298}$ of the phase in question, including it configurational entropy, if any. The essential limitation of this method is reached, when the experimental equilibrium data span only a short range of temperature. In such cases, unless the experimental control on the brackets is superb, no meaningful value of $\tilde{S}^o_{f,298}$ can be obtained. The point-by-point method may be taken recourse to in such cases, provided reliable $\tilde{S}^o_{f,298}$ data for the unkown phase is available.

5) Following extraction of thermodynamic data, it is a good practice to make sure that these are indeed consistent with the experimental phase equilibrium data from which they were obtained. This is best done by recalculating the equilibrium curve, using the retrieved thermodynamic data as input. Trivial though it may sound, a routine check would seem to be necessary in view of the fact that there are many cases of retrieved thermodynamic data in the literature not fulfilling this test of consistency. An example of such inconsistency is illustrated later (see Study Problems).

CALCULATION OF PHASE DIAGRAMS

Phase diagrams of geologic interest may be obtained by various method; these will be discussed in the following.

Linear Addition of ln*K* or ΔG^o_T

The fundamental basis of this procedure is derived by rearranging eq. (14) to

$$\ln K = -\Delta H^o_T/RT + \Delta S^o_T/R - \Delta V^o_{298,s}(P-1)/RT \qquad (24).$$

H, S, and V being state properties, they are additive; as such, ln*K*'s are also additive. Therefore, if an unknown reaction (x) can be expressed in terms of two (or more) reactions (y) and (z) whose ln*K*'s are known, then $\ln K_x$ can be obtained from the relation $\ln K_x = \nu_y \ln K_y + \nu_z \ln K_z$, where ν_y and ν_z are proper coefficients. This method has been immensely popular in geology, one recent example being its application to derive serpentine stability relations (Evans et al.,1976).

An analogous procedure is based on the relation

$$\Delta G^o_T = -RT\ln K - \Delta V^o_{298,s}(P-1) \qquad (13).$$

G is additive, because it is also a state function. Therefore, $\Delta G^o_{T,x}$ of an unknown reaction (x) may be derived by linearly combining $\Delta G^o_{T,y}$ and $\Delta G^o_{T,z}$ of the known reaction (y) and (z)(e.g. Froese,1976).

Thermodynamic Extrapolation of Equilibrium

For reasons of experimental or instrumental limitations, phase equilibria measurements are usually possible only over restricted ranges of pressure and temperature. If the equilibrium has been achieved at one point P_1T_1, it can be extrapolated to another point P_2T_2 by evaluation of a relation of the type -

$$\Delta G^{P_2}_{T_2} = 0 = -\int_{T_1}^{T_2} \Delta S_{f,s} dT + \int_{P_1}^{P_2} \Delta V_s dP + (G_{H_2O})^{P_2}_{T_2} - (G_{H_2O})^{P_1}_{T_1} \qquad (25).$$

In integrating this relation, both $\Delta S_{f,s}$ and ΔV_s are considered independent of P and T. An example of this type of study is the derivation of a self-consistent phase diagram for the system $CaO\text{-}Al_2O_3\text{-}SiO_2\text{-}H_2O$ (Wall and Essene,1972).

Direct Calculation from Thermodynamic Data

When a set of internally consistent thermodynamic data is available, phase diagrams can be computed by iterative or graphical solution of the eqs.(19) through (21), depending upon whether or not $\Delta C^o_{Pf,s}$ has to be considered explicitly. An example of the graphical evaluation of the relation (17), which is equivalent to eq.(19), is indicated in Fig.5. At equilibrium $\Delta G^P_T = 0$; therefore, from eq.(17) it follows that

$$\Delta G^o_{298,s} - \int_{298}^{T} \Delta S^o_{f,s}(T)dT + \Delta V^o_{298,s}(P-1) + \nu\tilde{G}^o_{f,T,H_2O} = -\nu RT\ln f^*_{H_2O} \quad (26).$$

Both sides of the eq.(26) has been numerically solved at 460°C and plotted in Figure 5 as a function of pressure; the point of intersection at 6950 bars indicates the equilibrium P_{H_2O} of the reaction (22) at 460°C.

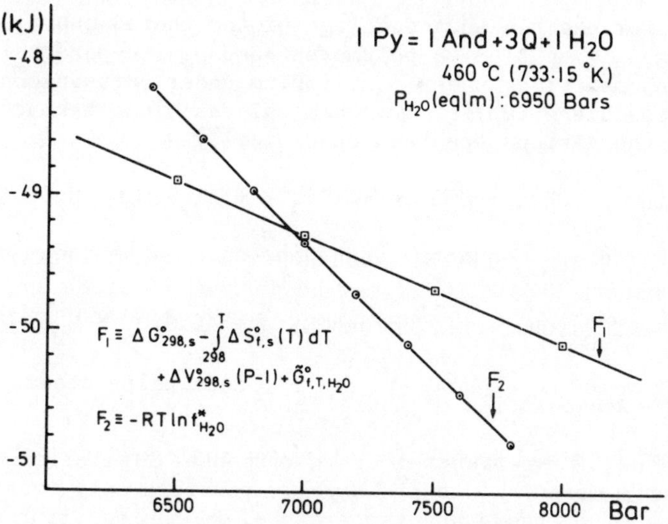

Figure 5. Calculation of the equilibrium P_{H_2O} of the reaction 1 pyrophyllite = 1 andalusite + 3 quartz + 1 H_2O (eq. 22) at 460°C by graphical solution of the eq.(26). Further explanation in the text.

An extensive example of the evaluation of the eq.(21) is provided by the derivation of the phase relations of margarite in the system CaO-Al_2O_3-SiO_2-H_2O (Chatterjee,1976). In this case, the relation (21) was further extended to take into account the effect of variable compositions in the fluid phase expressed as the activity of H_2O, a_{H_2O}.

Commonly, each of the above methods of calculation ignores the uncertainties of experimental phase equilibria or of thermodynamic data. Nevertheless, geologically meaningful results have been obtained (e.g. Chatterjee,1976; Evans et al.,1976), implying that satisfactory calculations can be made despite apparently large uncertainties in the input data. As Anderson (1976) points out, this might be an outcome of the fact that the standard errors of the input data are not necessarily independent, but rather, they are highly correlated.

DEHYDRATION EQUILIBRIA INVOLVING CRYSTALLINE SOLUTIONS

The basic principles used above to calculate phase diagrams among pure solids and H_2O can be extended to include crystalline solutions.

Consider for example an assemblage of the phases quartz + alkali feldspar$_{ss}$ + an Al_2SiO_5 polymorph + muscovite-paragonite$_{ss}$ + *pure* H_2O gas coexisting under equilibrium under certain conditions. This assemblage implies *two* chemical reactions taking place between the various species. Specifically,

$$\underset{\text{muscovite}}{KAl_2(AlSi_3O_{10})(OH)_2} + \underset{\text{quartz}}{SiO_2} = \underset{\text{sanidine}}{KAlSi_3O_8} + \underset{\text{andalusite}}{Al_2SiO_5} + \underset{\text{steam}}{H_2O} \quad (27),$$

$$\underset{\text{paragonite}}{NaAl_2(AlSi_3O_{10})(OH)_2} + \underset{\text{quartz}}{SiO_2} = \underset{\text{albite}}{NaAlSi_3O_8} + \underset{\text{andalusite}}{Al_2SiO_5} + \underset{\text{steam}}{H_2O} \quad (28).$$

At equilibrium, we shall have *two* relations analogous to eq.(4) -

$$\sum_i \nu_i^{(27)} \bar{G}_i^{(27)} = 0 \quad (29),$$

$$\sum_i \nu_i^{(28)} \bar{G}_i^{(28)} = 0 \quad (30).$$

The equilibrium constants of these reactions can be expressed as

$$\ln K_{(27)} = \ln a_{Kf}^{fd} - \ln a_{Ms}^{mi} + \ln f_{H_2O}^{*} \quad (31),$$

$$\ln K_{(28)} = \ln a_{Ab}^{fd} - \ln a_{Pg}^{mi} + \ln f_{H_2O}^{*} \quad (32),$$

with a_{Kf}^{fd} referring to the activity of the species $KAlSi_3O_8$ in feldspar crystalline solution, a_{Pg}^{mi} to the activity of the species $NaAl_2(AlSi_3O_{10})(OH)_2$ in mica crystalline solution etc. Substituting eq.(31) and (32) into eq.(13) -

$$\Delta G^{o}_{T(27)} = - RT(\ln a^{fd}_{Kf} - \ln a^{mi}_{Ms} + \ln f^{*}_{H_2O}) - \Delta V^{o}_{298,s(27)}(P-1) \qquad (33),$$

$$\Delta G^{o}_{T(28)} = - RT(\ln a^{fd}_{Ab} - \ln a^{mi}_{Pg} + \ln f^{*}_{H_2O}) - \Delta V^{o}_{298,s(28)}(P-1) \qquad (34).$$

Splitting up the activities, a_i into mole fractions X_i and activity coefficients γ_i, and recalling that for the binary feldspar and mica crystalline solutions $X^{fd}_{Ab} = (1-X^{fd}_{Kf})$ and $X^{mi}_{Pg} = (1-X^{mi}_{Ms})$, we derive

$$\Delta G^{o}_{T(27)} = - RT\,\{\ln X^{fd}_{Kf} + \ln \gamma^{fd}_{Kf} - \ln X^{mi}_{Ms} - \ln \gamma^{mi}_{Ms}\} - RT\ln f^{*}_{H_2O} - \Delta V^{o}_{298,s(27)}(P-1) \qquad (35),$$

$$\Delta G^{o}_{T(28)} = - RT\,\{\ln(1-X^{fd}_{Kf}) + \ln \gamma^{fd}_{Ab} - \ln(1-X^{mi}_{Ms}) - \ln \gamma^{mi}_{Pg}\} - RT\ln f^{*}_{H_2O} - \Delta V^{o}_{298,s(28)}(P-1) \qquad (36).$$

All the relevant data in eqs.(35) and (36), *except* the activity coefficients, are known from phase equilibria experiments on decomposition of pure muscovite and pure paragonite in the presence of quartz. Furthermore, by assuming a solution model, the pertinent activity coefficients can be derived for a given pressure and temperature as functions of compositions from the experimentally established binary solvii. Substituting these activity coefficients in eqs.(35) and (36), we are left with only two unknowns, X^{fd}_{Kf} and X^{mi}_{Ms}. The eqs.(35) and (36) can be solved *simultaneously* by iteration at any given P and T for the two unknowns, X^{fd}_{Kf} and X^{mi}_{Ms}. Examples of such simultaneous solutions and the resulting phase diagrams are given by Chatterjee and Froese (1975).

ACKNOWLEDGMENTS

The author wishes to thank D.G.Fraser for the invitation to give this review lecture; R.A.Robie for providing some valuable informations and data prior to publication; H.Halbach for assisting with some of the computations; K.Abraham, E.Froese, W.Maresch, R.C.Newton and D.R.Wones for commenting on a previous version of the manuscript. Thanks are also due to the Deutsche Forschungsgemeinschaft, Bonn-Bad Godesberg, for supporting this study by the grant Ch 46/5.

REFERENCES

Anderson, G.M. Geochim.Cosmochim.Acta (in press)(1976)
Burnham,C.W., Holloway,J.R. and Davis,N.F. Geolog.Soc.Am., Spec.Paper 132, 96p. (1969)
Chatterjee,N.D. Contrib.Mineral. and Petrol., 27, 244-257(1970)
Chatterjee,N.D. Fortschr. Mineral., 52, 47-60 (1975)
Chatterjee,N.D. Am. Mineralogist, 61, 699-709 (1976)
Chatterjee,N.D. and Johannes, W. Contrib. Mineral. and Petrol., 48, 89-114 (1974)
Chatterjee,N.D. and Froese, E. Am.Mineralogist, 60, 985-993, (1975)
Chayes, F. Am.Mineralogist, 53, 359-371 (1968)
Eugster,H.P. and Wones,D.R. J.Petrology, 3, 82-125 (1962)
Evans, B.W., Johannes, W., Oterdoom,H. and Trommsdorff, V. Schweiz. Mineral. Petrogr. Mittl., 56, 79-93 (1976)
Fisher, J.R. and Zen, E-an. Am. J. Sci., 270, 297-314 (1971)
Froese, E. Geolog. Survey Canada Paper 75-43, 37p. (1976)
Haas, H. and Holdaway, M.J. Am. J. Sci. 273, 449-464 (1973)
King, E.G. and Weller, W.W. U.S.Bur.Mines, Rep.Inv., 7369, 6p. (1970)
Kuskov, O.L. Geochem. Internatl., 10, 406-412 (1973)
Openshaw,R.E., Hemingway, B.S., Robie, R.A., Waldbaum, D.R. and Krupka, K.M. U.S.Geolog. Survey J. Research, 4, 195-204 (1976)
Orville P.M. and Greenwood, H.J. Am. J. Sci. 263, 678-683(1965)
Robie, R.A. and Stout, J.W. J.Phys.Chem. 67, 2252-2256 (1963)
Robie, R.A. and Waldbaum, D.R. U.S.Geolog.Survey Bull., 1259, 256 p. (1968)
Taylor, L.A. and Bell, P.M. Carnegie Inst. Washington Year Book 69, 193-194 (1970)
Wall, V. J. and Essene, E. J. Geolog. Soc. Am. Abstr. Progr., 700 (1972)
Weisbrod, A. Bull. Soc. Fr. Minéral. Cristallogr., 91, 444-452 (1968)
Wones, D.R. Geochim. Cosmochim. Acta 31, 2248-2253 (1967)
Zen, E-an Am. Mineralogist 54, 1592-1606 (1969)
Zen, E-an Am. Mineralogist 57, 524-553 (1972)

STUDY PROBLEM: DEHYDRATION EQUILIBRIA

	Corundum	Diaspore
$H^o_{f,298}$(KJ/mol)	- 1675.274	
$S^o_{f,298}$(J/^{o}K.mol)	- 313.189	- 263.341
V^o_{298} (cm^3)	25.575	17.760
$G^o_{f,298}$(KJ/mol)	- 1581.895	

For the univariant equilibrium curve of the reaction

$$2\ AlOOH = 1\ Al_2O_3 + 1\ H_2O$$

diaspore corundum steam

Haas (Am. Mineralogist 57, 1375-1385, 1972) gave the following experimental P_{H_2O} - T data:

P_{H_2O} (bars)	T^{o}C brackets
1750	389 - 408
2400	397 - 411
3500	412 - 426
4800	419 - 436
7000	455 - 468

Utilizing the thermodynamic and experimental data given above,

(1) derive the standard thermodynamic data for diaspore, using both the curve-fitting technique (eq.21) as well as the point-by-point method (eq.20);

(2) Calculate the P_{H_2O} - T curve of the above equilibrium, using both the sets of retrieved thermodynamic data;

(3) Compare your calculated curve with the original experimental brackets and comment on your reslts.

ANSWERS TO THE STUDY PROBLEMS: CHATTERJEE (DEHYDRATION EQUILIBRIA)

(1) The results of derivation of standard thermodynamic data for diaspore are displayed in Figure A. The *curve-fitting method* yielded $H^o_{f,298}$ = - 1002.629 KJ/mol, $S^o_{f,298}$ = - 269.231 J/°K.mol and $G^o_{f,298}$ = - 922.358 KJ/mol. By contrast, the *point-by-point calculation,* using the calorimetrically derived value of $S^o_{f,298}$ of - 263.341 J/°K.mol, gave an average $G^o_{f,298}$ = - 920.018 KJ/mol and $H^o_{f,298}$ = - 998.533 KJ/mol.

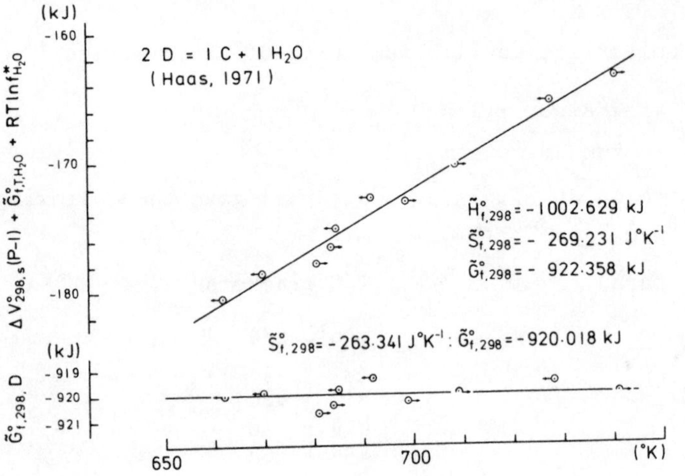

Figure A. Derivation of the standard thermodynamic data for diaspore based on the experimental phase equilibrium study of the reaction 2 diaspore = 1 corundum + 1 H_2O by Haas (1972). Curve-fitting method above; point-by-point method below.

(2) *Recalculation* of the equilibrium curve of the reaction 2 diaspore = 1 corundum + 1 H_2O, using both the sets of *retrieved* data for diaspore in conjunction with those known for corundum and H_2O, lead to the results indicated in Figure B.

(3) In comparing the results, the following points appear worthy of note:

(i) The derived $\tilde{S}^o_{f,298}$ value for diaspore obtained by the curve-fitting method does not check very well with that known from calorimetric work. This disparity also shows up in the observed trend (Fig. A) of the individual $\tilde{G}^o_{f,298}$ values for diaspore in the point-by-point calculation. As a direct consequence of this, the $\tilde{H}^o_{f,298}$ and $\tilde{G}^o_{f,298}$ values of diaspore retrieved by the curve-fitting and the point-by-point methods also differ mutually.

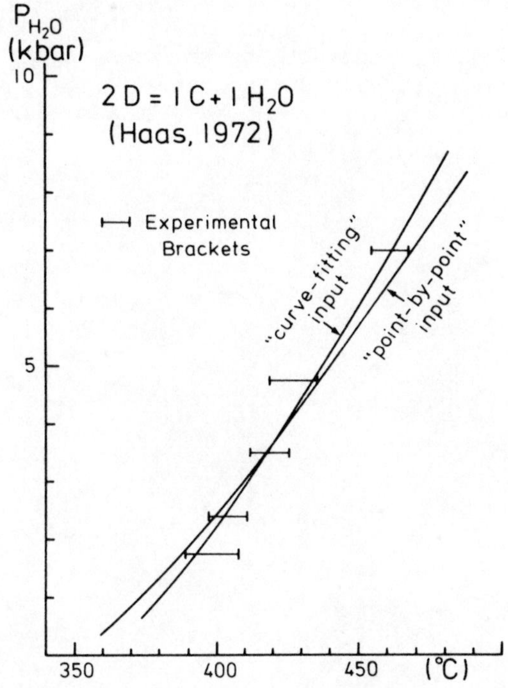

Figure B. The P_{H_2O} - T equilibrium curve of the reaction 2 diaspore = 1 corundum + 1 H_2O, recalculated using the extracted thermodynamic data for diaspore. The experimental reversal brackets by Haas (1972) are displayed for comparison.

(ii) The calculated equilibrium curve of the reaction 2 diaspore = 1 corundum + 1 H_2O, utilizing the curve-fitting input, passes through all the experimental brackets, while that based on the point-by-point data fails to cross the lowermost and the uppermost experimental reversal brackets. This is indicative of the fact that the calorimetric value of $\tilde{S}^o_{f,298}$ for diaspore is *not* compatible with the experimental phase equilibria data. It remains to be seen if the present failure to take into account $\Delta C^o_{pf,s} = f(T)$ data (not available at present) proves to be the reason of this ambiguity.

FUGACITY AND ACTIVITY OF MOLECULAR SPECIES IN SUPERCRITICAL FLUIDS

John R. Holloway

Division of Geochemistry
Department of Chemistry
Arizona State University
Tempe, Arizona 85281

INTRODUCTION

Naturally occuring gases (fluids) in metamorphic and igneous systems are known to be mixtures of several volatile species, usually including H_2O, CO_2, CO, H_2, CH_4, H_2S or SO_2 in major or minor amounts. The thermodynamic interpretation of experimental results, the thermodynamic calculation of reactions, and the accurate determination of fluid compositions recorded by natural assemblages require knowledge of the thermodynamic properties of the geologically important fluid-phase species. Because very few accurate experimental determinations of these properties exist, it seems necessary to rely on an equation of state which can be used to calculate thermodynamic properties of fluid mixtures in geologically important pressure-temperature regions.

This chapter covers the following topics: (1) a review of existing experimental data; (2) a brief discussion of the molecular dynamics of fluids which is used to provide a theoretical rationalization of; (3) equations of state of fluids, in particular; (4) the modified Redlich-Kwong equation. The Redlich-Kwong equation is described in detail and is used to calculate fugacities and activities of species in H_2O-CO_2 mixtures, and mixtures of CO, CO_2, H_2O, CH_4, and H_2 coexisting with graphite. That it is necessary to consider non-ideal mixing in geological calculations involving fluids is amply demonstrated in the final section of this chapter.

D. G. Fraser (ed.), Thermodynamics in Geology, 161-181.

SCOPE AND DEFINITIONS

This chapter deals exclusively with molecular fluids in the supercritical region. The term fluid is preferred over gas or vapor because for pressure-temperature conditions close to any reasonable geotherm, the density of mixtures of CO_2, H_2O, etc. is much closer to a liquid than a gas. The critical temperature for those mixtures is usually not above about 400°C. Ionized species are neglected for two reasons: the equations of state used cannot adequately describe ionic fluids; and the degree of dissociation of aqueous electrolytes is small at temperatures above about 500°C. The methods presented in this chapter for the calculation of activity and fugacity coefficients are thus restricted to temperatures above about 500°C for most geologically important fluids.

Relationships and abbreviations used in this chapter are given below. Derivations of the relationships can be found in standard textbooks on thermodynamics.

The <u>fugacity</u> of a pure substance (f°) is related to its molar volume (V) by:

$$RT\ell n(f°/Pl) = \int_{Pl}^{P} VdP \qquad (1)$$

where Pl is choosen at a sufficiently low pressure (invariably one bar or less) that the substance behaves as an ideal gas. Note that equation (1) is restricted to constant T. The fugacity of species i in a mixture (f_i) is related to the partial molar volume of the species ($\bar{V}_i$) by:

$$RT\ell n(f_i/Pl) = \int_{Pl}^{P} \bar{V}_i dP \qquad (2)$$

The activity of species i (a_i) is defined by:

$$a_i = f_i/f_i^\circ \qquad (3)$$

where all quantities are referred to the same P and T and f_i and f_i° have the same standard state.

The <u>mole fraction</u> (X_i) of species i in a mixture is given by:

$$X_i = \frac{n_i}{n_1+n_2+\ldots} \qquad (4)$$

where n_i represents the number of moles of i.

The <u>activity coefficient</u> (γ_i) is defined by:

$$\gamma_i = a_i/X_i = f_i/X_i f_i^\circ \qquad (5)$$

The <u>fugacity</u> <u>coefficient</u> (ϕ_i) is defined by:

$$\phi_i = f_i/PX_i \qquad (6)$$

where P represents total pressure. In the case of a pure substance, $X_i \equiv 1.0$, $f_i \equiv f_i^\circ$ and equation (6) becomes:

$$\phi = f^\circ/P \qquad (6a)$$

The partial molar free energy, or chemical potential (μ_i) of a substance can be calculated at P and T from:

$$\mu_i(P,T) = \Delta G^\circ(1\ \text{bar},\ T) + RT\ell n(X_i P \phi_i) \qquad (7)$$

or

$$\mu_i(P,T) = \Delta G^\circ(1\ \text{bar},\ T) + RT\ell n(f_i^\circ X_i \gamma_i) \qquad (8)$$

Equation (7) or (8) together with:

$$\Delta G^\circ(1\ \text{bar},\ T) = \Delta G^\circ(1\ \text{bar},\ 298^\circ K) + \int_{298}^{T} S dT \qquad (9)$$

where S is the molar entropy, enables calculation of all thermodynamic properties of a molecular gas species at P and T, provided that ϕ_i and γ_i are known. A primary purpose of this chapter is to examine our ability to calculate those parameters up to pressures of 40 Kbar in the range 500° to 2000°C.

REVIEW OF AVAILABLE EXPERIMENTAL DATA

The classical method of obtaining fugacity and activity coefficients of gases is by experimental measurement of the molar volume as a function of P and T (known as the P-V-T method) followed by application of equation (1) or (2). Unfortunately, the P-V-T method is difficult, expensive and time consuming when attempted at pressures above one Kbar and temperatures above 500°C. For those reasons very few measurements are available even for one-component (pure) substances. The P-T regions for which experimental P-V-T data are available for systems of geologic interest are listed in Table 1.

To extract the fugacity or activity of a species in a two component system (such as CO_2-H_2O) using P-V-T methods requires measurement of several different compositions along the binary join. That is because the partial molar volume ($\bar{V}_i$) in equation (2) can only be obtained by taking the slope of the mean molar volume <u>vs</u> composition curve. Because the P-V-T measurements must

TABLE 1. Summary of P-V-T data for geologically important molecules.

Molecule	Pressure range (Kbar)	Temperature range (C°)	Source
H_2	0.1 - 1.8	200 - 600	1
N_2	0.7 - 4.9	200 - 1000	2
CO_2	2 - 10	100 - 1000	3
CO_2	1 - 10	177 - 977	4
H_2O	0 - 1.0	0 - 1300	5
H_2O	0.1 - 10	20 - 1000	6

1. Presnall (1969); 2. Malbrunot and Vodar (1973); 3. Shmonov and Shmulovich (1974); 4. C.W. Burnham and V.J. Wall, written communication; 5. Keenan et al. (1969); 6. Burnham et al. (1969).

be differentiated, the precision of the experiment must be considerably greater than is necessary for the calculation of the fugacity of pure fluids. The required precision, together with the need for measurement of several compositions for each join, has resulted in few determinations of binary systems.

MOLECULAR DYNAMICS

There is an extensive body of literature covering the nature of physical interactions between molecules. The field of study is known as molecular dynamics, and it relies heavily on theories of statistical mechanics.

The purpose of this section is to present only those aspects of molecular dynamics which are useful in providing a theoretical rationalization for the semiempirical Redlich-Kwong equation of state and for the algebraic form of the combining rules for mixtures. No attempt is made to derive the relations presented in this section. For a rigorous treatment the reader should consult the published works on the subject, such as Prausnitz (1969) or Hirschfelder et al. (1964), which have supplied most of the information summarized in this section.

Intermolecular Forces

The forces between an isolated pair of molecules can be separated into the repulsive force which prevents the molecules from merging together, and the attractive force, which among other things, is responsible for holding molecules together in liquids. The total potential energy (Γ) in a system of two molecules can be written as:

$$\Gamma(\text{total}) = \Gamma(\text{repulsive}) + \Gamma(\text{attractive}) \tag{10}$$

This was first proposed by G. Mie in 1903 and is referred to as the Mie equation. A schematic representation of the potential function is shown in Fig. 2. The repulsive potential is not well understood but it is clear that it increases very rapidly as the intermolecular separation becomes small. The repulsive potential is commonly reresented as

$$\Gamma(\text{repulsive}) = \frac{A}{r^n} \qquad (11)$$

where A is a positive constant and n is usually taken to be between 8 and 16. Perhaps the most popular form of the Mie equation is that known as the Lennard-Jones potential:

$$\Gamma(\text{total}) = 4\varepsilon[(\sigma/r)^{12} - (\sigma/r)^{6}] \qquad (12)$$

where ε is the negative of the energy at the equilibrium distance between the molecules and σ is a distance parameter equal to the intermolecular separation at zero potential energy. (see Fig. 1)

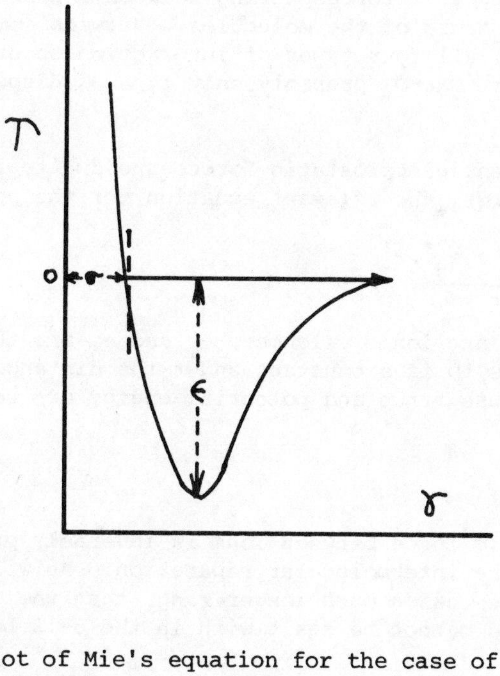

Figure 1. Plot of Mie's equation for the case of a 6-12 Lennard-Jones potential. Note the positions of ε and σ. The potential energy Γ and the intermolecular separation r are in arbitrary units.

In the above equation, the repulsive potential is represented by the twelfth power term and the attractive potential by the sixth power term, consequently the equation is known as a 6-12 Lennard-

Jones potential. There is little theoretical justification for the value of the exponent in the repulsive term; however, a range in values of (±2) has little effect on the total energy at ordinary molecular separations. There is considerable justification for the value of the attractive exponent and that will be summarized in the following paragraphs where consideration will be given to the most common types of attractive forces between molecules.

The types of attractive intermolecular forces considered can be classified as follows:

1. Electrostatic forces: arrise from charged particles (ions), permanent dipoles and permanent quadrapoles.
2. Induction forces between a permanent multipole and an induced dipole.
3. Forces of attraction between non-polar molecules (referred to as dispersion forces, also called induced dipole - instantaneous dipole forces).
4. Chemical forces causing association and complex formation.

The particular type of forces acting between a pair of molecules depends on the nature of the molecules. Between some molecules, such as H_2O-CO_2, all four types of interaction occur. Among other molecules such as CH_4-O_2 probably only type 3 (dispersion) forces occur.

The strongest electrostatic forces are due to ionic species of opposite charge, the relevant equation for the potential energy is

$$\Gamma_{ij} = \frac{Z_i Z_j e_i e_j}{Dr} \tag{13}$$

where Z_i and Z_j are ionic valences, e_i and e_j are the integral electric charges, D is a constant and r the distance separating the ions. Because force and potential energy are related by the expression:

$$F = - \frac{d\Gamma}{dr} \tag{14}$$

the electrostatic force between ions is inversely proportional to the square of the intermolecular separation. As will be seen later, that force has a much longer range than the other intermolecular forces and cannot be dealt with in the 6-12 Lennard-Jones equation.

Many of the geologically important fluid phase species are molecules which posses permanent dipole moments. Those moments are caused by the asymmetric distribution of electrons about the atomic nuclei. The dipole moments for several molecules are listed in Table 2 where it can be seen that HCl, H_2S, NH_3, SO_2, H_2O and HF all have significant dipole moments. The potential energy

TABLE 2. Selected properties of molecules

Molecule	ω	α	μ	Q
A_r	0	16.3	0	0
H_2	---	7.9	0	0.63
CH_4	0.013	26.0	0	0
O_2	0.021	16.0	0	---
N_2	0.040	17.0	0	1.5
CO	0.049	19.5	0.10	---
C_2H_4	0.085	42.6	0	---
CO_2	0.225	26.5	0	4.1
HCl	---	26.3	1.04	---
H_2S	---	37.8	(0.92)	---
NH_3	0.250	22.6	1.47	1.0
SO_2	---	37.2	1.63	---
H_2O	0.334	15.9	1.83	2.0
HF	---	24.6	1.91	---

Sources: Prausnitz (1969), Moelwyn-Hughes (1961), Hirschfelder, et al. (1964).

Note: a '---' indicates data not available, ω is the accentric factor, α the polarizability, μ the dipole moment, and Q the quadrupole moment.

between two dipoles will depend on the orientation of the dipoles with respect to each other. The lowest energy configuration (and consequently most probable) occurs when the molecules are aligned + - + -, etc., but that alignment is disturbed by thermal agitation of the molecules, which increases with temperature. Consequently the potential energy between dipoles is temperature dependent. The relevant potential energy equation for dipoles averaged over all orientations is:

$$\Gamma_{ij} = -\frac{2}{3}\frac{\mu_i^2\mu_j^2}{r^6 KT} \tag{15}$$

where K is the Boltzmann constant and T is absolute temperature. Note that the energy varies inversely with the sixth power of intermolecular separation, exactly as in the 6-12 Lennard-Jones potential.

In dipoles, the electric charge can be thought of as being concentrated at two separate points in the molecule. Quadrupole moments are found in molecules in which four separate points of charge exist. The simplest quadrupole is linear, such as CO_2. Experimental measurement of quadrupole moments is difficult and values are not available for many molecules. Values are listed in Table 2 for several molecules, where it can be seen that many molecules with dipole moments also have quadrupole moments. In such cases potential energy is determined mainly by the dipole moment and the quadrupole effect may be ignored. CO_2 is the only geologically important inorganic species in which the quadrupole contribution is important, but it is probably also important in hydrocarbons such as C_2H_4. The average potential energy for dipole-quadrupole interactions is inversely proportional to the eighth power of r and for quadrupole-quadrupole interactions it is inversely proportional to the tenth power of r. Consequently neither interaction is described accurately by the 6-12 Lennard-Jones potential. As in the case of the dipole-dipole interactions, the potential energy of interactions between quadrupoles is temperature dependent.

Many molecules have electric charge distributions which are spherically symmetrical and hence do not posses permanent dipole moments. But when those molecules are in the presence of polar molecules, the electric field of the polar molecule causes the electrons in the non-polar molecule to be displaced from their ordinary positions. That displacement creates an induced dipole moment μ_i which is proportional to the electric field strength E:

$$\mu^i = \alpha E \tag{16}$$

where α, the proportionality constant, is the polarizability, which measures the ease with which the molecules electrons can be displaced by an electric field. All molecules, including polar molecules, are polarizable to some extent, average polarizabilities are given in Table 2. The potential energy between a permanent dipole (j) and an induced dipole (i) is given by:

$$\Gamma_{ij} = -\frac{\alpha_i \mu_j^2}{r^6} \tag{17}$$

from which it can be seen that the interaction is not temperature dependent and is exactly described by the 6-12 Lennard-Jones potential.

The existance of crystalline forms of noble gases, such as argon, proves that attractive forces must exist between non-polar molecules. In 1930 F. London showed that a collection of non-polar molecules only lack polarity when viewed over a period of time. An instantaneous photograph of the molecules would show a distortion of the electrons relative to the nucleus sufficient to cause a temporary dipole moment. That instantaneous dipole moment causes induced dipoles in surrounding molecules which results in an attractive force between the non-polar molecules. The potential energy of the interaction was shown by London to be:

$$\Gamma = -\frac{3}{2}\frac{\alpha_i\alpha_j}{r^6}\left(\frac{h\nu_j h\nu_i}{h\nu_i+h\nu_j}\right) \quad (18)$$

Where the ν_i and ν_j terms are characteristic frequencies of the molecules. Because there is a relationship between that characteristic frequency and the index of refraction of the gas, the instantaneous dipole-induced dipole forces are called dispersion forces. From equation (19) it can be seen that the potential energy is not temperature dependent and that it follows the inverse sixth-power form.

We have now examined the three types of interaction between molecules which can easily be generalized. The fourth type - that caused by chemical bonding forces between molecules cannot be generalized and must be considered separately for each pair of molecules. That will be done in the following section for fluid mixtures of geological interest. At this point it is useful to assess the relative magnitude of the contributions from the electronic, induction, and dispersion forces for some typical molecules. The relative contributions for CO, HCl, NH_3 and H_2O are shown in Table 3. It can be seen that induction forces are generally not significant and that dispersion forces are important even for molecules of relatively high polarity such as NH_3. Of course in non-polar molecules only dispersion forces are present. Because the potentials between permanent dipoles, induced dipoles and instantaneous-induced dipoles are all described by the inverse sixth power relation, and because those are the most important contributions to the attractive forces between all neutral molecules, there is considerable reason to believe that one form of an equation of state would describe all one-component (pure) gases. That is one of the justifications for the corresponding state hypothesis.

TABLE 3. Relative magnitudes of intermolecular forces between identical molecules[1]

	Relative forces		
Molecule	Dipole	Induction	Dispersion
CO	.002	.039	64.3
HCl	24.1	6.14	107
NH_3	82.6	9.77	70.5
H_2O	203	10.8	38.1

[1]Adapted from Prausnitz (1969) p. 64.

Corresponding State Theory

The classical, or macroscopic, theory of corresponding states was proposed by van der Waals after he observed that, at the critical point of a fluid, the quantity PV/RT was very nearly the same for many different gases. The macroscopic theory of corresponding states can be stated as follows: If the reduced variables Pr, Tr and Vr are formed by dividing P, T and V by the respective critical values Pc, Tc and Vc, then any equation of state which describes the P-V-T relations of one fluid will also describe any other fluid. This principle has been used by Newton (1935) and Hougen and Watson (1947) to construct corresponding state charts for fugacity coefficients of pure gases. In its general form the classical theory states that the three parameters Tc, Pc and Vc define a universal function for fluids. But because the theory also states that the quantity PcVc/RTc is a constant, only two of the three parameters are independent so only Tc and Vc are needed to specify the properties of a fluid.

We have seen above that the 6-12 Lennard-Jones potential closely approximates intermolecular forces for many molecules. Equation (12) can be made dimensionless by dividing Γ by ε. This results in a universal function in which the dimensionless potenial is a function of the dimensionless distance of separation between the molecules, r/σ. The energy parameter ε_i and the distance parameter σ_i are characteristic values for a given molecule. This is a microscopic theory of corresponding states. It is related to the macroscopic theory through the critical properties of a fluid. Because the critical temperature is a measure of the kinetic energy of fluids in a common physical state, there should be a simple proportionality between the energy parameter ε_i and the critical temperature Tc. Because the critical volume reflects molecular size, there should also be a simple proportionality between σ_i and the cube root of Vc. For simple non-polar molecules which can be described by the 6-12 Lennard-Jones potential, the proportionalities have been found to be:

$$\varepsilon = (0.77K)\cdot Tc \; , \quad \sigma = 8.41\times 10^{-9}(Vc)^{1/3} \qquad (19)$$

where K is Boltzmann's constant. The microscopic theory of corresponding states should prove useful in geology because, unlike the macroscopic theory, it can be used to calculate transport properties of fluids as well as equilibrium properties.

The two parameter corresponding state theory is limited to 'simple' molecules which have highly symmetric force fields. The non-polar molecules Ar, CH_4, O_2, N_2 and CO fall into the simple class. Polar molecules show deviations from simple two-parameter corresponding state theory. A useful measure of the extent of deviation is the accentric factor proposed by K.S. Pitzer, who defined it as:

$$\omega = -\log_{10}(Ps/Pc) - 1 \text{ at } T/Tc = 0.7 \qquad (20)$$

where Ps is the saturation pressure of the liquid. The accentric factor is defined so that it is zero for ideally simple molecules. Values of ω are given in Table 1. It has been found that although molecules with large accentric factors do not obey corresponding state equations based on simple molecules, they can be grouped together on the basis of their accentric factors. Thus a corresponding state equation which accurately describes H_2O will also describe HF, etc. We will return to this topic in a following section.

Forces in Molecular Mixtures

The simplest mixtures are those involving only non-polar molecules. Dispersion forces are the only attractive forces which act on such mixtures and we have seen that they are not temperature dependent. The equation for potential energy (equation 18) can be rewritten in terms of the first ionization potential as:

$$\Gamma_{ij} = -\frac{3}{2}\frac{\alpha_i\alpha_j}{r^6}\left(\frac{I_iI_j}{I_i+I_h}\right) \qquad (21)$$

Where I_i and I_j are the first ionization potentials of molecules i and j. Because the above equation is more sensitive to the polarizabilities (α), and because the ionization potentials for all species of interest here vary by less than ±15%, we can make the approximation that

$$\frac{I_iI_j}{I_i+I_j} \simeq \frac{\bar{I}}{2} = k \qquad (22)$$

Where $\bar{I}$ is an average first ionization constant for all species, and k is a constant. Now $\Gamma_{ij} \simeq k\alpha_i\alpha_j/r^6$, $\Gamma_{ii} \simeq k\alpha_i^{\,2}/r^6$, and $\Gamma_{ji} \simeq k\alpha_j^{\,2}/r^6$ and because $\alpha_i\alpha_j = (\alpha_i^{\,2}\alpha_j^{\,2})^{\frac{1}{2}}$,

$$\Gamma_{ij} \simeq (\Gamma_{ii}\Gamma_{jj})^{\frac{1}{2}} \qquad (23)$$

The above equation gives an approximate theoretical basis for the "geometric mean" rule which is often applied to mixtures of non-polar gases. Because a mixture of a polar and a non-polar gas can result only in induction and dispersion forces, and because the dispersion forces are an order of magnitude larger than the induction forces, equation (23) should also be a good approximation for mixtures of polar and non-polar gases.

However, in mixtures of polar gases, such as H_2O-NH_3 or H_2O-HCl, the dipole forces are of about the same magnitude as the dispersion forces and equation (23) is probably not valid.

The mixing rules for the repulsive potential result from the assumption that there is a close relationship between molecular "size" and the repulsive potential. A useful measure of molecular size is the distance parameter σ described above. If a hard-sphere model for molecular interaction is assumed, it can be shown that for molecules i and j:

$$\sigma_{ij} = \tfrac{1}{2}(\sigma_i + \sigma_j) \tag{24}$$

Equations (23) and (24) have been found to be good approximations for a variety of mixtures, especially mixtures of molecules having nearly the same size and dipole moment.

THE REDLICH-KWONG EQUATION OF STATE

In this section the modified Redlich-Kwong equation will be described in detail. Of the many other equations of state which have been proposed, they either do not adequately represent available data or they contain too many adjustable parameters for the present purpose.

van der Waals equation. This equation is discussed here because of the physical significance of its parameters and because of its similarity to the Redlich-Kwong equation. The van der Waals equation is:

$$P = \frac{RT}{V-b} - \frac{a}{V^2} \tag{25}$$

Where the constant a is a measure of the cohesion between the molecules and is thus related to the attractive potential energy term of the Mie equation; and b is a measure of the volume of the molecules, and is thus related to the repulsive potential term of the Mie equation. Although the van der Waals equation is useful over limited ranges of P and T, it fits experimental data poorly over extended ranges.

Redlich-Kwong Equation. This equation, proposed by Redlich and Kwong (1949) has two adjustable parameters and is convenient to use as a corresponding state equation. It is:

$$P = \frac{RT}{V-b} - \frac{a}{(V^2+bV)(T)^{\frac{1}{2}}} \tag{26}$$

The constants a and b have the same physical significance as in the van der Waals equation. Redlich and Kwong (1949) showed that a and b parameters calculated by the corresponding state method resulted in good fits to experimental data for simple, non-polar molecules such as N_2, O_2, CO and CH_4. For the corresponding state representation of a and b, Redlich and Kwong (1949) use:

$$a = \frac{0.4278 Tc^{2.5} R^2}{Pc} \quad , \quad b = \frac{0.0867 Tc}{R\,Pc} \qquad (27)$$

Where R is the gas constant (= 83.12 cm^3 bar/deg mole). Because of the large quantum-mechanical effects in fluids of low molecular weight, such as H_2, the measured values of Tc and Pc must be adjusted. For H_2 the values suggested by Newton (1935) give good agreement with experimental results. For molecules which possess large dipole (or quadrupole) moments, the molecular dynamics arguments in the previous section suggest that corresponding state theory can not be used to calculate accurate values of a and b. In addition, equation (15) suggests that the a parameter should have a temperature dependence proportional to $1/T^2$ in the case of dipole moments (it can also be shown to hold for quadrupole moments). The Redlich-Kwong equation was modified by de Santis, et al. to take the above factors into account for H_2O, CO_2 and mixtures of them with non-polar molecules. The modified equation will be referred to as the MRK in this chapter. For H_2O and CO_2, de Santis, et al. (1974) calculated the temperature dependence of a from experimental data up to 800°C. They write the expression for a as:

$$a(T) = a° + a_1(T) \qquad (28)$$

Where the temperature independent constant a° represents dispersion forces and $a_1(T)$ represents intermolecular forces due to hydrogen bonds, permanent dipoles, and quadrupoles. The values of $a_1(T)$ for H_2O given by de Santis et al. (1974) do not accurately represent the H_2O data at pressures above 2 Kbar., so I derived a new expression of $a_1(T)$ for water using the data of Burnham, et al. (1969) up to 900°C and the extrapolated data of Holloway, et al. (1971) from 900° - 1300°C. For both CO_2 and H_2O I extrapolated the $a_1(T)$ function to 1800°C using the $1/T^2$ proportionality, and the limiting value of a_o. The variation of a with T is not a smooth function in the vicinity of the critical temperature and I did not fit the analytical function to temperatures below 400°C for H_2O.

Redlich and Kwong (1949) derived the following expression from equation (26) for the fugacity coefficient of a pure fluid:

$$\ln\phi = Z - 1 - \ln(Z - BP) - (A^2/B)\ln(1 + BP/Z) \qquad (29)$$

where $B = b/(RT)$, $A^2 = a/(R^2T^{2.5})$ (30)

A numerical technique for the solution of equation (29) is given by Edmister (1968).

The fit of the MRK to available data for pure fluids is summarized in Table 4, as are the equation for a(T) and the de Santis, et al. (1974) values of a° and b. It is noteworthy that the MRK provides a good fit to the H_2O data with only five adjustable parameters, none of which is a function of pressure. Because of the lack of pressure dependent parameters in the MRK, extrapolations to much higher pressures should be possible. Indeed the functional form of the equation was choosen by Redlich and Kwong (1949) to furnish good approximations at high pressure.

Redlich and Kwong demonstrated that their equation could be used to calculate the activity of species in mixtures of simple, non-polar molecules. These mixing rules they proposed are based on the molecular dynamics principles discussed in the previous section. For mixtures of n species, they can be generalized as:

$$b = \sum_{j}^{n} X_i b_i \quad , \quad a = \sum_{i}^{n} \sum_{j}^{n} X_i X_j a_{ij} \tag{31}$$

TABLE 4. Comparison of values calculated by the MRK equation with literature values.

Species	Variable	P-range (Kbar)	T-range (°C)	Relative Deviation	Source
H_2	f	0.1-3	25-1000	±2% maximum	1
H_2O	f	0.5-10	20-1000	±1.3% (2)	2
CO_2	f	1.0-10	177-977	±5% maximum	3
CO_2	V	2.0-10	100-1000	±7% maximum	4

CO_2: $a° = 46\times10^6$, $b = 29.7$
$a(T) = 73.03\times10^6 - 71400.\ T + 21.57\ T^2$

H_2O: $a° = 35\times10^6$, $b = 14.6$
$a(T) = 166.8\times10^6 - 193080.\ T + 186.4\ T^2 - 0.071288\ T^3$

For the a(T) function, T is in degrees celsius.

Source: 1. Shaw and Wones (1964); 2. Burnham, et al. (1969); 3. C. Wayne Burnham and Victor J. Wall, written communication; 4. Shmonov and Shmulovich (1974).

with

$$a_{ij} = (a_i a_j)^{\frac{1}{2}} \tag{32}$$

The above mixing rules are not necessarily valid for mixtures of polar molecules. For mixtures of H_2O and CO_2, de Santis, et al. (1974) proposed that the a_{ij} term be given as:

$$a_{H_2O-CO_2} = (a^{\circ}_{H_2O} a^{\circ}_{CO_2})^{\frac{1}{2}} + \tfrac{1}{2}R^2T^{5/2}K \tag{33}$$

Where K is an equilibrium constant describing the observed complex formation between H_2O and CO_2 and is given by:

$$\ell n K = -11.07 + \frac{5953}{T} - \frac{2746\times10^3}{T^2} + \frac{464.6\times10^6}{T^3} \tag{34}$$

Where T is in Kelvin. For mixtures of H_2O or CO_2 with simple, non-polar molecules the cross term is given by:

$$a_{12} = (a^{\circ}_1 a_2)^{\frac{1}{2}} \tag{35}$$

Where a°_1 refers to either H_2O or CO_2. The justification for using the temperature independent a° parameter in mixtures of polar and non-polar fluids was given in the molecular dynamics section of this chapter.

The moderately polar species HCl, H_2S, NH_3, SO_2 and HF will require a(T) functions to adequately describe their thermodynamic properties. Also, mixtures of those species with each other or with CO_2 or H_2O may require formulation of expressions such as equations (33) and (34) to accurately represent their partial molar properties.

Using the above mixing relationships, the fugacity of species i in a supercritical fluid mixture is given by Redlich and Kwong (1949) as:

$$\ell n\phi_i = \frac{B_i}{B}(Z-1) - \ell n(Z-BP) - \left(\frac{2A_i}{A} - \frac{B_i}{B}\right)(A^2/B)\ell n(1+BP/Z) \tag{36}$$

Where B_i and A_i are calculated from equations (30) and A and B are calculated by first calculating a and b from the mixing rules [equations (31), (32), (33), and (35)] and then using equations (30). Note that the values of Z, A and B in equation (36) are calculated for the mixture in question and are thus composition dependent, and different from the corresponding parameters in equation (29) for a pure fluid. For a binary system, equation (36) yields activity-composition relations which are, in general, not symmetric. Equation (36) fits experimental data very well for several systems involving simple, non-polar molecules (Redlich and Kwong, 1949; Prausnitz, 1969). Very few experimental data are available to test the mixing model for compositions, pressures, and temperatures of petrologic interest. Shaw (1963) reported activity-composition relations for H_2-H_2O mixtures at 800 bars and 700°C. The MRK yields activities that are systematically lower than those calculated from Shaw's (1967) equation, but the curve calculated using the MRK falls within the probable error of his experiments (Shaw, 1963). For H_2O-CO_2 mixtures, values of the compressibility factor calculated using the MRK fit the Franck and Tödheide (1959) data with an average deviation of 6% relative. Activity coefficients calculated from the MRK fit those calculated by Greenwood (1973) to within ±5% relative. Calculations made using the MRK also predict a topology for the two-phase region in the H_2O-CO_2 system that is in qualitative agreement with the topology depicted by Tödheide and Franck (1963). Eggler, Kushiro, and Holloway (1976) show that MRK fugacities provide a much better fit to the experimentally determined decarbonation reaction enstatite + magnesite = forsterite + CO_2 at 25 Kbar (H_2O-CO_2 fluid) than does the ideal mixing model.

The available data suggest that the MRK can be used to estimate the fugacity of CO_2, H_2O, H_2, CO and CH_4 in multispecies fluids at temperatures from about 450° to 1800°C and at pressures from 0.5 to about 40 Kbar. Figure 2 shows activity-composition relations for binary mixtures of H_2O and CO_2 for several pressures at 700°C. Note that although the deviations from ideality increase with increasing pressure, they approach an asymptotic limit at very high pressure. The behavior on other isotherms is similar, although at constant pressure nonideality becomes more pronounced at lower temperatures.

Large positive deviations of H_2O and CO_2 from ideality at small mole fractions will have a considerable effect on calculated equilibria involving dehydration-decarbonation reactions. The reaction tremolite + 3 calcite + 2 quartz = 5 diopside + 3 CO_2 + H_2O will serve as an example (Skippen, 1971, Fig. 9). In an isobaric temperature-fluid composition projection, the stability of the tremolite + calcite + quartz assemblage is expanded to higher mole fractions of <u>both</u> CO_2 and H_2O if activity values for H_2O and CO_2 are calculated using the MRK instead of an ideal mixing model. This expansion is a general phenomenon for those reactions in which H_2O and CO_2 appear together on the same side of the reaction. For those reactions in which H_2O and CO_2 appear on opposite sides, the effect of nonideal mixing cannot be so easily predicted and should be calculated for specific cases.

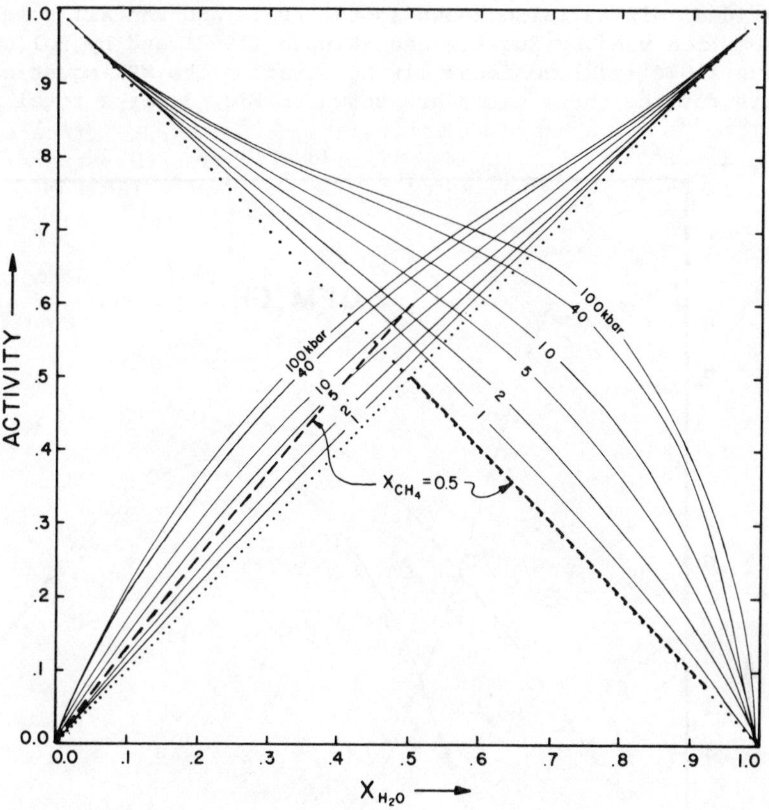

Figure 2. Activity-composition relations for H_2O and CO_2 at 700° calculated using the MRK equation. Light solid lines represent the 1, 2, 5, 10, 40 and 100 Kbar isobars for the system CO_2-H_2O. Dashed lines represent CO_2 and H_2O in mixtures at 2 Kbar with X_{CH_4} fixed at 0.5.

The effect of dilution of H_2O-CO_2 mixtures by a third species is also shown in Fig. 2 for a fluid at a total pressure of 2 Kbar containing 50 mole % CH_4. Note that addition of methane causes CO_2 to behave more ideally and H_2O much less ideally because CH_4 is chemically very similar to CO_2 but quite dissimilar to H_2O. Consequently, for a geological system in the two-phase field of aqueous liquid and CO_2-rich vapor, methane would partition strongly into the vapor.

To illustrate the effect of nonideal mixing in fluids of geological interest, the fluid composition in equilibrium with graphite in the C-O-H system was calculated using the equations derived by French (1966). Calculations were carried out for three cases: (1) ideal gas behavior, the assumption used by French (1966); (2) ideal mixing (also known as the Lewis and Randall rule), the assumption used by Eugster and Skippen (1967) and by Holloway and Reese (1974); (3) nonideal mixing based on the MRK equation. Results for the three cases are shown in Fig. 3 for a total pressure

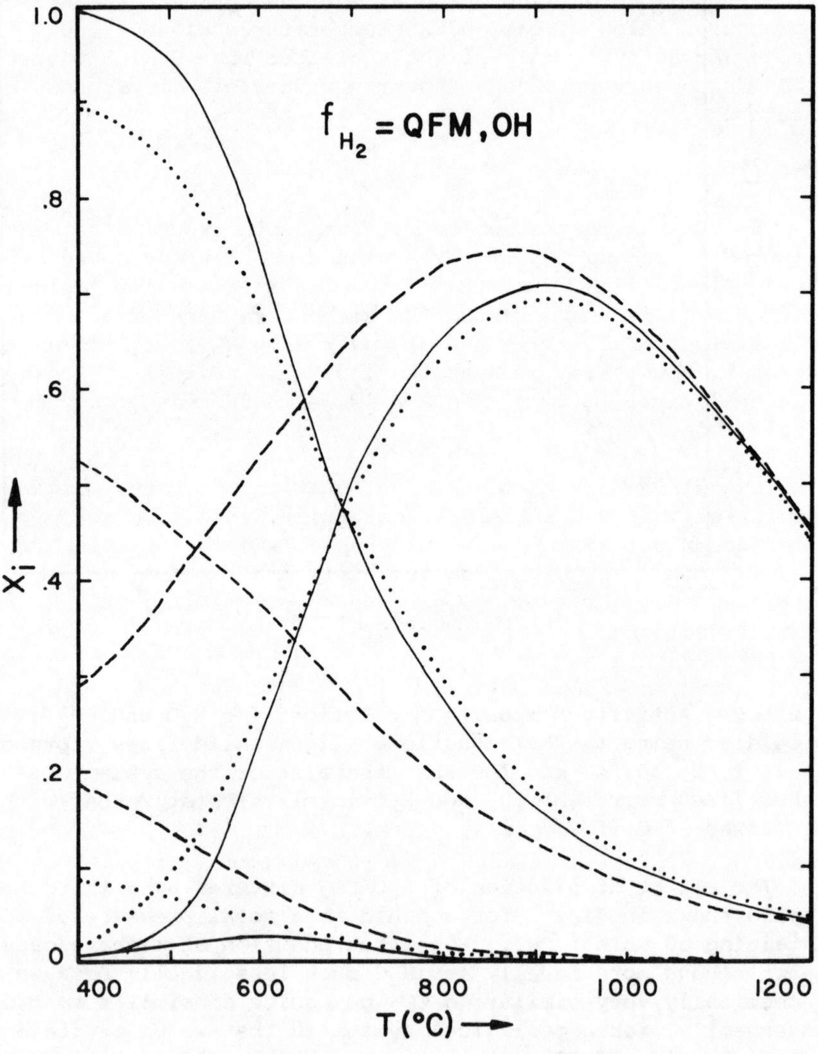

Figure 3. Comparison of calculated concentrations of CO_2, CH_4 and H_2O in fluids in equilibrium with graphite in the C-O-H system at 2 Kbar total pressure. Dashed lines calculated using ideal gas assumption, dotted lines calculated using ideal mixing assumption, and solid lines calculated using the MRK equation.

of 2 Kbar. The general appearance of plots at other pressures is quite similar. It can be seen that at temperatures above about 600°C, there are only small differences between the ideal mixing and the MRK calculations. Calculations at other pressures and hydrogen fugacities suggest that if the hydrogen fugacity is lower than that fixed by the assemblage quartz + fayalite + magnetite + H_2O (QFM[OH]), methane will be an abundant species only at pressures below 2 kbar and at temperatures of 400°-600°C. If hydrogen fugacity is only slightly greater than QFM(OH), however, methane becomes very abundant over a wide range of pressure and temperature.

CONCLUSIONS

The MRK has been shown to fit available experimental data for pure fluids and for mixtures with good accuracy. Use of the MRK to calculate the activity of species in fluids of geologic importance is far superior to the common practice of assuming ideal mixing. The MRK is useable over a much wider P-T range than covered by the tables of Ryzhenko and Malinin (1971). Because it is analytical in form, the MRK is well adapted to computer calculations.

It would appear that a simple equation of state, combined with insights from the field of molecular dynamics, can provide a good description of the thermodynamic behavior of supercritical molecular fluids. Only a comparatively small number of experimental measurements are necessary to supply missing parameters for most geologically important molecular species. However, there is a large gap in our knowledge of fluids in the two phase region and of mixtures of non-polar and ionic species.

AKCNOWLEDGEMENTS

This work was partially supported by National Science Foundation grant DES 72-01357A01 and by the Geophysical Laboratory, Carnegie Institution of Washington through the good offices of Dr. H. S. Yoder, Jr.

REFERENCES

Burnham, C.W., Holloway, J.R. and Davis, N.F. Amer. Jour. Sci., 267A, 70-95 (1969).

Burnham, C.W., Holloway, J.R. and Davis, N.F. Geol. Soc. Am. Sp. Paper 132, 96 pp (1969).

de Santis, R., Breedvelde, G.F.J. and Prausnitz, J.M. Ind. Eng. Chem. Process Des. Dev. 13, 374-377 (1974).

Edmister, W.C. Hydrocarbon Process. 47, 239-244 (1968).

Eggler, D.H., Kushiro, I. and Holloway, J.R. Carnegie Inst. Wash. Year Book 75, in press (1976).

Eugster, H.P. and Skippen, G.B. in Researches in Geochemistry 2, John Wiley and Sons, Inc., 492-520 (1967).
Fisher, J.R. and E-an Zen Am. Jour. Sci., 270, 297-314 (1971).
Franck, E.U. and K. Tödheide A. Phys. Chem., 37, 232-243 (1959).
French, B.M. Rev. Geophys. 4, 223-253 (1966).
Greenwood, H.J. Am. Jour. Sci., 267A, 191-208 (1969).
Greenwood, H.J. Am. Jour. Sci., 273, 561-571 (1973).
Hirschfelder, J.O., Curtiss, C.F. and Bird, R.B. Molecular Theory of Gases and Liquids, John Wiley and Sons, Inc. 1249 pp (1964).
Holloway, J.R., Eggler, D.H. and Davis, N.F. Geol. Soc. Am. Bull. 82, 2639-2642 (1971).
Holloway, J.R. and Reese, R.L. Am. Mineral. 59, 587-597 (1974).
Hougen, O.A. and Watson, K.M. Chemical Process Principles, John Wiley and Sons (1947).
Keenan, J.H., Keyes, F.G., Hill, P.G. and Moore, J.G. Steam Tables, John Wiley, 162 pp (1969).
Malbrunot, P. and Vodar, B. Physica, 66, 351-363 (1973).
Moelwyn-Hughes, E.A. Physical Chemistry, Pergamon Press, 1334 pp (1961).
Newton, R.H. Ind. Eng. Chem., 27, 302-306 (1935).
Prausnitz, J.M. Molecular Thermodynamics of Fluid-Phase Equilibria, Prentice-Hall, Inc., 523 pp (1969).
Presnall, D.C. Jour. Geophys. Res., 74, 6026-6033 (1969).
Redlich, O. and Kwong, J.N.S. Chem. Rev., 44, 233-244 (1949).
Richardson, J.M., Arons, A.B. and Halverson, R.R., Jour. Chem. Phys., 15, 785 (1947).
Ryzhenko, B.N. and Malinin, S.D. Geochem. Internat. 8, 562-574 (1971).
Shaw, H.R. Science, 139 1220-1222 (1963).
Shaw, R.H. in Researches in Geochemistry 2, John Wiley and Sons, Inc., 521-541 (1967).
Shaw, H.R. and Wones, D.R. Am. Jour. Sci., 262, 918-929 (1964).
Shmonov, V.M. and Shmulovich, K.I. Doklady Akad. Nauk SSSR, Earth Sci. Sec., 217, 206-209.
Skippen, G.B. Jour. Geol., 79, 457-481 (1971).
Tödheide, K. and Franck, E.U. A. Phys. Chem., 37, 387-401 (1963).

STUDY PROBLEMS

1. Calculate the subsolidus decarbonation of the assemblage Enstatite + Dolomite (= Diopside + Forsterite + CO_2) at 25 kbar assuming:

   a. Ideal gas
   b. Modified Redlich-kwong

2. Calculate the above decarbonation at 25 kb if $X^{Fl}_{CO_2} = 0.5$ ($X_{H_2O} = 0.5$) assuming:

   a. Lewis and Randall rule
   b. Modified Redlich-kwong

3. Choose a set of values for P, T and f_{O_2} and calculate the molefractions of the molecular species and the atomic composition for a point on the graphite surface in the C-O-H system assuming:

   a. Ideal gas law
   b. Lewis and Randall mixing of real gases
   c. Modified Redlich-kwong

COMPOSITIONS AND THERMODYNAMICS OF METAMORPHIC SOLUTIONS

Hans P. Eugster

Department of Earth and Planetary Sciences
The Johns Hopkins University
Baltimore, Maryland 21218

INTRODUCTION

The composition of the supercritical fluid phase which is in equilibrium with metamorphic or igneous mineral assemblages can be assessed either through fluid inclusion studies (Touret, 1971; Roedder, 1972; Poty <u>et al</u>., 1974), or by considering the nature and composition of the minerals themselves. In this review we restrict ourselves to the latter approach. It is based on a combination of buffer and exchange reactions and fugacity indicators in conjunction with laboratory calibrations. A simple buffer reaction is a reaction between a group of solids and a single gas species. When all participants in the reaction are present, the chemical potential of the gas species remains constant at a given P and T, regardless of any changes in bulk composition, provided the compositions of the solids remain constant. Hematite + magnetite, quartz + calcite + wollastonite, gypsum + anhydrite, pyrite + pyrrhotite, are commonly occurring buffers for oxygen, CO_2, H_2O and S_2 respectively:

$$6\ Fe_2O_3 \rightleftharpoons 4\ Fe_3O_4 + O_2 \qquad (K_1)_{P,T} = f(O_2) \qquad (1)$$

$$CaCO_3 + SiO_2 \rightleftharpoons CaSiO_3 + CO_2 \qquad (K_2)_{P,T} = f(CO_2) \qquad (2)$$

$$CaSO_4 \cdot 2\ H_2O \rightleftharpoons CaSO_4 + 2\ H_2O \qquad (K_3)_{P,T} = f(H_2O)^2 \qquad (3)$$

$$2\ FeS_2 \rightleftharpoons 2\ FeS + S_2 \qquad (K_4)_{P,T} = f(S_2)\ . \qquad (4)$$

D. G. Fraser (ed.), Thermodynamics in Geology, 183-202. *All Rights Reserved.*

Equilibrium constants, determined in the laboratory or calculated from free energy data, are usually expressed by an Arrhenius equation, modified to take into account the effect of the difference in molar volumes between reactant and product solids (the C term, Eugster and Wones, 1962):

$$\log K = -\frac{A}{T} + B + \frac{C(P-P^\circ)}{T} \quad (5)$$

where T is in °K and P° is the reference pressure. A complex buffer reaction involves two or more gas species. In this case, the product or ratio of the chemical potentials, as expressed by the equilibrium constant, are buffered with respect to bulk composition changes. Exchange reactions are reactions which occur between solids which can contain variable ratios of two or more gas species. Common examples are OH-HF exchanges in micas or amphiboles, such as:

$$KMg_3AlSi_3O_{10}(OH)_2 + 2\ HF \rightleftharpoons KMg_3AlSi_3O_{10}F_2 + 2\ H_2O\ . \quad (6)$$

Exchange reactions define the ratios of the chemical potentials of the exchanged gas species, but they do not buffer them. Fugacity indicators are solid solutions with compositions varying regularly with changes in fugacity, so that a composition measurement amounts to a fugacity measurement. Examples are wüstite (Eugster, 1959) or pyrrhotite (Toulmin and Barton, 1964, see fig. 9). In addition to containing stable gases, the metamorphic fluid also contains acids and bases, salts and ionic species. Many of these are derived from solution and precipitation reactions of minerals the fluid is in contact with. The concentrations or activities of the neutral species can also be evaluated from buffer and exchange reactions, but for the ionized species we need in addition the relevant ionization constants. In the latter part of this review we will present some of the preliminary data available as well as the experimental methods for obtaining them.

O-H GASES

The most important compositional system is the system O-H, containing as principal species H_2, O_2 and H_2O, which are related by the reaction

$$H_2 + \tfrac{1}{2} O_2 \rightleftharpoons H_2O \qquad (K_7)_T = \frac{f(H_2O)}{f(H_2)\cdot f(O_2)^{1/2}} \ . \quad (7)$$

K_7 is a constant dependent only on temperature (Robie and Waldbaum, 1968, p. 114). Of the three variables, $f(H_2O)$, $f(H_2)$ and $f(O_2)$, two are independently variable. Even after $f(H_2O)$ is specified, $f(O_2)^{1/2}$ and $f(H_2)$ can still vary, but the product is constant, that is they must vary antithetically. This inverse relationship is very important in rocks, because it means that diffusion of hydrogen is equivalent to diffusion of oxygen in the opposite direction. $f(H_2)$ is always very much larger than $f(O_2)$ and redox reactions are accomplished by hydrogen diffusion (Eugster, 1959). Presumably, hydrogen also diffuses much more rapidly through rock fabrics than H_2O, thus changing $f(O_2)$, unless it is buffered, since $f(O_2)$ is a measure of the H_2O/H_2 ratio.

The next important condition to be fulfilled is the gas pressure equation:

$$P(gas) = P(H_2O) + P(H_2) + P(O_2) = \frac{f(H_2O)}{\gamma(H_2O)} + \frac{f(H_2)}{\gamma(H_2)} + \frac{f(O_2)}{\gamma(O_2)} . \quad (8)$$

Fugacity coefficients $(\gamma)_i$, as a function of P and T, have been determined for the most important gases (Burnham <u>et al</u>., 1969; Presnall, 1969) and some gas mixtures (Shaw, 1967; Greenwood, 1973). Where they are not known, they can be approximated from corresponding states plots (Hougen and Watson, 1946; Ryzhenko and Malinin, 1971) or equations of state (Redlich and Kwong, 1949). Where data are lacking, gas mixtures are treated as ideal mixtures of real gases. In order to be able to solve for the individual fugacities, we need one additional constraint. We may either specify the bulk composition, or we must define one of the individual fugacities or a fugacity ratio. Let us say the fluid has the bulk composition of pure H_2O, that is hydrogen is twice as abundant as oxygen. Hence

$$P(H_2) = 2\ P(O_2) . \quad (9)$$

Combining eqs. (7), (8), and (9) we can solve for the individual fugacities. A solution for 2 kb is shown in fig. 1, labeled "pure water".

For natural assemblages it is more reasonable to search for buffer assemblages which define one of the fugacities than to assume a fixed fluid composition. H_2O and H_2 buffers are not common in nature, quartz + muscovite + K-feldspar + sillimanite and biotite + K-feldspar + magnetite being respective examples (Day, 1973; Wones and Eugster, 1965). The latter buffer is illustrated in fig. 2. On the other hand, it is often possible to define $f(O_2)$ quite closely using a number of oxygen barometers available, such as oxide pairs (Eugster, 1959; Buddington and Lindsley, 1964; Huebner, 1971; see figs. 3 and 4), pairs of iron-bearing silicates or oxygen sensors (Sato, 1971, 1972). Having evaluated $f(O_2)$, $f(H_2O)$ and $f(H_2)$ can be calculated from (7) and (8) (eqs. 6 and 7 in Eugster and Wones, 1962). The results for a number of oxygen buffers are also shown in fig. 1. A third method, used extensively by experimentalists (Eugster, 1957; Eugster and Wones, 1962; Huebner, 1971), is based on fixing $f(H_2)$ through an osmotic membrane of Pt, AgPd or AuPd. $f(H_2)$ is fixed externally by an oxygen buffer + H_2O, and equilibrates across

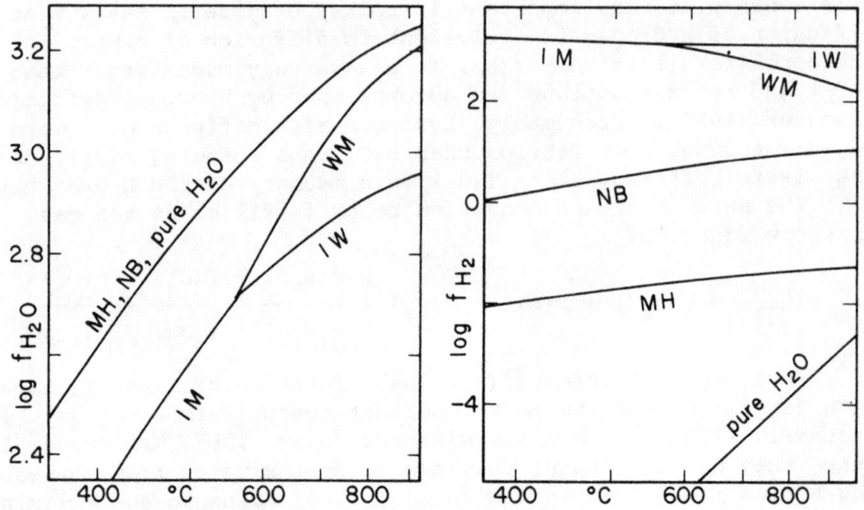

Figure 1. Fugacities of H_2O and H_2 at P(gas) = 2000 bars. IW: Fe + FeO, IM: Fe + Fe_3O_4, WM: FeO + Fe_3O_4, NB: Ni + NiO, MH: Fe_3O_4 + Fe_2O_3. Pure H_2O is a gas phase with H:O :: 2:1.

the membrane. That $f(H_2)$ value together with eqs. (7) and (8) allows calculation of $f(H_2O)$ and $f(O_2)$ in the internal system. It should be emphasized that the fluid compositions in the external and internal systems need not be the same (Eugster and Skippen, 1967; Huebner, 1971).

In summary, most metamorphic mineral assemblages are not buffered with respect to H_2O or H_2, but $f(O_2)$ is usually either buffered or defined (cases 2 and 4 of Eugster and Skippen, 1967, p. 518. The latter case corresponds to the intrinsic oxygen fugacity of Sato, 1972). To be able to evaluate $f(O_2)$, it is necessary first to estimate T from any available geothermometer, such as solid solutions or isotopic ratios. Next we need to put limits on the fluid pressure. This is usually accomplished by estimating P(load) from overburden or the mineral assemblages, yielding an extreme estimate for P(fluid) = P(load). Eqs. (7) and (8) will give $f(H_2O)$ and $f(H_2)$ with the aid of the fugacity coefficients. Assuming Raoult's law, we can then calculate the approximate gas composition from $X_i = f_i/f_i^*$ where X_i is the mole fraction of i and f_i the fugacity of i in the mixture and f_i^* is the fugacity of pure i at the same P and T. If P(fluid) < P(load), the gas composition can normally not be obtained, unless an H_2O or H_2 buffer is present in the mineral assemblage.

Because $f(O_2)$ of crustal environments is always very low, O-H fluids are essentially mixtures of H_2O and H_2. Normally, $f(O_2)$ lies between MH and FMQ. The former dictates a fluid which is essentially pure H_2O, containing no more than 0.001% H_2,

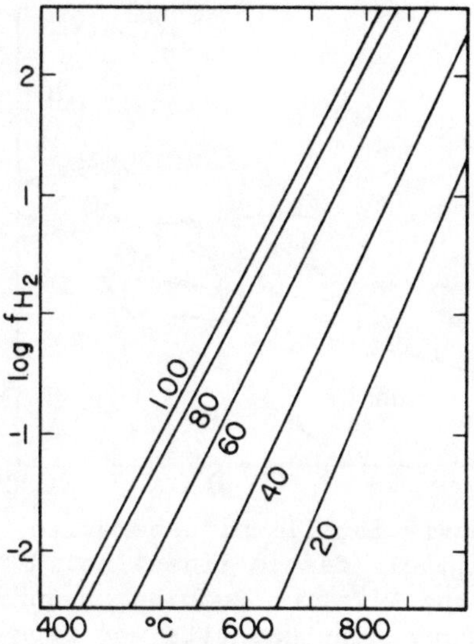

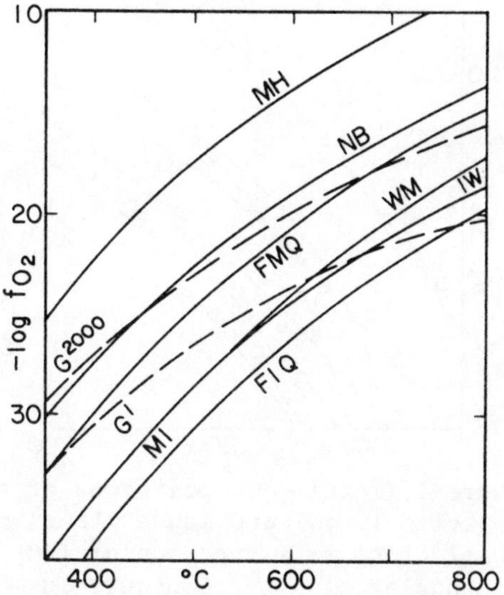

Figure 2 (left). The assemblages biotite + K-feldspar + magnetite as hydrogen buffers (after Wones and Eugster, 1965). Contours are for mol percent $KFe_3AlSi_3O_{10}(OH)_2$ in biotite.

Figure 3 (right). Oxygen fugacities of the standard oxygen buffers. They are, in addition to those of fig. 1, FIQ: Fe_2SiO_4 + Fe + SiO_2 , FMQ: Fe_2SiO_4 + Fe_3O_4 + SiO_2. The graphite curves G^1 (1 bar) and G^{2000} (2000 bars) are drawn for the pure system C-O. Data from Huebner (1971), Haas and Robie (1973), French and Eugster (1965).

while fluids equilibrated with FMQ contain up to 1.6% H_2, depending upon T (Eugster, 1972). Hydrogen becomes the dominant species for the more reducing buffers like WM, FIQ. It is interesting to note that although $f(O_2)$ decreases with T for every buffer, the H_2/H_2O ratio does not vary much over considerable temperature ranges. For instance, cooling FMQ from 1023 to 723°C changes $\log f(H_2)/f(H_2O)$ only from -1.65 to -1.69. The petrologic significance of this fact is obvious. It has been shown (Carmichael and Nichols, 1967) that igneous assemblages follow anhydrous buffer curves during their cooling through the subsolidus region, that is, paths of essentially constant fluid composition and constant Fe/Mg ratios of the solids. Hydrous Fe-Mg silicates, on the other hand, have constant Fe/Mg contours which are much flatter in a $f(O_2)$-T diagram and hence cooling at constant fluid composition leads to strong iron enrichment in these silicates (see fig. 5). In such silicates, constant Fe/Mg ratios during cooling can be achieved only by substantial preferential losses of hydrogen.

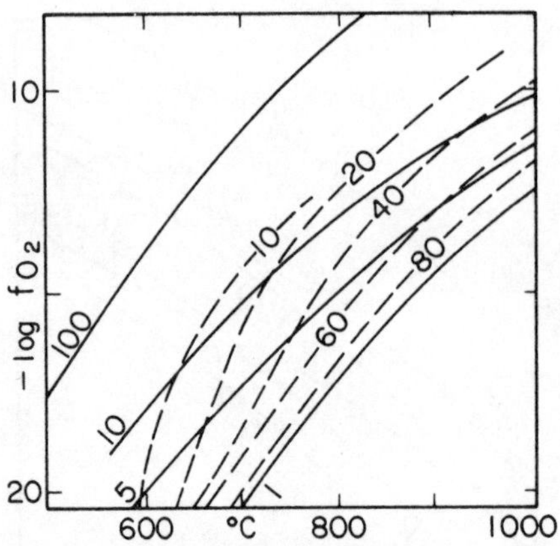

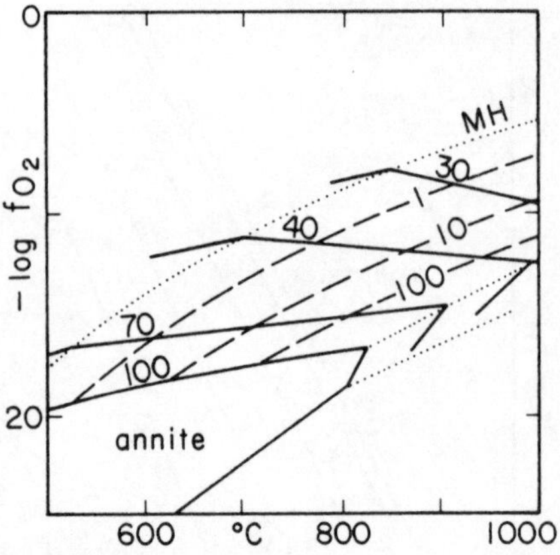

Figure 4 (left). Compositions of coexisting ilmenite-hematite (solid lines) and magnetite-ulvospinel (dashed lines) solid solutions as a function of $f(O_2)$ and T (after Buddington and Lindsley, 1964). Contours in mol per cent hematite and ulvospinel, respectively.

Figure 5 (right). Stability limits with respect to oxidation, of biotites (solid lines) and olivines (dashed lines) at 2070 bars (after Wones and Eugster, 1965). Contours are in mol per cent annite and fayalite, respectively. Note the oblique intersection of the contours.

C-O GASES

There is complete analogy between the C-O and O-H systems. The reaction

$$CO + \tfrac{1}{2}\, O_2 \rightleftharpoons CO_2 \qquad (K_{10})_T = \frac{f(CO_2)}{f(CO)\cdot f(O_2)^{\frac{1}{2}}} \qquad (10)$$

combined with the P(gas) equation

$$P(gas) = P(CO_2) + P(CO) = \frac{f(CO_2)}{\gamma(CO_2)} + \frac{f(CO)}{\gamma(CO)} \qquad (11)$$

governs the gas compositions. Again, an additional parameter must be specified, such as $f(O_2)$, to allow us to calculate individual fugacities. The relevant equations and solutions are given in French and Eugster (1965). There is another way of achieving invariance and one which geologically is very significant: the presence of graphite. It adds a third, independent equation, which can be written either involving CO or CO_2

$$C + O_2 \rightleftharpoons CO_2 \qquad (K_{12})_{P,T} = \frac{f(CO_2)}{f(O_2)} \quad . \qquad (12)$$

Because graphite is consumed or produced during the reaction, its volume change must be taken into account. The three fugacities can be calculated from eqs. (10), (11) and (12), for given values of P(gas) and T, hence graphite acts as an oxygen buffer in the pure system C-O. Because of the large volume change, the $f(O_2)$ values are strongly pressure dependent (see fig. 3). Towards higher temperatures, graphite becomes progressively more reducing with respect to the iron oxides, that is graphite, mostly derived from organic material, becomes an important reducing agent in high grade metamorphic rocks (Eugster, 1972), by reactions such as

$$6\,Fe_2O_3 + C \longrightarrow 4\,Fe_3O_4 + CO_2 \quad . \tag{13}$$

If other gases are added, such as H_2, graphite ceases to be an oxygen buffer, because the gas composition does not remain invariant. However, graphite always controls the CO_2/CO ratio, regardless of bulk composition, because of

$$C + CO_2 \rightleftharpoons 2\,CO \qquad (K_{14})_{P,T} = \frac{f(CO)^2}{f(CO_2)} \tag{14}$$

which is a combination of eqs. (10) and (12).

C-O-H GASES

This system has been treated in detail by French (1966), Skippen (1967) and Eugster and Skippen (1967). Bulk compositions for selected $f(O_2)$'s have been tabulated by Deines et al. (1974). The principal gas species are H_2O, H_2, O_2, CO_2, CO and CH_4. The system can be discussed most conveniently in two parts, one representing oxidized and one reduced environments. In oxidized environments, the only abundant species are H_2O and CO_2. Such mixtures are most commonly found during the metamorphism of impure limestones, which involves reactions between carbonates and hydrous or anhydrous silicates. An extensive literature has accumulated on this subject during the last few years. Laboratory calibrations are summarized in Skippen (1971, 1974), Skippen et al. (1974), Slaughter et al. (1975), while a recent field study was reported by Moore et al. (1976). Fig. 6 is a typical mole-fraction diagram containing a number of invariant reactions. Mineral phases participating in such reactions specify and buffer, for a given temperature and pressure, the composition of the fluid, here expressed as mole fraction of CO_2. An example would be the assemblage quartz + dolomite + tremolite + calcite, related by the reaction

$$8\,Q + 5\,Do + H_2O \rightleftharpoons Tr + 3\,Cc + 7\,CO_2 \quad . \tag{15}$$

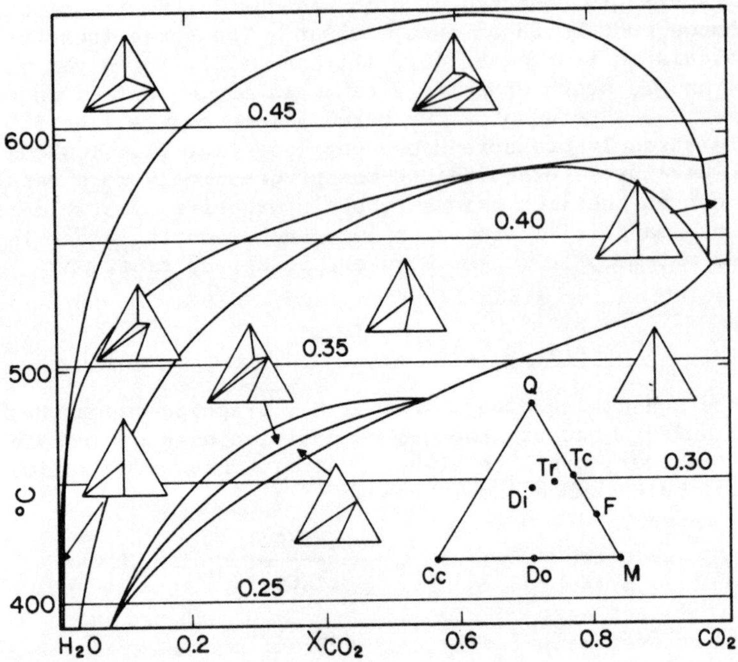

Figure 6. Mineral assemblages of siliceous carbonates at 3000 bars pressure as a function of temperature and fluid composition (from Skippen, 1975). $X(CO_2)$ is the mol fraction of CO2. Q: SiO_2, Cc: $CaCO_3$, Do: $CaMg(CO_3)_2$, M: $MgCO_3$, Di: $CaMgSi_2O_6$, Tr: $Ca_2Mg_5Si_7O_{22}(OH)_2$, Tc: $Mg_3Si_4O_{10}(OH)_2$, F: Mg_2SiO_4. Equilibria calculated for activities of $MgCO_3$ and $CaCO_3$ specified by the calcite-dolomite immiscibility gap. Horizontal contours are activities of $MgCO_3$.

Such buffered assemblages are quite common in nature and make contact aureoles of impure carbonate rocks one of the best areas to study the evolution of metamorphic fluids, because fluid compositions are essentially controlled by the mineral reactions. Progressive metamorphism initially leads to a rapid enrichment in CO_2, because the talc and tremolite reactions consume H_2O and release CO_2. At higher temperatures, water enrichment can take place by the decomposition of tremolite to diopside and/or forsterite. Changes in the compositions of the solids will affect the fluid ratios, and in fig. 6 the compositional changes of coexisting calcite and dolomite as a function of temperature have been taken into account.

In the presence of graphite, H_2O and CO_2 are still important, but CH_4, CO and H_2 also become significant. French (1966) and Skippen (1967) have analyzed the C-O-H system in equilibrium with graphite. The relevant equations are (7), (12), (14), (16) and (17).

$$C + 2\,H_2 \rightleftharpoons CH_4 \qquad (K_{16})_{P,T} = \frac{f(CH_4)}{f(H_2)^2} \tag{16}$$

$$P(gas) = \frac{f(H_2O)}{\gamma(H_2O)} + \frac{f(H_2)}{\gamma(H_2)} + \frac{f(CO_2)}{\gamma(CO_2)} + \frac{f(CO)}{\gamma(CO)} + \frac{f(CH_4)}{\gamma(CH_4)} + \frac{f(O_2)}{\gamma(O_2)} \;. \tag{17}$$

Again, one degree of freedom remains and has to be removed either by stipulating a bulk composition, a fixed $f(H_2)$ or $f(O_2)$. The explicit solution for fixed $f(O_2)$ in terms of $f(H_2O)$, can be obtained from

$$f(H_2O)^2\left[\frac{K_{16}}{(K_7)^2\cdot f(O_2)\cdot\gamma(CH_4)}\right] + f(H_2O)\left[\frac{1}{\gamma(H_2O)} + \frac{1}{K_7\cdot f(O_2)^{\frac{1}{2}}\cdot\gamma(H_2)}\right]$$

$$+\left[\frac{f(O_2)\cdot K_{12}}{\gamma(CO_2)} + \frac{[f(O_2)\cdot K_{12}\cdot K_{14}]^{\frac{1}{2}}}{\gamma(CO)} - P\right] = 0 \;. \tag{18}$$

Methane is the dominant species in equilibrium with graphite up to intermediate metamorphic grade (see fig. 7). Upon rising temperature it is replaced by H_2O, CO_2 and CO, in that order. This effect is mirrored beautifully in the fluid inclusion compositions of quartz crystals across the Alps (Poty et al., 1974). It is convenient to show the compositions of C-O-H gases in equilibrium with graphite in terms of partial pressure diagrams (see fig. 8). At a fixed gas pressure, any point in the diagram represents two fixed fugacities or partial pressures and hence, in the presence of graphite, the temperature must also be fixed. In other words, it is possible to draw isotherms on the graphite surface. We can also contour the gas pressures for standard oxygen buffers + graphite. Fig. 8 demonstrates again that pure $H_2O + CO_2$ is approached most closely at high T and/or high $f(O_2)$. Values below the graphite surface are not accessible unless gases outside the system C-O-H are present. Similarly, any fluid not in equilibrium with graphite and not fully oxidized to $H_2O + CO_2$ must lie between the graphite surface and the $P(gas) = P(H_2O) + P(CO_2)$ line.

C-O-H-S GASES

Perhaps the most important four-component gas system for crustal rocks is the system C-O-H-S, because of the common coexistence of silicates, carbonates and sulfides. Addition of one component means that an additional restriction is needed to specify the system. Usually this is achieved either by the presence of a sulfide buffer, such as pyrite and pyrrhotite (see fig. 9), by the participants of a sulfide-silicate reaction,

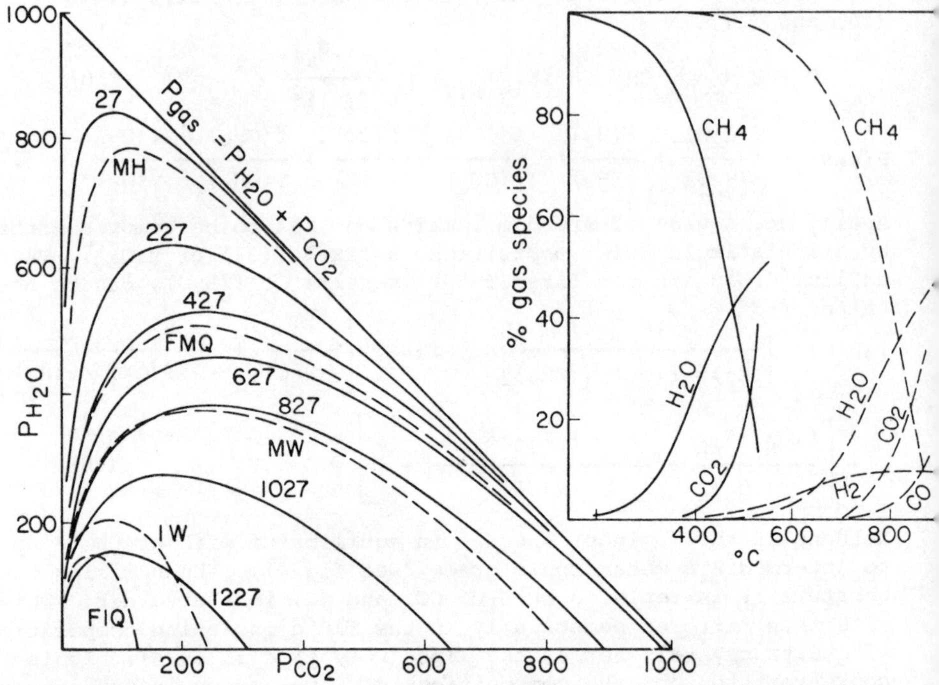

Figure 7 (right). Abundance of gas species in a fluid at 1000 bars in equilibrium with graphite + fayalite + magnetite + quartz (solid) and graphite + magnetite + wüstite (dashed). From French (1966). No fugacity coefficient corrections.

Figure 8 (left). Partial pressures of CO_2 and H_2O of a fluid in equilibrium with graphite at 1000 bars. From French (1966). No fugacity coefficient corrections. The dashed curves are the traces of the standard oxygen buffers (see fig. 3) on the graphite surface. Contours in °C.

such as

$$\underset{\text{annite}}{2\ KFe_3AlSi_3O_{10}(OH)_2} + 12\ H_2S + 3\ O_2 \rightleftharpoons \underset{\text{K-feldspar}}{2\ KAlSi_3O_8} + \underset{\text{pyrite}}{6\ FeS_2} + 14\ H_2O$$

or by a phase which defines $f(S_2)$ (intrinsic sulfur fugacity), such as pyrrhotite (see fig. 9). Gas compositions for eq. (19) have been calculated by Eugster and Skippen (1967). Gerlach and Nordlie (1975) studied the C-O-H-S system in detail with regards to magmatic gases. For the sulfur-rich portions, cooling at constant bulk composition follows a path which is oxidizing with respect to the standard oxygen buffers. A similar

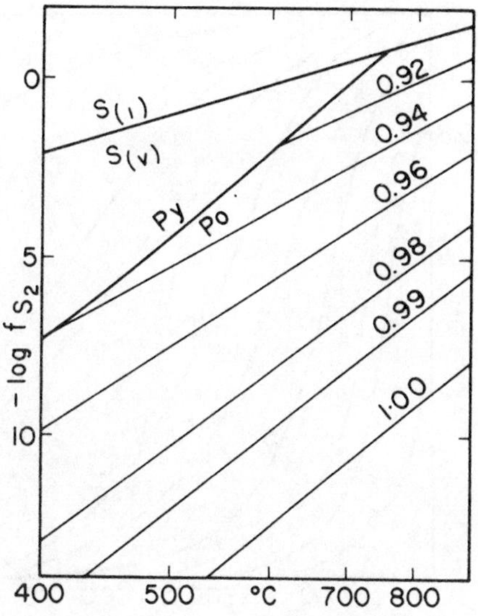

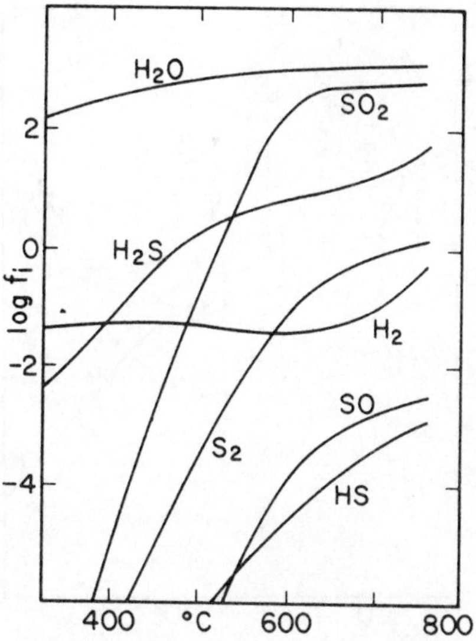

Figure 9 (left). Composition of pyrrhotite as a function of $f(S_2)$ and T (after Toulmin and Barton, 1964). S(L) and S(V) are sulfur liquid and vapor, Po: pyrrhotite, Py: Pyrite. Contours are in mole fraction FeS.

Figure 10 (right). Fugacities of gas species in a fluid in equilibrium with magnetite + pyrrhotite + pyrite at 2000 bars (from Eugster and Skippen, 1967).

observation was made by Sato and Wright (1966) for the gases from the drill holes in the Makaopuhi lava lake. Fig. 10 presents the results of calculations for the system H-O-S, where $f(O_2)$ and $f(S_2)$ are buffered internally by common mineral assemblages. The most abundant gas species are H_2O, H_2S and SO_2, with SO_2 becoming dominant towards the higher temperatures. There now are experimental methods available to study mineral reactions with such gases at high pressures. Fig. 11 shows a preliminary calibration of the SO_2 buffer, which is patterned after the HCl buffer of Frantz and Eugster (1973). A gas of composition H-O-S has a variance of 4, hence, in addition to P and T, two intensive parameters must be specified. In Fig. 11 this was done by using Ag + Ag_2S to fix $f(S_2)$, while $f(H_2)$ was controlled externally by an oxygen buffer + H_2O.

H-O-F AND H-O-N GASES

Fluorine can substitute for OH in many hydrous silicates such as amphiboles, micas, humites, topaz and this substitution has a pronounced effect on the properties and stability of such phases. Methods for calibrating mineral-gas reactions involving

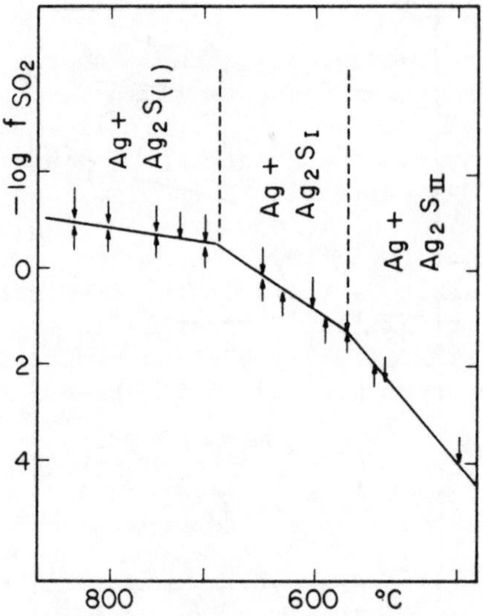

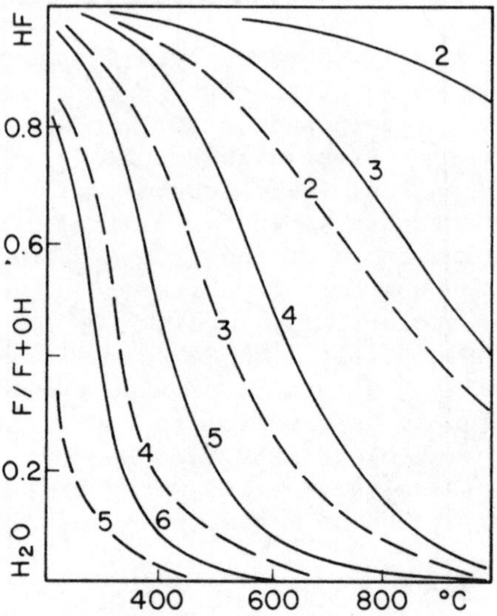

Figure 11 (left). Fugacity of SO_2 of the Ag + Ag_2S buffer at 2000 bars. The hydrogen fugacity is controlled externally by magnetite + hematite. Arrows indicate individual experiments. Unpublished data of Rudert and Eugster.

Figure 12 (right). Compositions of fluids in equilibrium with phlogopite (solid curves) and annite (dashed curves), from Munoz and Ludington (1974). The contours are labeled for log $f(H_2O)/f(HF)$.

fluorine were presented by Munoz and Eugster (1969), employing fluorine buffers, such as $CaCO_3$ + CaF_2 + $CaSiO_3$. To assess the abundance of fluoride species in igneous and metamorphic fluids, exchange reactions such as those in phlogopite (eq. 6) are useful. Munoz and Ludington (1974) have calibrated this reaction for biotites (see fig. 12), while Stormer and Carmichael (1971) used phlogopite-apatite pairs. Methods for calibrating H-O-N gases were presented by Hallam and Eugster (1974, 1976), using Cr + CrN as a nitrogen buffer, while Holloway and Reese (1974) proposed procedures for generating N_2-CO_2-H_2O fluids for hydrothermal experiments.

Other gas compositions can be treated in a similar manner. It is important to define the number of components C to be included for a specific metamorphic or igneous fluid. For a given P and T, C-1 intensive parameters, such as individual fugacities or fugacity ratios, must be evaluated before the simultaneous equations can be solved. For experimental systems, $f(H_2)$ is normally fixed by using hydrogen membranes in conjunction with oxygen buffers. For natural assemblages, simple or complex buffers or fugacity indicators must be used. If the necessary minerals and mineral assemblages are missing, this approach fails.

ACIDS AND BASES

All minerals, even the most refractory silicates, are soluble to a certain extent in a supercritical aqueous fluid (Morey and Hesselgesser, 1951). Hence igneous and metamorphic fluids must be treated as solutions and not just as mixtures of gases (Helgeson, 1967; Eugster, 1970). Very few true solubility constants are known for minerals, the most complete data available being those of Anderson and Burnham (1965, 1967) on quartz and corundum and of Currie (1968) on albite. The solution chemistry of geologically important supercritical fluids is still in its infancy, although Helgeson and Kirkham (1974, 1976) have made a valiant effort to provide the tools for predicting the behavior of aqueous electrolytes at high P and T.

Hemley (1959) and Hemley and Jones (1964) studied several acid-salt reactions which are important for establishing the compositions of metamorphic solutions, such as the muscovite-feldspar equilibrium

$$KAl_3Si_3O_{10}(OH)_2 + 6\ SiO_2 + 2\ K^+ \rightleftharpoons 3\ KAlSi_3O_8 + 2\ H^+ . \qquad (20)$$

From fluid inclusion studies (Roedder, 1972) we know that chloride is usually the principle anion, hence for the P-T range where acids and salts are associated, we should write

$$KAl_3Si_3O_{10}(OH)_2 + 6\ SiO_2 + 2\ KCl \rightleftharpoons 3\ KAlSi_3O_8 + 2\ HCl. \qquad (21)$$

Dissociation constants for KCl and HCl are listed in Helgeson and Kirkham (1976). From this information it is possible to predict the compositions of solutions in equilibrium with feldspar, micas and aluminosilicates (Eugster, 1970; Shade, 1974; Wintsch, 1975; for Russian references see Burt, 1976). Examples are shown in figs. 13 and 14. Frantz and Eugster (1973) developed a method for studying acid-salt reactions, whereby the fugacity of HCl is controlled at P and T. It is based on fixing $f(Cl_2)$ internally by using Ag + AgCl, and imposing a fixed $f(H_2)$ externally. This controls $f(HCl)$ through the reaction

$$\tfrac{1}{2}\ H_2 + \tfrac{1}{2}\ Cl_2 \rightleftharpoons HCl \qquad (K_{22})_T = \frac{f(HCl)}{f(H_2)^{\frac{1}{2}} \cdot f(Cl_2)^{\frac{1}{2}}} . \qquad (22)$$

The calibration of this buffer has made it possible to determine the dissociation constant of HCl

$$HCl \rightleftharpoons H^+ + Cl^- \qquad (K_{23})_T = \frac{a(H^+) \cdot a(Cl^-)}{a(HCl)} \qquad (23)$$

and to measure free energies and dissociation constants of acids and salts in supercritical aqueous fluids with a minimum of quench problems. Using this approach, the solubility of magnetite was measured by Chou and Eugster (in press; see fig. 15). The solubility increases gradually with decreasing temperature,

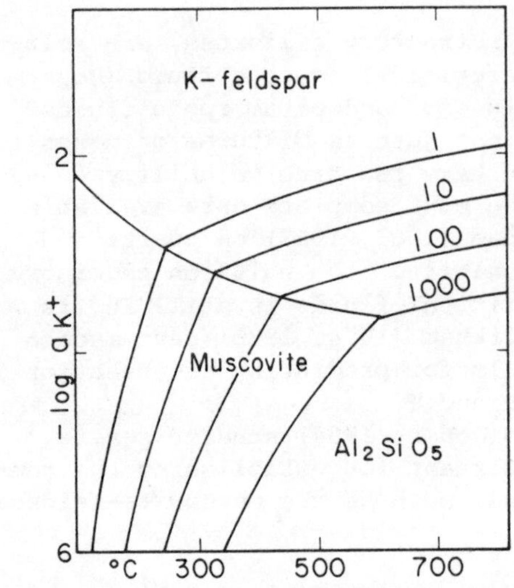

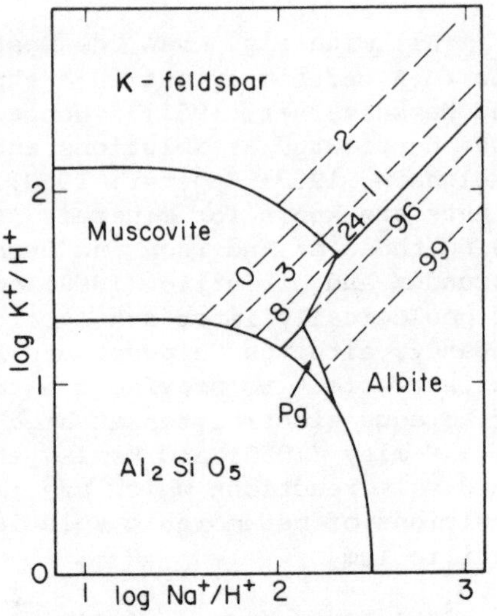

Figure 13 (left). Stability of aluminosilicates as a function of potassium activity at a constant H^+ activity of 10^{-7}. Contours are for $f(H_2O)$ in bars. From Eugster (1970).

Figure 14 (right). Stability of aluminosilicates as a function of potassium and sodium activities at 500°C, 1000 bars. After Wintsch (1975). Contours are labeled for mol per cent albite and paragonite, respectively. Pg: paragonite.

Based on determinations of the chloride and total iron concentrations after quench, the solubility curve is thought to represent the reaction

$$Fe_3O_4 + 6\ HCl + H_2 \rightleftharpoons 3\ FeCl_2 + 4\ H_2O \qquad (24)$$

from which the free energy of formation of $FeCl_2$(aq.) at 2 kb and T can be calculated. This then allows us to define the equilibrium constant for

$$Fe_2O_3 + 4\ HCl + H_2 \rightleftharpoons 2\ FeCl_2 + 3\ H_2O \qquad (25)$$

Reactions (24) and (25) are evaluated in fig. 16 for a constant molality of HCl. The contours refer to the molality of associated $FeCl_2$ at P and T. At lower temperatures, $FeCl_2$ dissociates, according to

$$FeCl_2 \rightleftharpoons Fe^{++} + 2\ Cl^- \qquad (26)$$

and the solubility of magnetite is governed by

$$Fe_3O_4 + 6\ H^+ + H_2 \longrightarrow 3\ Fe^{++} + 4\ H_2O \qquad (27)$$

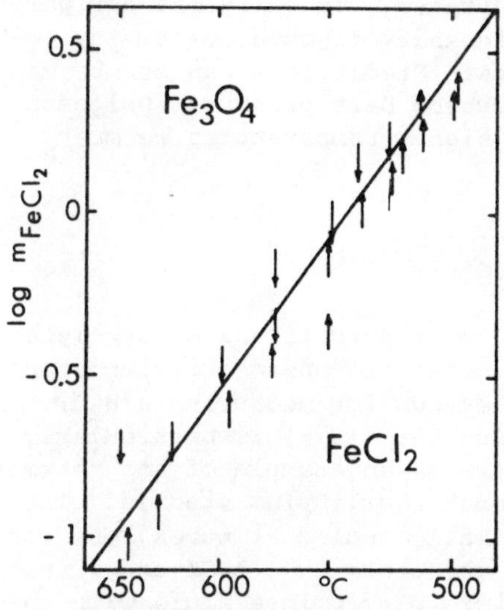

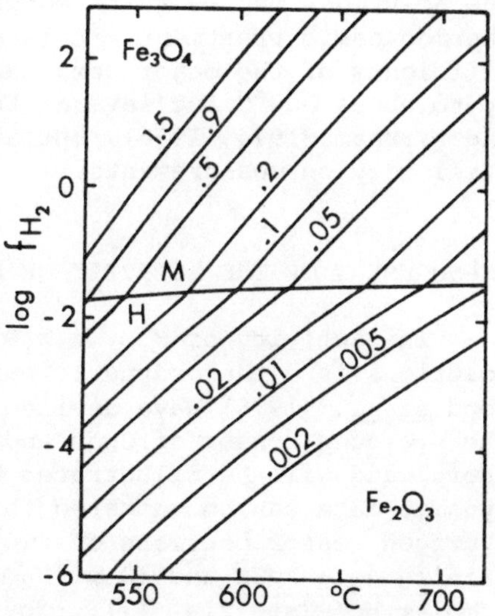

Figure 15 (left). Solubility of magnetite in a chloride fluid at 2000 bars (Chou and Eugster, in press). The fugacity of HCl is controlled by Ag + AgCl internal and MH external (Frantz and Eugster, 1973). Arrows are individual experiments.
Figure 16 (right). Solubilities of magnetite and hematite as a function of H_2 fugacity at a constant molality of HCl of 0.03 and 2000 bars pressure (from Chou and Eugster, in press). Contours are labeled for molality of $FeCl_2$.

and the total amount of iron in solution, $m(FeCl_2) + m(Fe^{++})$ must be calculated from the $m(FeCl_2)$ values, equations (23), (26) and the electrical neutrality condition:

$$2\ m(Fe^{++}) + m(H^+) = m(Cl^-) + m(OH^-)\ . \tag{28}$$

In calculations of ionic equilibria, the electrical neutrality equation plays the rolè of the P(gas) equation, eqs. (8), (11), and (17). Data such as those presented in fig. 15 and 16 can be used to define the compositional parameters of ore-forming fluids which precipitate magnetite and hematite. From the thermodynamic parameters extracted from such data, it is now possible to predict the compositions of the supercritical solutions in equilibrium with many metamorphic and igneous assemblages. This is the first step towards mapping compositional gradients in the fluid for specific field occurrences. Data for the system SiO_2-MgO-H_2O-HCl at 1 and 2 kb have been presented by Frantz (1973). Gunter (1974) has extended the original work of Hemley (1959) and has included $MgCl_2$ and $CaCl_2$ in addition to KCl.

Judging from the dissociation constants available, above about 500°C the solutes in the supercritical fluid are largely associated molecules. Fluid inclusion studies indicate that the solutions may be quite concentrated. To carry out a rigorous thermodynamic treatment, it is necessary to obtain activity coefficients of the major components. Predictions can now be made up to about 500°C and several thousand bars pressure (Helgeson and Kirkham, 1974, 1976), but at higher temperatures we must still rely on measurements.

MEASUREMENT OF THE ACTIVITY OF H_2O

The activity of H_2O, $a(H_2O)$, of supercritical electrolyte solutions is a thermodynamic parameter of fundamental importance. Wood et al. (1975) have devised methods for measuring $a(H_2O)$. The hydrogen sensor of Chou and Eugster (1976) is particularly useful and will be illustrated here as an example of how thermodynamic data can be gathered through equilibrium studies. The hydrogen sensor consists of two small, sealed Pt tubes, one containing Ag + AgCl and 10 µl H_2O, the other Ag + AgCl and 10 µl 3 molar HCl (see fig. 17). They are placed in a fluid with unknown $f(H_2)$. After quench, pH and m(Cl) are measured in both tubes, each defining an m(HCl) and thus also an $f(H_2)$ value, approached from opposite sides, bracketing the equilibrium value. The relations necessary for calculating $f(H_2)$ are, in addition to eqs. (7), (22) and (23)

$$Ag + \tfrac{1}{2} Cl_2 \rightleftharpoons AgCl \qquad (K_{29})_{P,T} = f(Cl_2)^{-\frac{1}{2}} \tag{29}$$

$$P(gas) = P(H_2O) + P(HCl) + P(H_2) + P(Cl_2) \tag{30}$$

$$f(H_2)ext \equiv f(H_2)int \tag{31}$$

$$H_2O \rightleftharpoons H^+ + OH^- \qquad (K_{32})_T = \frac{a(H^+).a(OH^-)}{a(H_2O)} \tag{32}$$

$$m(H^+) = m(Cl^-) + m(OH^-) \quad . \tag{33}$$

Assuming that $m(OH^-)$ is small, and that all $(HCl)_{P,T}$ dissociates upon quench, we have from (33):

$$m(H^+)1,25° = m(Cl^-)1,25° = m(HCl)_{P,T} + m(Cl^-)_{P,T} \quad . \tag{34}$$

Using eqs. (23) and (34) and either a quench pH or chloride measurement, m(HCl)P,T can be calculated, provided the activity coefficient quotient in (23) is near unity.

$$m(HCl)^2{}_{P,T} - m(HCl)_{P,T}\,[2m(H^+)1,25° + K_{23}\,] + m(H^+)^2 1,25° = 0. \tag{35}$$

Assuming Raoult's law to hold, we have:

$$m(HCl) = X(HCl) \cdot 55.5 = \frac{f(HCl) \cdot 55.5}{f^*(HCl)} = K_{36} \cdot f(HCl) \tag{36}$$

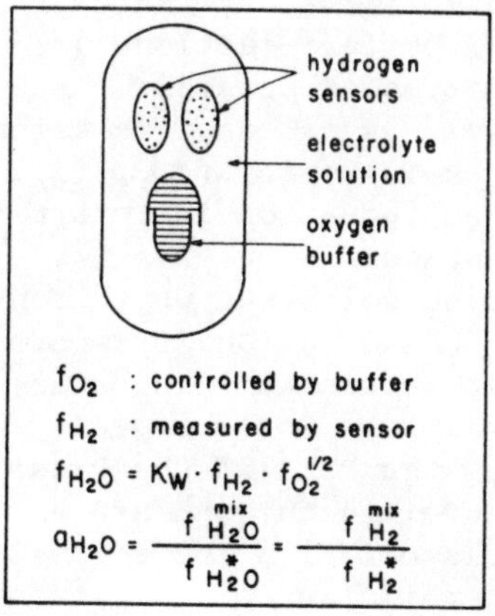

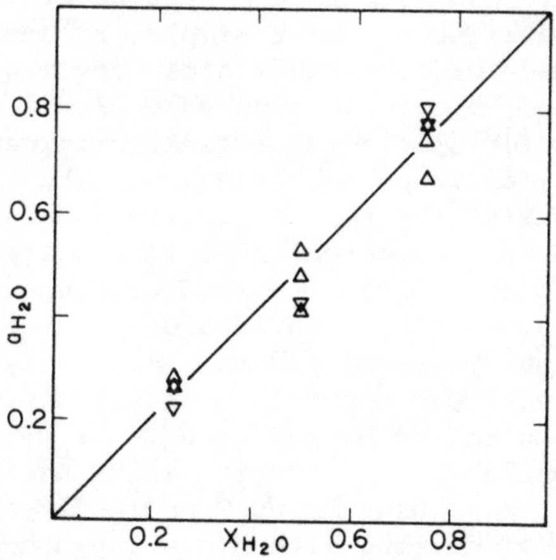

Figure 17 (left). The hydrogen sensor of Chou and Eugster (1976). Hydrogen sensors are in sealed Pt tubes, the oxygen buffer in an open tube and the electrolyte in a sealed gold tube. K(w) is K_7 and f_i^* the fugacity of pure i at P and T.

Figure 18 (right). Activity versus mol fraction of H_2O in KCl-H_2O solutions at 700°C, 2000 bars. Unpublished data of Wood et al. (in press). For individual measurements, direction towards equilibrium is shown by the apex of each triangle.

where f^*(HCl) is the fugacity of pure HCl at P and T. Combining eqs. (22), (29) and (36)

$$m(HCl)_{P,T} = \frac{K_{22} \cdot K_{36}}{K_{29}} \cdot f(H_2)^{\frac{1}{2}}_{P,T} \qquad (37)$$

defines $f(H_2)$ in the fluid.

The hydrogen sensors in conjunction with an oxygen buffer makes it possible to measure $a(H_2O)$ in electrolyte solutions at high P and T. The preliminary results for KCl-H_2O at 700°C and 2 kb pressure are shown in fig. 18. KCl-H_2O solutions are essentially ideal under these conditions. Development of the hydrogen sensor has also made it possible to measure diffusion of hydrogen through Pt at high P and T. It is hoped that these techniques can be extended to measure H_2 and H_2O diffusion through natural rock fabrics.

SUMMARY AND CONCLUSIONS

The supercritical fluid in equilibrium with mineral assemblages has been treated first as a mixture of uncharged gas species and then as an electrolyte solution. Although the procedures are general, most examples have been drawn from metamorphic rocks.

Pressure and temperature must be evaluated first, by whatever means available. Next simple or complex buffer reactions are searched for to establish as many fugacities as possible. Fugacity indicators are also valuable, as well as exchange reactions, the latter specifying fugacity ratios. A value for $f(O_2)$ is usually assigned most readily. The P(gas) equation is important for solving the set of simultaneous equations.

Natural supercritical fluids are also solutions which contain acids, salts, ions and complexes. Dissociation is important below about 500°C, while above that temperature the fluids are essentially associated and relatively simple. For charged species, the electrical neutrality equation plays the role of the P(gas) equation for uncharged gases. Experimental techniques are now available for studying acid-base, dissociation and solubility reactions at high P and T, markedly extending our previous capability of dealing with hydration-dehydration, carbonation-decarbonation and redox reactions. The igneous and metamorphic fluid phases are now treated as true electrolyte solutions and their thermodynamic parameters are determined from the point of view of solution chemistry, using well-established approaches modified appropriately for the high pressure-high temperature environment.

REFERENCES

Anderson, G.M. and Burnham, C.W., Am. J. Sci., 263, 494-511, 1965.
_____, Am. J. Sci., 267, 12-27, 1967.
Buddington, A.F. and Lindsley, D.H., J. Petrol., 5, 310-357, 1964.
Burnham, C.W., Holloway, J.R. and Davis, N.F., Geol. Soc. Am. Spec. Pap. 132, 96 p.
Burt, D.M., Econ. Geol., 71, 665-671, 1976.
Carmichael, I.S.E., and Nicholls, J., J. Geophys. Res., 72, 4665-4687, 1967.
Chou, I. and Eugster, H.P., E⊕S, 57, 340, 1976.
_____, Econ. Geol., in press.
Currie, K.L., Am. J. Sci., 266, 321-341, 1968.
Day, H.W., Am. Min., 58, 255-262, 1973.
Deines, P., Nafziger, R.H., Ulmer, G. and Woermann, E., Penn. State Bull. Earth Sci., 88, 129 p., 1974.
Eugster, H.P., J. Chem. Phys., 26, 1760, 1957.
_____, in Abelson, ed., Researches in Geochemistry, John Wiley, 397-426, 1959.
_____, Fortschr. Min., 47, 106-123, 1970.
_____, 24th Int. Geol. Congr., 10, 3-11, 1972.
_____, and Skippen, G.B., in Abelson, ed., Researches in Geochemistry, v. 2, John Wiley, 492-520, 1967.
_____, and Wones, D.R., J. Petrol., 3, 82-125, 1962.
Frantz, J.D., Ph.D. Thesis, Johns Hopkins Univ., Baltimore, 59 p., 1973.
_____, and Eugster, H.P., Am. J. Sci., 273, 268-286, 1973.

French, B.M., Rev. Geophys. 4, 223-253, 1966.
_____, and Eugster, H.P., J. Geophys. Res., 70, 1529-1539, 1965.
Gerlach, T.M. and Nordlie, B.E., Am. J. Sci., 275, 353-410, 1975.
Greenwood, H.J., Am. J. Sci., 273, 561-571, 1973.
Gunter, W.D., Ph.D. Thesis, Johns Hopkins Univ., Baltimore, 195 p., 1974.
Haas, J.L., and Robie, R.A., Trans. Am. Geophys. Union, 54, 483, 1973.
Hallam, M. and Eugster, H.P., E⊕S, 55, 452, 1974.
_____, Contr. Min. Petr., in press.
Helgeson, H.C., in Abelson, ed., Researches in Geochemistry, v. 2, John Wiley, 362-404, 1967.
_____, and Kirkham, D.H., Am. J. Sci., 274, 1089-1261, 1974.
_____, Am. J. Sci., 276, 97-240, 1976.
Hemley, J.J., Am. J. Sci., 257, 241-270, 1959.
_____, and Jones, W.R., Econ. Geol. 59, 538-569, 1964.
Hougen, O.A. and Watson, D.R., Chem. process principles charts, John Wiley, 219 p., 1946.
Huebner, J.J., in Ulmer, ed., Research techniques for high pressure and temperature, Springer-Verlag, 123-177, 1971.
Moore, J.N. and Kerrick, D.M., Am. J. Sci., 276, 502-524, 1976.
Morey, G.W. and Hesselgesser, J.M., Econ. Geol., 46, 821-835, 1951.
Munoz, J.L. and Eugster, H.P., Am. Min. 54, 943-959, 1969.
_____, and Ludington, S.D., Am. J. Sci., 274, 396-413, 1974.
Poty, B.P., Stalder, H.A. and Weisbrod, A.M., Schweiz. Min. Petr. Mitt., 54, 717-752, 1974.
Presnall, D.C., J. Geophys. Res., 74, 6026-6033, 1969.
Redlich, O. and Kwong, J.N.S., Chem. Rev. 44, 233-244, 1949.
Robie, R.A. and Waldbaum, D.R., U.S. Geol. Surv. Bull. 1259, 256 p., 1968.
Roedder, E., U.S. Geol. Surv. Prof. Paper, 440 JJ, 164 p., 1972.
Rudert, V. and Eugster, H.P., unpublished data.
Ryzhenko, B.N. and Malinin, S.D., Geoch. Intern. 8, 562-574, 1971.
Sato, M., in Ulmer, ed., Research techniques for high pressure and high temperature, Springer-Verlag, 43-99, 1971.
_____, Geol. Soc. Am. Memoir, 135, 289-307, 1972.
_____, and Wright, T.L., Science, 153, 1103-1105, 1966.
Shade, J.W., Econ. Geol., 69, 218-228, 1974.
Shaw, H.R., in Abelson, ed., Researches in Geochemistry, v. 2, John Wiley, 521-541, 1967.
Skippen, G.B., Ph.D. Thesis, Johns Hopkins Univ., Baltimore, 251 p., 1967.
_____, J. Geol., 79, 457-481, 1971.
_____, Am. J. Sci., 274, 487-509, 1974.
_____, Fortschr. Min., 52, 75-99, 1975.
_____, and Hutcheon, I., Can. Min., 12, 327-333, 1974.
Slaughter, J., Kerrick, D.M. and Wall, V.J., Am. J. Sci., 275, 143-162, 1975.
Stormer, J.C. and Carmichael, I.S.E., Contr. Min. Petr., 31, 121-131, 1971.
Toulmin, P. and Barton, P.B., Geoch. Cosmoch. Acta, 28, 641-671, 1964.

Touret, J., Lithos, 4, 423-436, 1971.
Wintsch, R.P., J. Petrol., 16, 57-79, 1975.
Wones, D.R. and Eugster, H.P., Am. Min., 50, 1228-1272, 1965.
Wood, J.R., Chou, I. and Gunter, W.D., Geol. Soc. Am. Abstr. 7, 1321, 1975.

SUGGESTED PROBLEMS

Problem 1: Calculate the fugacities of H_2O, H_2 and O_2 of a gas phase containing twice as much hydrogen as oxygen ("pure" H_2O) between 600 and 1200 °K at 2 kb pressure. Compare the $f(O_2)$ values with those of an O-H gas in equilibrium with Fe_3O_4 + Fe_2O_3.

Problem 2: Evaluate the effect of making fugacity coefficient corrections on figure 7.

Problem 3: A marble contains the assemblage phlogopite + tremolite + dolomite + calcite + quartz. Accessory minerals are pyrrhotite, magnetite and graphite. The phlogopite contains 20% F-phlogopite and 20% annite. The $MgCO_3$ activity in dolomite is 0.35 and the pyrrhotite has a FeS mole fraction of 0.95. Estimate the gas composition, assuming all phases are in equilibrium.

THE SIGNIFICANCE OF FLUID INCLUSIONS IN METAMORPHIC ROCKS

Jacques Touret

Département des Sciences de la Terre,
Univ. Paris 7, L.A. CNRS 196(*)

INTRODUCTION

At high magnification, quartz and many other non-opaque minerals show the presence of small (a few µm) inclusions filled with various fluids. These are mainly H_2O and CO_2, and more rarely hydrocarbons. The basic rules for their recognition and interpretation are almost as old as petrography, as they were clearly formulated by Sorby in 1858, but only in recent times have decisive advances in technology (microthermometry) and in the knowledge of mineral reactions helped to make it possible to understand their importance as possible samples of the volatile phase during many petrogenetic processes.

Fluid inclusion techniques have been developed mainly by students of ore deposits (Smith, 1953; Roedder, 1967a, 1972), but they are now applied to many geological problems and especially to the study of metamorphic rocks for a number of reasons:

1. Most metamorphic rocks result from dehydration or decarbonation reactions which involve a fluid phase (heterogenous equilibria). The many methods of investigation of these reactions and the high degree of precision reached now by experimental petrology place strict limits to the possible compositions of the fluid phase, which can be compared with the actual content of the fluid inclusions. As will be seen later, there are so many causes of error (leakage, "necking down", reaction with the host mineral, etc.) that no fluid inclusion can be considered "a priori" as

(*) Postal address: Lab. Pétrographie, Muséum Hist. Naturelle, 61, rue de Buffon, 75005 Paris, France.

D. G. Fraser (ed.), Thermodynamics in Geology, 203-227.

representative of a large amount of volatiles. It is always necessary to check and compare them with a model composition to be certain that some limiting conditions are fulfilled.

2. "Hydrothermal" mineral veins or segregations occur in all metamorphic terrains. Since such veins contain large, sometimes idiomorphic crystals, they are of course much better suited for the study of fluid inclusions than massive rocks. Very improperly, they are often neglected by petrologists, but their mineralogy may be much more complex and interesting than is commonly assumed; in many cases, they contain metamorphic minerals (zeolites, aluminosilicates, etc.) which indicate the same metamorphic grade as the surrounding rock.

3. Except in the case of high grade anatectic rocks, metamorphic rocks do not involve the complex and poorly understood interactions between fluids and magmas. It is nevertheless evident that methods discussed in this paper apply also to any magmatic rock which has crystallized in the presence of fluids.

During the past ten years a number of papers have shown the interest of fluid inclusion studies in many different domains: Metamorphism of the Western Alps (Poty, 1969; Poty et al., 1974), hydrothermal metamorphism of the oceanic crust (Jehl, 1975; Jehl et al., 1976), pegmatite evolution in low pressure type Hercynian metamorphism (Weisbrod and Poty, 1975), granulite facies and related rocks (Touret, 1974; Hollister and Burruss, 1976). The results of these works are summarized elsewhere (Weisbrod et al., 1976). This paper will review some basic principles of the study of fluid inclusions in rocks. For reasons of space, only some aspects and techniques (notably microthermometry) will be discussed in detail and for a more general presentation the interested reader must consult the classical papers of Roedder, notably Roedder, 1967a and 1972. Fundamental notions on fluid systems are also given in F.G. Smith (1953) and Ypma (1963).

OBSERVATION BY CONVENTIONAL TECHNIQUES (PETROGRAPHIC MICROSCOPY)

The study of fluid inclusions starts with their observation under the microscope which, done carefully, may yield a surprising amount of information. Millions of fluid inclusions may occur in a single thin section of hydrothermal quartz in a high grade metamorphic rock, but their size rarely exceeds a few tens of microns in segregation and a few microns in massive rocks. High magnification ("normal" light) and doubly polished thick (up to 0.5 mm) plates, which allow 3-dimensional observation, are therefore necessary, but with some training many fluid inclusions can also be observed in "normal" thin sections (x 50 or immersion objective, condenser slightly lower than for conoscopy; <u>no</u> crossed nicol).

Among the many features which can be recorded, three are especially important (Fig. 1): Morphology (size, shape, etc.), mode of occurrence and content at room temperature (T_{lab}).

Morphology

The dimensions and the shape of the fluid inclusion cavity are the most obvious but the least informative features. Many small inclusions tend to have a minimal energy configuration, i.e. a "negative crystal" shape. But this has no immediate significance (Ypma, 1963; Roedder, 1967a), even if some very general trends can be deduced from the study of many fluid inclusions, for example CO_2 and "secondary" inclusions (see below) are more commonly negative crystal shaped than H_2O and "primary" inclusions. The size also is not meaningful: except in the case of metastability (Roedder, 1967), a small inclusion may be more representative than a big one, which is more liable to leakage or secondary alteration. Only the lower limit is important: below a few microns, fluid inclusions can hardly be observed and they cannot be studied properly by present day optics. As cavities in minerals grade continuously into crystal dislocations as shown by electron microscopy (Green, 1972), there is indeed a wide field which cannot be studied at present.

Finally, the most interesting morphological features are those which indicate that the following two basic assumptions have not been fulfilled (of course, their absence does not indicate that they have been!). These are:

a) The fluid trapped in the fluid inclusions is representative of the volatiles which did exist in a relatively large domain at the time of the trapping. This is not the case when a tubular inclusion is fragmented into smaller ones during slow cooling (necking down phenomenon, Roedder, 1967a; see section 3 and Fig. 3).

b) No modification of the fluid phase has occurred after the trapping ("leakage" problem). There are many indications that leakage is the exception more than the rule, especially if quartz is the host mineral (Roedder and Skinner, 1968). Some inclusions however show small cracks (Fig.1.6) or more typically a star like shape with arrays of small secondary inclusions (Fig. 1.7) which indicate that natural decrepitation has occurred either during cooling as the internal pressure exceeded the strength of the host mineral, or during a reheating event (Lemmlein and Kliya, 1954). Examples of this phenomenon are known in many hydrothermal dykes and on a regional scale, in some high grade polymetamorphic domains (Touret, 1974).

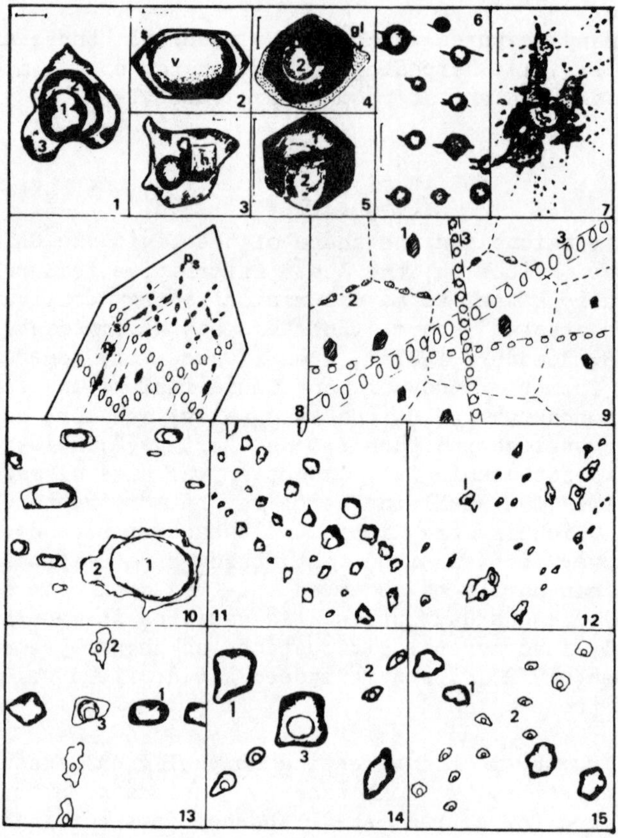

Fig. 1. Fluid inclusion morphology and chronology. Scale bar = 10 µm. All are taken at T_{lab}.

1-7: Different types of isolated (single) inclusions

1. Three fluid phases: (1) $CO_2(g)$; (2) $CO_2(l)$; (3) $H_2O(l)$. Fluid inclusions in porphyry coppers (after Roedder, 1972, Plate 4). These occur in distinct cracks but in the same crystal and possibly represent contemporaneous fluid immiscibility (cf. Fig. 4).

2. Negative crystals containing a low density fluid (large vapour bubble rich in CO_2; small amount of low salinity H_2O liquid) s: possibly specularite.

3. Very dense brine with large halite crystal (h) and several other solid phases (anhydrite, specularite? chalcopyrite? etc.).

4. & 5. Carbonic (high density CO_2 ± hydrocarbons) inclusions (Touret, 1974). Typical of mantle derivatives and deep-seated crustal rocks (granulite facies).

4: in olivine from an ultrabasic nodule in basalt (Roedder, 1972, Pl.4, 6) gl: basaltic glass; 1: H_2O; 2. CO_2(g).

5: negative crystal in orthoclase from Itrongay pegmatite, Madagascar (Touret, 1974). 1. CO_2(l); 2. CO_2(g); s. unknown solid, possibly graphite or bitumen.

6. & 7. Shape change from temporary overheating.

6: Microfissure carbonic inclusions from Bournac granulites (Bilal and Touret, 1974).

7. "Exploded" carbonic inclusion in the Ansignan charnockite (Touret, 1974).

<u>8-9</u>: "Primary" and "secondary" inclusions

8. In idiomorphic crystals. p = primary; s = secondary; ps = pseudosecondary.

9. In massive rocks. Chronology is defined with respect to grain boundaries. 1. Isolated inclusions in grains; 2. inclusions along grain boundaries; 3. lines of inclusions along healed cracks.

<u>10-12</u>: "Homogeneous" fluid in secondary inclusions. (Hoefs and Touret, 1975)

10. (1) CO_2(l); (2) H_2O.

11. High density CO_2 ($T_h = 10 \pm 5^{\circ}C$).

12. H_2O:CO_2 constant ratio V_{gas}:V_{liq}.

<u>13-15</u>: Chronology of secondary inclusions (see Pagel, 1975)

13. Intersection of planes of CO_2(1) and H_2O(2) inclusions. H_2O inclusion (3) at intersection indicates that (1) and (2) were not contemporaneous, but it does not give any indication of the time relationships.

14. Plane of former high-density CO_2 inclusions (1) is intersected by a set of low density ones (2). The density of (1) is lowered with conservation of the cavity at the intersection (3).

15. Late inclusions (2) crossing former ones (1) with obliteration of cavities of (1).

Mode of occurrence. Primary/secondary inclusions, a "faux problème"

Among fluid inclusions, a classical distinction exists between "primary" (fluids have been trapped along growing faces or edges of the host crystal) and "secondary" ones (they are formed by the healing of an open crack in already crystallized material) (Fig. 1.8). N.P. Ermakov (1949) has added the term "pseudosecondary" for inclusions which have the appearance of secondary - but the significance (related to the growth of the host crystal) of primary ones: the healing of the crack occurs during crystal growth (Fig. 1.8).

Features that may help to distinguish primary and secondary inclusions have been listed by many authors, notably Roedder (1967a) (see also Lemmlein, 1929; Ermakov, 1950; Deicha, 1955, Ypma, 1963; Poty, 1969), but few are unambiguous: primary inclusions occur in solitary or crystallographically induced positions, typically along growth surfaces of the host mineral, while secondary ones are regularly disposed along recrystallized (healed) microcracks or fractures which may crosscut crystal boundaries. As growth planes are never visible in massive rock-forming minerals, true primary inclusions are, with a few exceptions (Berglund and Touret, 1976), never seen in metamorphic rocks. Strictly speaking, all their inclusions must therefore be considered as secondary, but careful observation nevertheless makes it possible to establish in many cases a crude chronology between relatively "primary" inclusions (Hollister, 1973) (Fig. 1.9) occurring alone or in small groups scattered within the host mineral, and true "secondary" ones, regularly disposed along healed microcracks. It is also often possible to distinguish between early cracks and later ones which are much less recrystallized and may sometimes grade into open fissures and, using simple criteria derived from the immediate observation, to propose a partial chronology of the different fissures (Fig. 1.13-15).

The fact that most metamorphic inclusions are secondary has led to the diffuse opinion that they are useless. Notwithstanding the evidence that they are an intrinsic part of the rock and have consequently to be studied, I see the problem in an entirely different way: primary and secondary inclusions are exactly alike; neither is "a priori" secure as any is liable to have suffered one of the many processes which may alter the representativity of the enclosed fluid, such as selective trapping, leakage, necking down or reaction with the host mineral. Therefore any fluid inclusion data must be compared with results derived independently by other methods. This is the point at which we are now, as many of these data become available in metamorphic petrology from experimentation or theoretical models. Only when some limiting conditions have been fulfilled may additional data be obtained and it comes then as a happy surprise to discover that some "late" inclusions may be in fact remarkably close to the peak of metamorphism, e.g. quartz segregations in regional metamorphism (Rich, 1975) or CO_2 inclusions in granulite facies rocks (Touret, 1974).

Composition at T lab

Without any sophisticated instrumentation, some most important types of fluid inclusion are easily recognized under the petrographic microscope (for a complete and detailed description of the criteria see Roedder, 1972). Most common fluids may be classified as aqueous: H_2O (Fig. 1.2), H_2O + NaCl (Fig. 1.3), H_2O + NaCl + many other daughter minerals (14 in Volhynia pegmatite, each a different species, Lyakhov, 1967), H_2O + immiscible hydrocarbons and oil, carbonic (CO_2 ± dissolved hydrocarbons, Fig. 1.4 and 5) and composite (H_2O + CO_2, Fig. 1.1). Similar in appearance to some of these fluid inclusions are glassy (Clochiatti, 1975) and empty low density gas inclusions, which must be checked by microthermometry (see below).

From our experience, it is possible to estimate roughly the relative abundance of different types (notably H_2O and CO_2) especially in coarse grained homogeneous rocks (granites, charnockites, etc.). We normally estimate the percentage in the field of view of a microscope with the help of percentage charts common in sedimentary petrography and we take the average of 10-15 fields for a single thin section. This method is evidently applicable only if fluid inclusion composition varies markedly in the investigated samples, but in several instances (South Norway quartzites (Touret, 1972), Kleivatn granite (South Norway (Touret, 1974 and Konnerup-Madsen, in press)), it has helped to establish an amazing correspondence between fluid inclusion composition and mineral isograds. For CO_2, the results are consistent with crushing stage determinations (Al Khatib and Touret, 1973).

MICROTHERMOMETRY: T_f and T_h

Since the time of Sorby, microthermometry (temperature of phase changes under the microscope) has always been the specific method used to study fluid inclusions, initially by heating (Smith, 1953; Ermakov, 1950) and more recently by heating and cooling (Roedder, 1962; Ypma, 1963; Poty, 1969). The method has many advantages - it is non-destructive, easy to perform, requires relatively simple instrumentation and allows the possibility of studying single inclusions, but also has some limitations - it is time consuming, difficult to study small inclusions and to obtain accurate results[1]. As decisive technological advances have been made recently, notably by B. Poty (Poty et al., 1976), microthermometry may now be performed on a routine basis in the temperature range -180 to +600°C and should become a standard method for most petrographic investigations.

Interpretation of Microthermometric Data

Metastability being a major problem in the closed and often concentrated system of the fluid inclusion (Roedder, 1967b), only temperatures of disappearance of an already existing phase can be interpreted: the temperature of nucleation of the same phase may differ by several tens of degrees.

Among all temperatures which may be recorded, two are especially important and will be discussed in some detail (Fig. 2). These are the temperature of fusion (T_f = solid → liquid) and the temperature of homogenization (T_h = disappearance of the meniscus between liquid and vapour). Also, temperatures of dissolution of the daughter minerals (notably halite), often recorded as T_s, are interesting, but since their interpretation is considered in Weisbrod et al. (1976), it will not be repeated here (see also Touray, 1970).

T_f: For a one component system, T_f, where three phases coexist (solid, liquid, vapour), must correspond to the triple point of the system (H_2O: 0.01°C, CO_2: -56.6°C, CH_4:-182.5°C, etc.) (Fig. 2.1 and 2). Conversely, melting temperatures at the triple point are the best indication yet available of the purity of the fluid. But the measurement must be accurate (of the order of 0.1°C): for water, melting temperatures below 0°C indicate the presence of dissolved ions (cryometry) and they are commonly interpreted as "equivalent NaCl" relative to the $NaCl-H_2O$ system (Sourirajan and Kennedy, 1962). A more complete analysis is possible if the ratio of the different species (Na^+, K^+, Ca^{++}, etc.) has been determined by the "leaching" method (Roedder et al., 1963; Ohmoto, 1968). However, to the knowledge of the writer, this is only possible at present for idiomorphic quartz in alpine type clefts (Poty et al., 1974). For aqueous fluids again, melting temperatures slightly above 0°C indicate the formation of gas hydrates (clathrates, e.g. $CO_2.5.75\ H_2O$, see Takenouchi and

(1) Microthermometric technique is discussed in detail by Roedder in the section on laboratory technique in a chapter "Fluid inclusion evidence on the genesis of ores in sedimentary and volcanic rocks" in a three volume set entitled "Ores in Sediments, Sedimentary and Volcanic Rocks" edited by K.H. Wolf, to be published by Elsevier Pub. Co.

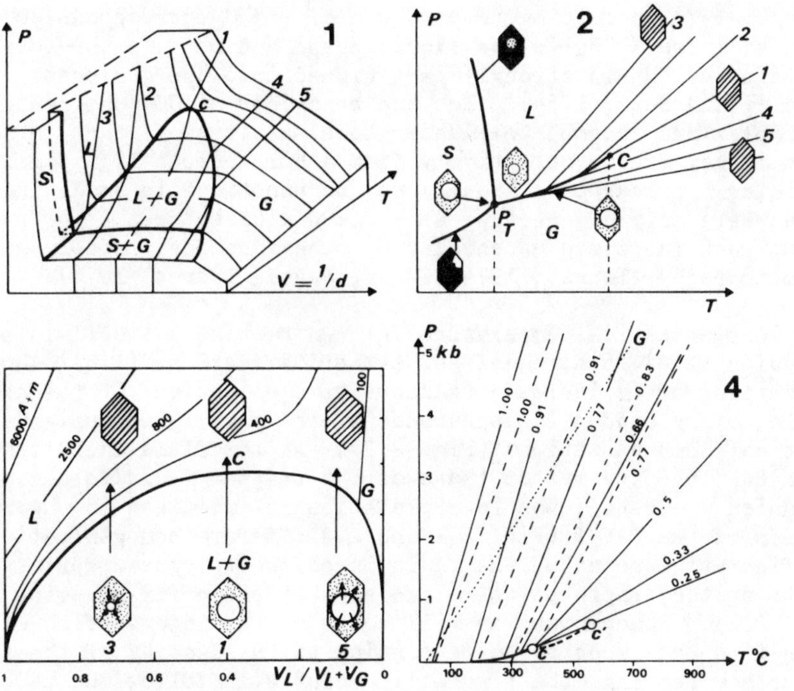

Fig. 2. Interpretation of microthermometric data: T_f and T_h

1. P-V-T of one component system. d = density; S = solid; L = liquid; G = gas (vapour); C = critical point; 1-5 = isochores (as in 2 and 3).

2. P-T section. P_T = Triple point. For the sketch inclusions, black = solid; stippled = liquid; white = gaseous; hatched = fluid.

3. T-V section for H_2O (after Roedder, 1967a), illustrating liquid (3), critical (1) and gaseous (5) homogenization. For water, as $d_{Liquid} = 1$, $d_{Gas} \neq 0$, $d_{Inclusion} = V_L/(V_L + V_G)$.

4. Isochores of H_2O (Solid lines - data from Kennedy, 1950) and H_2O + 30% NaCl (Dashed lines - data from Lemmlein and Klevtsov, 1956) Gt: "normal" geotherm (30°C/km). C = critical point (H_2O). C' = critical point (H_2O + 30% NaCl).

Kennedy, 1965; Roedder, 1972; Hollister and Burruss, 1976), which may be useful in helping to recognize extremely small quantities of gases dissolved in water (Poty, 1969; Leroy, 1971).

For CO_2, melting temperatures lower than -56.6°C indicate the presence of hydrocarbons, notably CH_4: After Hollister and

Burruss, 1976, a melting temperature of -65°C corresponds to about 45 moles % CH_4 in the liquid phase and more in the total inclusion as CH_4 is strongly partitioned into the gas phase (Donally and Katz, 1954). But the behaviour of CO_2 on freezing is complex and incompletely understood, as it seems that in some cases it may freeze selectively from a homogeneous liquid mixture. Remaining hydrocarbons (liquid) may be unnoticed in small inclusions where only CO_2 melting at -56.6°C will be seen. The purity of CO_2 must therefore be checked by other methods, notably gas chromatography (Touret, 1974; Cuney, Pagel, Touret, 1976).

Homogenization temperature T_h: On heating a two fluid phase inclusion (L+G) it homogenizes, either by decrease (liquid homogenization) or by increase (gaseous homogenization) of the gas bubble, or by sudden disappearance of the gas/liquid meniscus (critical homogenization) (Fig. 2.3). As the fluid inclusion is a constant density system (Mass of the trapped fluid/Volume of the inclusion = const.), the interpretation is immediate in the P-T section of the P-V-T (Fig. 2.1 and 2.2): Before homogenization, the figurative point of the fluid is on the two-phase curve which starts at the triple point P_T and ends at the critical point C (Fig. 2.2). Homogenization will occur at the intersection of this curve with the isochore corresponding to the density of the fluid. On further heating, the figurative point will follow the isochore, which for many systems may be taken in first approximation as straight lines (Fig. 2.2 and 2.4). The true temperature of trapping <u>differs</u> from T_h by the (unknown) amount followed on the isochore, which is often referred to in the specialized literature as the "pressure correction". This correction can hardly be neglected as it may reach several hundreds of degrees and experience shows that its determination - in terms of pressure - is one of the most important pieces of information which may be gained from fluid inclusion studies, at least in metamorphic rocks (Poty et al., 1974).

In conclusion: the interpretation of T_f and T_h may be summarized as follows: T_f indicates the chemical composition of the fluid and then if the P-V-T of the system have been determined experimentally, T_h gives its <u>density</u>, i.e. a first order relationship between P and T at the time of the trapping. Complete determination of P and T necessitates consequently the independent determination of one of these two unknowns (notably T, see 4.1) or the rather improbable discovery of two different fluids trapped simultaneously under the same P-T conditions. P-V-T data exist for a limited number of systems: H_2O (Kennedy, 1950; Burnham et al., 1970) (Fig. 2.4), H_2O + NaCl (Lemmlein and Klevtsov, 1956; Sourirajan and Kennedy, 1962) (Fig. 2.4), CO_2 (Kennedy, 1954), CH_4 (Douslin et al., 1964; Robertson and Bab, 1969) etc. Except for H_2O, they do not cover much of the geologically interesting range. A great amount of experimental work still remains to be done, simply to be able to interpret the already available fluid inclusion data.

Example: Determination of X_{CO_2} in a three phase inclusion (Fig. 1.1). In a three fluid phase inclusion (at T lab.), let V_1 = volume CO_2 (liquid+gas), V_2 = volume H_2O (both optically estimated under the microscope; T_f (H_2O) and T_h (CO_2) are measured by microthermometry. If T_f (CO_2) = -56.6°C, the fluid is approximated by the three-component system H_2O-CO_2-NaCl. Mutual miscibility of H_2O and CO_2 is low at T lab: traces of H_2O in CO_2 (L), 2.3 moles % of CO_2 in H_2O (L) (Wiebe and Gaddy, 1939). The quantity of H_2O in the gas bubble ("vapour pressure" of H_2O at T lab.) is also neglected.

If n = number of CO_2 moles, n_2 = number of H_2O moles, n_3 = number of NaCl moles,

$$X_{CO_2} = \frac{n_1}{n_1 + n_2 + n_3}$$

$$n_1 = V_1 \cdot \frac{d}{44} + \frac{2.3 \cdot V_2}{100 \cdot 18}$$; d = CO_2 density is derived from T_f CO_2 on P-V-T of CO_2 (Fig. 4.1)

$$n_2 = \frac{V_2}{18}$$; (density of liquid water = 1.0)

$$n_3 = n_2 \cdot x;$$ x is derived from T_f (H_2O) in the H_2O-NaCl system (Sourirajan and Kennedy, 1962) = equivalent NaCl

If X_{CO_2} is to be expressed in weight %, the above quantities must not be divided by gram formula weight of H_2O and CO_2.

Homogeneous and heterogeneous fluids. Miscibility/Immiscibility

It is impossible to determine T_f and T_h for all fluid inclusions occurring within a single thin section (many thousands!). Sets of similar inclusions must therefore be recognized. The basic (and ideal) criterion is of course chronology: a family of inclusions consists of all fluid inclusions trapped at the same time.

As primary fluid inclusions are exceedingly rare in rocks, contemporaneous inclusions can only be recognized if they belong to the same healed microcrack. This is often immediately noticeable (regular arrangement within the former plane of the crack, constancy of size and shape, similar crystallographic orientation of negative crystals, etc.). However, unfortunately the common occurrence of "necking-down" (Lemmlein and Kliya, 1960b; Roedder,

1967a) introduces a great additional complexity. If crack healing occurs in a gradient of decreasing temperature, the content of the fluid inclusion will be very variable, especially if the fluid was in the multiphase (L ± V ± S) region (Fig. 3.1 and 3.2).

In many cases, traces of necking down are directly visible as thin capillary tubes connecting several inclusions (Vacher, 1976, Fig. 6). Homogenization temperatures will then be very variable; they may be much higher than the maximum temperature reached by the fluid if a vapour bubble of a large inclusion has been trapped in a smaller one (Fig. 3.1). Also, any temperature may be recorded between this maximum and a minimum which corresponds theoretically to the complete healing of the crack. Homogenization temperature histograms are consequently flat without any well defined maximum.

If no necking down has taken place then the most important question becomes: Was the fluid homogeneous or heterogeneous (L + S, L + G or two immiscible fluids) at the time of trapping?

Homogeneity is indicated by the constancy of volume ratios : gas/liquid (Fig. 1.12), solid/liquid/gas (for a magnificent example, see Stalder, 1976, Fig. 1a), fluid 1 (CO_2)/fluid 2 (H_2O) (Fig. 1.10), etc. Microthermometric data are also constant (e.g. within 0.1°C in Stalder, op.cit.). In that case (see p.212) the homogenization temperature is the minimum temperature of trapping which may in fact be much higher (pressure correction).

The interpretation of heterogeneity is more complex: variability of the solid/liquid ratio may indicate that the fluid was oversaturated at the time of the trapping, but it may also be due to a selective leakage of the liquid (as in some fluorite or quartz crystals where the cavity is entirely occupied by a halite cube). Leakage may also explain the variability of gas/liquid ratio, which is otherwise evidence of boiling (especially "retrograde boiling" caused by isothermal pressure decrease, see Fig. 4.3). Most important is the case of two immiscible fluids (Fig. 3.2): two fluids A and B for instance (H_2O + NaCl) and (CO_2) occur in contemporaneous healed cracks or more typically in the same crack, but they homogenize at the same temperature (temperature of exsolution, Fig. 3.2). Rare composite inclusions (mechanical mixture of A and B) homogenize at a slightly higher temperature which theoretically fits the crest of the miscibility gap between A and B.

Necking-down of course complicates this ideal picture and, in the case of immiscible fluids, it makes the interpretation almost impossible. In practice we estimate that it can be neglected if values of T_h in a single crack do not differ by more

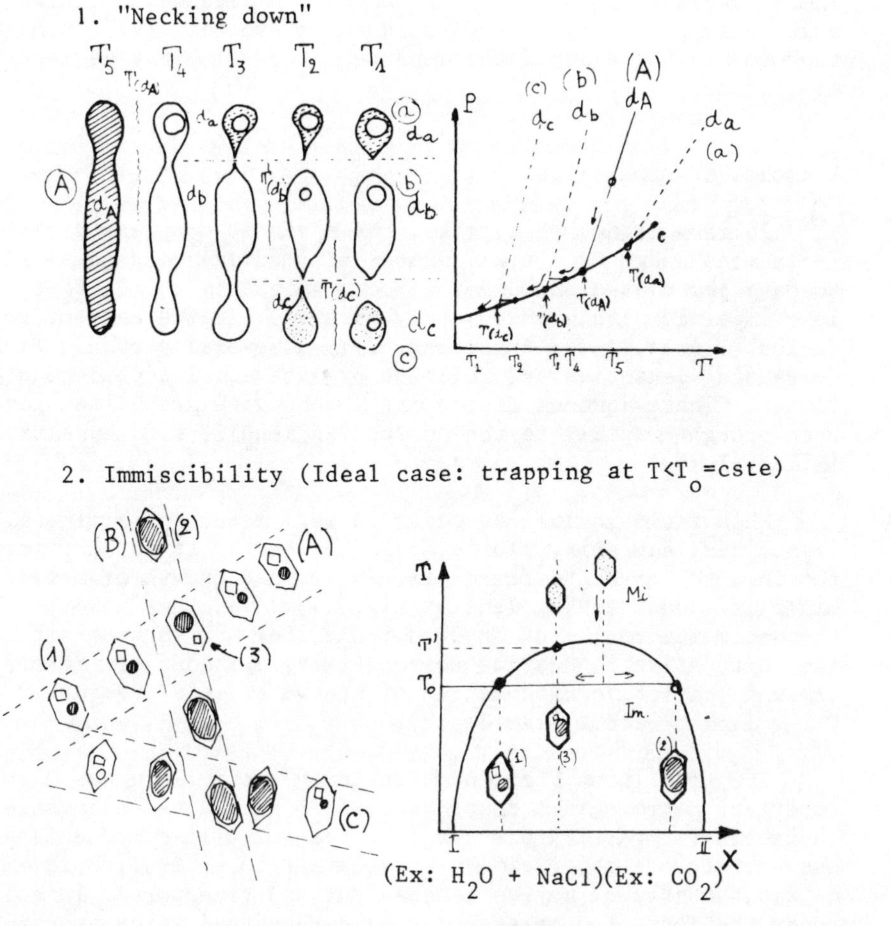

Fig. 3. Homogenization temperatures of heterogenous sets of inclusions.

3.1: "Necking-down" (after Roedder, 1967) and its interpretation on a P-T diagram. Under decreasing temperature ($T_5 \rightarrow T_1$), a one-phase inclusion A (density:d_A) evolves into three inclusions (densities: d_a, d_b, d_c respectively). As $d_a < d_A$ (due to the importance of the vapour phase) $T_h(a)$, homogenization temperature of $a(T(d_a))$ is higher than $T_h(A)$, while $T_h(c)$ ($=T(d_c)$) $< T_h(b) < T_h(A)$.

3.2: Immiscibility. Two immiscible fluids (1) and (2) occur in different ((A) and (B)) or in the same (C) healed crack. (1) and (2) homogenize at the same temperature T_o, while (3) (mixture of (1) and (2)) homogenizes at $T' > T_o$.Mi: domain of miscibility, Im: domain of immiscibility of the system.

than 10-20°C, the mean T_h being taken as the representative value. These considerations show also that the existence of a given fluid must be supported by several hundred determinations. In microthermometry the ease of the measurement is at least as important as its accuracy.

Examples of Application

In many metamorphic rocks, the apparent complexity of fluid inclusions suggests a small number of (mostly homogenous) fluids to have percolated through the rock, which may therefore be much less impervious than immediately assumed. In the case of granulite facies (Touret, 1974; Armstrong, 1975), several carbonic fluids of decreasing densities are followed by late and independent aqueous fluids. These aqueous fluids are absent when granulites have been brought rapidly to the surface as xenoliths in basaltic lavas (Bilal, 1976).

When fluid inclusions occur in many metamorphic minerals (zenolites, adularia, albite, calcite, etc.), it is sometimes possible to correlate fluid circulation and growth of metamorphic minerals. Jehl (1975; Jehl et al., 1976) was thus able to show that epimetamorphism of dredged tholeiites and serpentinites from the North Atlantic was due to repetitive episodes of microfracturing and convective circulation of sea water in a pressure range 0.4 - 1.5 kb (about 4 km depth).

The transition homogeneous/heterogeneous fluid has also important petrological consequences: it introduces a drastic change in the partial pressures of the volatile components and it has therefore a major effect on volatile/solid equilibria. Some metals, notably U, may be leached out and transported in solution under the form of complexes (uranylcarbonates) which are stable under a relatively high CO_2 pressure. If fluid immiscibility occurs, the initial fluid will separate into two phases, one richer in CO_2 (relatively low density-vapour dominant), the other richer in H_2O + dissolved salts. The gaseous CO_2 will leave the system, inducing the lowering of P_{CO_2}, uranyl complex destruction and U precipitation. Such a process has been demonstrated in some deposits from the French massif Central (Leroy and Poty, 1969; Leroy, 1971).

Fluid immiscibility is also widespread in porphyry copper deposits (Fig. 3.2), but it generally occurs over a large temperature range (Roedder, 1971; Denis, 1974) in low pressure metamorphism (Hercynian type) and in pegmatite genesis (Weisbrod and Poty, 1975). Important too is the transitory immiscibility ("retrograde boiling", Fig. 3.3) which may occur in binary systems

(H_2O + CO_2) at T constant under a release of the pressure.

Construction of Isochores in Simple Systems

Interpretation of microthermometric data (see above) shows the fundamental importance of isochores. Well-determined experimental isochores unfortunately exist only for a limited number of systems (H_2O, H_2O + NaCl, CO_2, CH_4, see p.212), and, except for H_2O, they cover but a very small part of the possible geological P-T conditions. Large extrapolations are therefore needed, based to a first approximation on the hypothesis that isochores are straight lines. This approximation is certainly valid for most one component systems in the supercritical domain. For CO_2 especially, extremely large extrapolations of experimental data (Fig. 4.1) are compatible with P-T estimates derived independently from the study of solid phases (Fig. 4.2) (see also Armstrong, 1975; Bilal, 1976; Touret, 1974).

Much remains to be done for the geologically most important system H_2O + CO_2 as only very few experimental points have been determined by Khitarov and Malinin (1956). Only crude approximations are possible based also on the "straight isochore" approximation which is unfortunately very dubious in this case (Ypma, 1963).

A first method is simply to add the partial pressure of H_2O and CO_2 (first estimate the partial density of H_2O and CO_2 by assuming that each component occupies the whole cavity at room temperature). But, as the mixture is not ideal (positive volume of mixing, Greenwood, 1973), the obtained pressure is lower than the real pressure. Khitarov and Malinin (1966) have experimentally determined some corrections in the region 20°C - 300°C, but they are far from sufficient, as is the present knowledge of fugacity coefficients (Rhyzenko and Volkov, 1971; Mel'nik, 1972; Barron, 1973). It is hoped that the precise knowledge of these coefficients will soon allow more accurate determinations over the whole range of P-T conditions which is needed (at least 800°C, 10 kb).

Another approximation, also based on the assumption (H_2O + CO_2) isochores = straight lines, is presented on Fig. 5 as the "two homogenization points method". The isochore is determined by two points A and B: A, at low temperature, i.e. within the miscibility gap of the system, corresponds to T_h CO_2 (Fig. 4.1) (the contribution to the pressure of H_2O liquid (vapour pressure near room temperature can be neglected). The temperature of total homogenization defines B as the intersection of T_h and the "critical" curve for a given X_{CO_2} (determined as in p.213). This curve is defined as the intersection of the immiscibility plane in

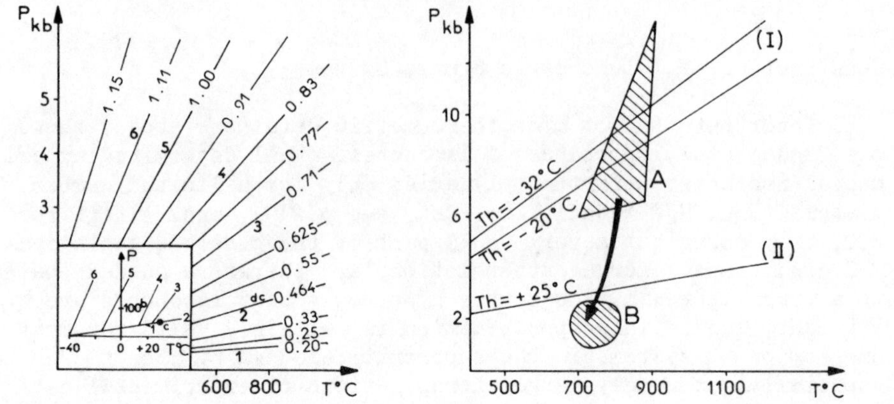

Fig. 4. Isochores for CO_2 fluids.

1. Experimental isochores for CO_2 (data from Kennedy, 1954, extrapolated by Armstrong, 1975). Number on the isochores=densities in g/cm^3.

2. Example of correspondence between solid phases and fluid inclusion data: Bournac granulites (see Bilal and Touret, 1976). Stability fields of critical minerals indicate a retromorphic evolution from field A to field B; fluid inclusions are divided into early fluids (I) and late fluids (II) on the basis of T_hCO_2 and of their occurrence in rocks.

the H_2O-CO_2 system and the isocompositional plane. Approximate as it is, this method is certainly better than another one, also based on the "straight line isochore hypothesis" which has been proposed by Naumov and Malinin (1968). In this, one point is determined by T_hCO_2, the other by the temperature of decrepitation for which the internal overpressure is taken as 850 atm. It is our experience that this last value is so variable (depending on the size of the inclusion, but also on its shape, content, etc.) that it can hardly be retained even as a first approximation.

OTHER ANALYTICAL METHODS

Many other analytical methods have been used or are potentially usable (for a complete listing, see Roedder, 1972 and

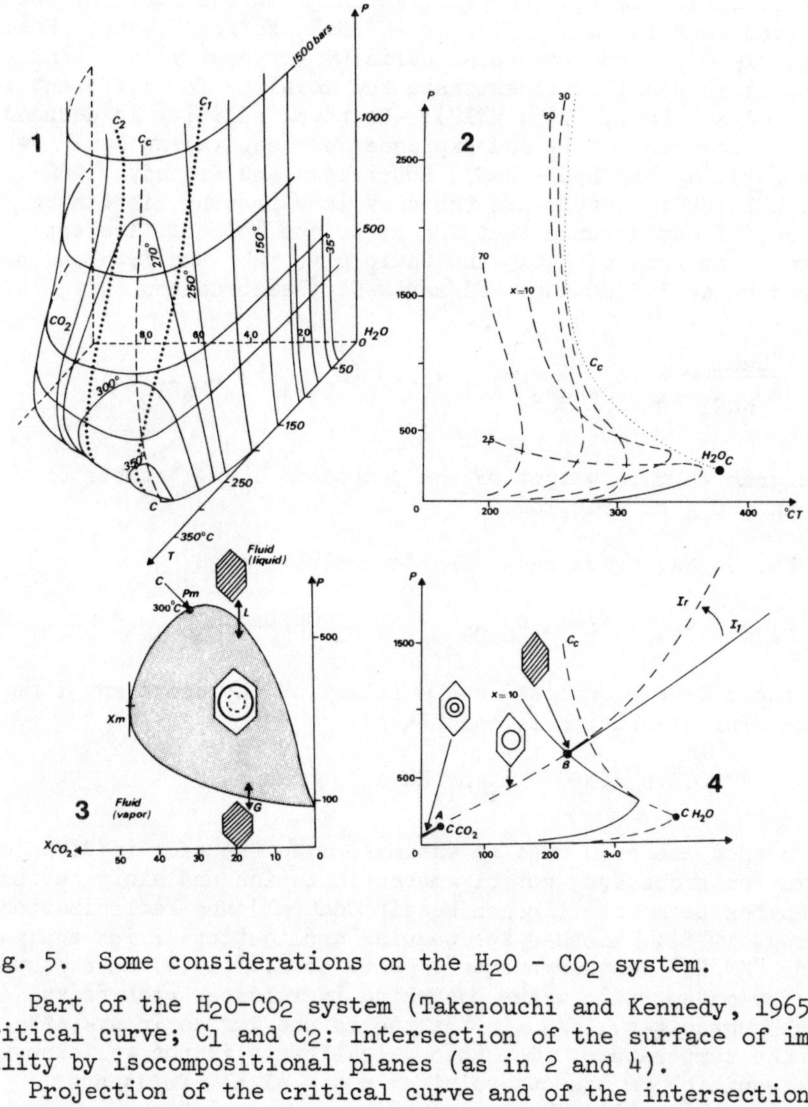

Fig. 5. Some considerations on the H_2O - CO_2 system.

1. Part of the H_2O-CO_2 system (Takenouchi and Kennedy, 1965) C_c: Critical curve; C_1 and C_2: Intersection of the surface of immiscibility by isocompositional planes (as in 2 and 4).
2. Projection of the critical curve and of the intersections of the immiscibility surface by planes of various X_{CO_2} on the P-T plane.
3. Section through the 300°C isotherm (I), illustrating the "retrograde boiling" phenomenon. X_m = maximum composition, P_m = maximum pressure for immiscibility to occur. C: section of the critical curve.
4. Construction of the isochores in the H_2O-CO_2 system by the "two homogenization point method" (see text). A: CO_2 homogenization, B: H_2O + CO_2 homogenization (for a given X_{CO_2}), I_r: trend of deviation for the real isochores (from Ypma, 1963). All X_{CO_2} in mole %.

Weisbrod et al., 1976) but few lead to a real interpretation. A major exception is ion analysis, which gives the ratio of the dissolved ions in aqueous fluids (K^+/Na^+, Na^+/Ca^{++}, etc. Roedder et al., 1963). From the total salinity (cryometry) and ionic ratio, it is possible to estimate the molality for different ions (Weisbrod and Poty, 1975, p.12): the total salinity is deduced from the lowering of T_f and expressed as "equivalent NaCl" (weight % = W_{NaCl}) (system H_2O - NaCl, Sourirajan and Kennedy, 1962). If NaCl, KCl, $CaCl_2$, $MgCl_2$ are the only ions present, it can be written, if one assumes that for up to one mole of divalent cations, one mole of $MgCl_2$ and $CaCl_2$ give the same freezing point depression as 1.5 mole of KCl and NaCl (Weisbrod and Poty, 1975):

$$\frac{W_{NaCl}}{MW_{NaCl}} = n_{NaCl} + n_{KCl} + 1.5\ (n_{CaCl_2} + n_{MgCl_2})$$

MW_i : gram formula weight of the component i, n_i: number of moles of i in 100 g of solution.

If K/Na, Ca/Na, Mg/Na have been determined, then:

$$n_{KCl} = n_{NaCl}.K/Na;\ n_{CaCl_2} = n_{NaCl}.Ca/Na;\ n_{MgCl_2} = n_{NaCl}.Mg/Na$$

From these four equations, the molality of the component i (m_i) in the fluid inclusion is calculated:

$$m_i = 1000\ n_i/(100 - \Sigma_i\ n_i\ MW_i)$$

This method has been used to estimate mass transfer in different alteration processes, notably muscovitization and albitization in the Mayres pegmatite (French Massif Central) and kaolinization (Charoy, 1975). Another spectacular application of ion analysis is the "K^+/Na^+ geothermometer". Poty et al. (1974) have shown that for metamorphic veins occurring in feldspar rich rocks (notably granites), K/Na in the inclusions varies in sympathy with the temperature. The theoretical basis is the well known ionic equilibrium aqueous solution - two alkali feldspars (Orville, 1963, 1972; Hemley, 1967).

$$\text{Orthoclase} + \text{Na} \rightleftarrows \text{Albite} + \text{K}$$

The equilibrium constant K = Na^+/K^+ = NaCl / KCl is essentially dependent on the temperature (at least in the range 300-600°C, 1-3 kb) or, if the following conditions are fulfilled: a) KCl and NaCl are the major dissolved species in the fluid; b) there is no calcium in the alkali feldspar and c) the newly-formed alkali feldspars grew as adularia and high albite, respectively. This

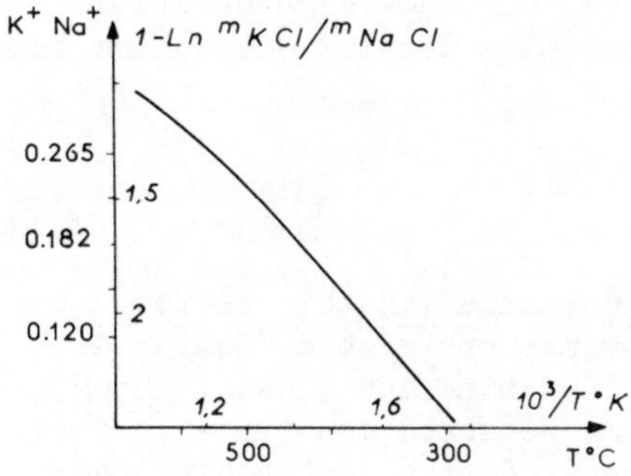

	K/Na	T°C	P(bar)
Pelvoux	0.106	335	1700
Mt. Blanc (av.12 det.)	0.161	410	2800
Aar (av. central zone)	0.179	430	2800
Gotthard	0.231	505	2700

Table 1. Estimated T and P of formation of some alpine fissures (Poty et al., 1974, p.735).

Fig. 6. Calibration curve of the K/Na thermometer (Poty et al., 1974).

is the case in the Western Alps and the results (Table 1, Poty et al., 1974), from a calibration curve based on the Orville and Hemley experimental data (op. cit.) (Fig. 6) are consistent indeed with most interpretations of alpine metamorphism: except for Pelvoux, constant pressure of about 3 kbar, temperature increasing regularly from the external domains towards the more internal one (Table 1).

CONCLUSION

During the history of modern geology, the interest of petrologists in fluid inclusions has been somewhat sinusoidal: some have praised them as wonder tools to solve all kinds of problems, while others have considered them as absolutely useless. I think that we have now sufficient knowledge of fluid behaviour to take inclusions for what they are: an intrinsic part of the rock, which has to be studied with all the other minerals. Their study must always be conducted together with the investigations on solid phases and their results internally checked by independent methods. Only then may new hypotheses arise, sometimes completely unexpected or thought to be impossible. The interpretation is also limited by the poor experimental data available and, when one considers the present sophistication of solid phase experimental petrology, it is strange to find that the behaviour of the most fundamental fluid system ($H_2O + CO_2$) is so badly known. Any further development of the method depends on better experimental data: P-V-T, dissociation and activity coefficients, behaviour of complex mixtures (notably hydrocarbons) at low temperature, etc. Many of these experiments are certainly within the capabilities of present day technology (notably the P-V-T measurements) and it is hoped that, in the not too distant future, our knowledge of fluid systems will match that of the much more complex - but also much better known - silicates.

ACKNOWLEDGEMENTS

Most of the ideas which are developed in this review have been timelessly debated by all the members of the "Equipe de recherches sur les équilibres fluides-minéraux", Nancy among which its co-leaders B. Poty and A. Weisbrod deserve a special mention. I have also greatly benefited from lectures at the University of Nancy by Dr. P. Poty, who reviewed critically the manuscript and from comments by G. Fisher and J. Konnerup-Madsen, while the drawings were made by O. Sonsini and J. Dion. The help of Don Fraser is gratefully acknowledged.

REFERENCES

Al Khatib, R. and Touret, J. Bull. Soc. Geol. France (1973).
Armstrong, E. 3rd Cycle Thesis, Nancy, 145 p. (1975).
Barron, L.M. Contr. Mineral. and Petrol., 39, 184 (1973).
Berglund, L. and Touret, J. Lithos, 9, 139-148 (1976).
Bilal, A. 3rd Cycle Thesis, Nancy, 160 p. (1976).
Bilal, A. and Touret, J. Bull. Soc. fr. Min. Crist., 99, 134-139 (1976).
Burnham, W., Holloway, J.R. and Davis, N.F. Geol. Soc. America, Sp. Paper 132.
Charoy, B. Petrologie, 1-4, 253-266 (1975).
Clochiatti, R. Mem. Soc. Geol. France, LIV, 122, 96 p. (1975).
Cuney, M., Pagel, M. and Touret, J. Bull. Soc. fr. Min. Crist., 99, 2/3, 169-177 (1976).
Deicha, G. Les lacunes des cristaux... 1 Vol., Masson, 126 p. (1955).
Denis, M. 3rd Cycle Thesis, Nancy, 146 p. (1974).
Donnally, H.G. and Katz, D.L. Ind. Eng. Chem., 46, 511-517 (1954).
Douslin, D.R., Harrison, R.H., Moore, R.T., McCullouch, J.P. J. Chem. Eng. Data, 9-1, 358-363 (1964).
Ermakov, N.P. L'vov Geol. Obsh. Sbornik, 3, 21-27 (1949).
Ermakov, N.P. Research on mineral forming solutions. Kharkov Univ. Press, 460 p. (1950).
Green, H. Nature, 238, 1-5 (1972).
Greenwood, H. Am. Jour. Sci., 273, 561-571 (1973).
Hemley, J.J. Abstr. Prog. Ann. Meet., Geol. Soc. Amer., 94-95 (1967).
Hoefs, J. and Touret, J. Contr. Mineral. and Petrol., 52, 165-174 (1975).

Hollister, L. Geol. Soc. Amer. An. Meet., Abst. 672 (1973).
Hollister, L. and Burruss, R.C. Geoch. Cosm. Act., 40, 163-175 (1976).
Jehl, V. Doc. Ing. Thesis, Nancy, 1 vol. (1975).
Jehl, V., Poty, B. and Weisbrod, A. An. Geoph. Un; Spring An. Meet. Prog., p.37 (1976).
Kennedy, G.C. Am. Jour. Sci., 248, 540-564 (1950).
Kennedy, G.C. Am. Jour. Sci., 252, 225- (1954).
Khitarov, N. and Malinin, C. Geochemistry Int., 3, 246-256 (1956).
Lemmlein, G.C. Zeitsch. Krist., 71, 237-256 (1929).
Lemmlein, G.C. and Klevtsov, P.V. Vses. Min. Ob. Zap., 85, 529-534 (1956).
Lemmlein, G.C. and Kliya, M.O. Ak. Nauk SSSR Dok., 94, 233-236 (1954).
Lemmlein, G.C. and Kliya, M.O. Int. Geol. Rev., 2-2, 120-124 (1960a).
Lemmlein, G.C. and Kliya, M.O. Int. Geol. Rev., 2-2, 125-128 (1960b).
Leroy, J. 3rd Cycle Thesis, Nancy, 87 p. (1971).
Leroy, J. and Poty, B. Miner. Deposita, 4, 395-400 (1969).
Liakhov, Yu V. Geochemistry Int., 4-3, 618-625 (1967).
Mel'nik, Yu P. Geokhimiya, 6, 654-662 (1972).
Naumov, V.B. and Malinin, S.D. Geochemistry Int., 5, 382-391 (1968).
Ohmoto, H. Ph.D. Thesis, Princeton, Pt. 2, 53 p. (1968).
Orville, P. Am. Jour. Sci., 261, 201-237 (1963).
Orville, P. Am. Jour. Sci., 27, 234-272 (1972).
Pagel, M. 3rd Cycle Thesis, Nancy (1975).
Poty, B. Sci. Terre, Nancy, Mem. 17, 151 p. (1969).
Poty, B., Leroy, J. and Jachimovicz, L. Bull. Soc. fr. Min. Crist., 99, 2/3, 182-186 (1976).
Poty, B., Stalder, H.A. and Weisbrod, A. Schweiz. Min. Petr. Mitt., 54-2/3, 717-752 (1974).
Rhyzenko, B.N. and Volko, P. Geokhimiya, 7, 760-773 (1971).
Rich, R.A. Ph.D. Thesis, Harvard Univ., 297 p. (1975).
Robertson, S.L. and Babb, S.E. J. Phys. Chem., 51, no. 4 (1969).
Roedder, E. Ch.12 in "Geochemistry of hydrothermal ore deposits", ed. H.L. Barnes, 515-574 (1967a).
Roedder, E. Science, 155, 1413-1417 (1967b).
Roedder, E. Econ. Geol., 66, 98-120 (1971).
Roedder, E. U.S.G.S. Prof. Pap., 440-JJ, 164 p. (1972).
Roedder, E., Ingram, B. and Hall, W.E. Econ. Geology, V.58, 353-374 (1963).
Smith, F.G. Physical Geochemistry, 1 vol. (1953).
Sorby, H.C. Geol. Soc. Lond. Quat. Jour., 14-I, 453-500 (1858).
Sourirajan, S. and Kennedy, G.C. Am. Jour. Sci., 260, 115-141 (1962).
Stalder, H.A. Bull. Soc. Fr. Min. Crist., 99-2/3, 80-84 (1976).
Takenouchi, S. and Kennedy, G.C. Jour. Geol., 73, 383-390 (1965).

Touray, J.C. Schweiz. Min. Petr. Mitt., 50-1 (1970).
Touret, J. An. Soc. Geol. Belgique, vol. P. Michot, 267-287 (1974)
Vacher, A. Bull. Soc. fr. Min. Crist., 99-2/3, 131-133 (1976).
Weisbrod, A., Poty, B. Petrologie, I, 1-16 and 89-102 (1975).
Weisbrod, A., Poty, B. and Touret, J. Bull. Soc. fr. Min. Crist., 99-2/3, 140-152 (1976).
Wiebe, R. and Gaddy, V.L. Jour. Amer. Chem. Soc., 50-61, 315-318, (1939).
Ypma, P. Thesis, Leiden, 212 p. (1963).

PROBLEMS

I. For the three fluid phase inclusions shown in Fig. 1.1, the following microthermometric results have been determined: (all T°C)

$T_f(H_2O) = -3.0$, $T_f(CO_2) = 56.6$, $T_h(CO_2)_{(L)} = 27.5$,

$T_{h(to)} = +300$

1. Give the composition of the fluid (weight %, mole %).
2. What is the pressure at 500°C: a) by adding the partial pressures of H_2O and CO_2 (ideal mixing). b) by assuming that the isochores in H_2O-CO_2 system are straight lines. Give the trend of deviation for the real system.

II. In a quartz crystal, fluid inclusions freeze at a constant -5°C. Chemical analysis of water leachates give the following results (in ppm): K : 0.328 Na : 5.72 Ca : 1.98 Mg : 0.092 Cl : 13.7.
What are the molalities for KCl, NaCl, $CaCl_2$, $MgCl_2$?
Are there other species present?

III. In some crystalline massif of the Alps, quartz bearing cavities with the same mineral association occur at the top (Alt. 4000 m) and at the basis (Alt. 1000 m) of the mountain. For inclusions at the top: $T_f = 0°C$, $T_h = 250°C$.

1. What is the homogenization temperature at the base, if one assumes that pressure is simply due to the weight of the overlying rocks (d = 2.7) (Sometimes, it works!).

2. Assuming a geothermal gradient of 35°C. km^{-1} for cavities at the top, what are the true P.T. conditions of formation?

3. What K^+/Na^+ is to be expected in the fluid if these estimates are correct?

IV. A quartz crystal contains two types of fluid inclusions, one filled with: H_2O (T_f = -7^oC, $T_{h(L)}$ = 220^oC), the other with CO_2 (T_f = -56.6^oC).

1. What is the "degree of filling" ($V_L/V_L + V_G$) of the aqueous inclusions at T_{Lab}.

2. If K^+/Na^+ = 0.154 in the aqueous inclusions, which homogenization temperature of CO_2 would indicate that the two sets of inclusions are contemporaneous? Discuss the possibility of this assumption and cite other arguments which would confirm it. If there are some three fluid phase inclusions, what will be their total homogenization temperatures?

SOLUTIONS TO PROBLEMS

I.1. In Fig. 1.1, $V_{CO_2}(L+G)$ is estimated to be equal to V_{H_2O} (rough approximation). Hence (p.), $V_1 = V_2 = 0.5$.

$T_f(H_2O) = -3^oC \rightarrow$ 7% NaCl (weight), 2 mols % NaCl

$T_h(CO_2)1 = +27.5 \rightarrow d_{CO_2} = 0.70$ g.cm^{-3}.

From the formulae p.

mol %: $X_{CO_2} = 23.3$, $X_{H_2O} = 75.2$, $X_{NaCl} = 1.5$

weight %: $X_{CO_2} = 41.4$, $X_{H_2O} = 54.8$, $X_{NaCl} = 3.8$

2.a First estimate the apparent densities d* of H_2O and CO_2 (as if each occupies alone the cavity)

$d^*_{H_2O} = d_{H_2O} \times \frac{1}{2} = 0.5$ g.cm^{-3}

$d^*_{H_2O} = 0.70 \times \frac{1}{2} = 0.35$ g.cm^{-3}

At 500^oC, the partial pressures are estimated from the respective sets of isochores (Fig. 2.4 and Fig. 4.1):

P_{H_2O} = 800 bar, P_{CO_2} = 500 bar, P_{tot} = 1300 bar

b On Fig. 5.2, the "straight isochore" joins the point 27, 5^oC, 50 bar (T_hCO_2) (practically the origin) to the point T = 300^oC, X_{CO_2} = 22.4 (between C_c and X = 30) (see Fig. 5.4). Extrapolation to 500^oC gives P $\neq$ 1200 bar, remarkably close to a, but reflecting merely the same basic hypothesis ($H_2O + CO_2$ = ideal gas).

II. Calculated molalities: [KCl] = 0.023, [NaCl] = 0.671, [$CaCl_2$] = 0.131, [$MgCl_2$] = 0.010 (An. 176, Weisbrod and Poty, 1975, p.11). Charge balance (Σ cations/Σ anions) = 0.94; this slight deviation from 1 indicates that some other species are present, but in minor amounts.

III.1) Same mineral association = constant temperature, so the variation of T_h will be related to the variation of pressure. Assuming $P_{fluid} = P_{lithost}$ and an average rock density of 2.7, 3 km corresponds to $P_{lithost}$ = 810 bar. On the H_2O isochores (Fig. 2.4) $T_h(250^oC) \rightarrow d_{H_2O} = 0.78$ g.cm^{-3}. To find $T_{h_{Hw}}$ at the base, trace on the 0.78 isochore, at a "reasonable" temperature (between 300°C - 400°C), a segment parallel to the pressure axis = 810 bar. This gives the required isochore (d = 0.87 g.cm^{-3}) → T_h = 200°C. Note that the choice of temperature is arbitrary, but that it does not bring a significant error as the isochores are nearly parallel. The last sentence means that most determinations in the Alps, especially from Mt. Blanc (Poty, 1965) do not support this model (the density is independent of the altitude). One exception is the Pelvoux massif (Poty, unpublished data).

2) T = 450°C, P = 2800 bars.

3) 0.190 (Fig. 6).

IV. 1) If one neglects the mass of H_2O gas, as d_{H_2O} L = 1, the "degree of filling" = density of the inclusion.

$T_f = -7^oC \rightarrow 10$ % NaCl (weight) = 3 mol % NaCl

From the isochores of H_2O (the correction for NaCl can be neglected)

T_h 220° → d = 0.83 g.cm^{-3}. Hence: degree of filling = 0.83.

2) Assuming the validity of the K^+/Na^+ geothermometer (Fig. 6),

$K^+/Na^+ = 0.154 \rightarrow T = 400^oC$, which on the 0.83 isochore corresponds to P = 2500 bars.

Supercritical CO_2 in the same P,T conditions has a density of 0.86 g.cm^{-3} (note the very similar densities), whose T_h is + 10°C (Fig. 4.1). A necessary condition for both sets of inclusions to be contemporaneous is therefore T_h CO_2 = +10°C. Further confirmation relates to "immiscibility criteria", as immiscibility between H_2O and CO_2 (possible because of the NaCl current of H_2O) is almost the only possible hypothesis. T_h of 3 fluid phase inclusion would correspond to the 2500 bar section of the miscibility gap in the (NaCl)-H_2O-CO_2 system (unfortunately not yet determined experimentally).

THE STABILITY OF PHLOGOPITE IN THE PRESENCE OF QUARTZ AND DIOPSIDE

David R. Wones and Franklin C.W. Dodge

U.S. Geological Survey, Reston, VA 22092 and
Menlo Park, CA 94025

The stability of phlogopite in the presence of quartz and diopside is of interest in the interpretation of biotite-bearing assemblages in granitic rocks and of biotite-clinopyroxene-amphibole assemblages in metamorphic rocks. This study, begun to compliment the studies of Luth (1967), Schairer (1954), and Shaw (1963), is now being abandoned and the results to date are presented herein. Numerous persons have contributed to the study. Starting materials were contributed by W.T. Pecora, J.F. Schairer, and H.S. Yoder. Laboratory assistance was provided by J.V. Chernosky, C.J. Duffy, H.R. Shaw, and J. Whitney. N. Chatterjee, W.C. Luth, and D.A. Hewitt discussed many aspects of the theoretical and applied aspects of the study. J.S. Huebner and M.E. Woodruff were particularly helpful in the acquisition of x-ray powder intensity data. J.V. Chernosky provided library access and working space at the Univ. of Maine, Orono, so that this manuscript could be finished in the midst of an active field season. Critical reviews by J.S. Huebner, R.A. Robie, D.B. Stewart and E-an Zen helped to improve the manuscript.

EXPERIMENTAL METHODS

Starting materials. Glasses or their anhydrous crystalline equivalents (cristobalite, enstatite, and sanidine) were provided by J.F. Schairer and H.S. Yoder from the study of Schairer (1954). The compositions used are given in Table 1. Gel starting materials were prepared following the methods of Wones (1967) and Shaw (1963). Gem quality quartz crystals from Brazil were provided by W.T. Pecora. Synthetic phlogopite, tremolite, sanidine and diopside were crystallized hydrothermally, then mixed with the quartz to make up the compositions "PQ" and "TRK" given in Table 1. The ternary glasses (or crystalline equivalents) were crystallized in

D. G. Fraser (ed.), Thermodynamics in Geology, 229-247.

Table 1. Anhydrous Starting Materials Used in This Study

Symbol	Mole percent $KAlO_2$	MgO	SiO_2	CaO
Qz	0.0	0.0	100.0	
En	0.0	50.0	50.0	
Fo	0.0	66.7	33.3	
Ks	50.0	0.0	50.0	
Lc	33.3	0.0	66.7	
Sa	25.0	0.0	75.0	
Ph	14.3	42.9	42.9	
PQ	10.0	30.0	60.0	
A3	8.2	24.7	67.1	
A9	7.7	16.6	75.7	
A10	11.5	8.9	79.6	
A11	14.6	9.1	76.3	
A15	5.4	16.7	77.9	
A25	4.4	13.4	82.2	
A36	12.3	13.2	74.5	
A38	20.2	4.7	75.1	
TRK	5.3	26.3	57.9	10.5

the presence of an H_2O-rich gas to form starting assemblages. Most experiments were carried out in sealed capsules of Au, containing 30 to 35 mg of solids and 5 to 10 mg of distilled, deionized H_2O. No intentional experiments were made in an H_2O-unsaturated region.

A problem was encountered in the synthesis of tremolite. All synthesized material contained 1-2% quartz and diopside. N. Chatterjee (personal comm. 1971) has made a rigorous study of this problem and concluded that tremolite synthesized at 750° contains 5 to 10 mol percent $Mg_7Si_8O_{22}(OH)_2$. A survey of over 1200 amphibole analyses (Deer et al. 1963; Leake, 1968) produced only nine analyses in which the calculated structural formula contained between 1.90 and 2.10 Ca, less than 2.10 "X" cations, and greater than 7.80 Si. Boyd (1959) and Troll and Gilbert (1972) also describe the difficulties of tremolite synthesis. It appears that, at temperatures above 700°C, stoichiometric $Ca_2Mg_5Si_8O_{22}(OH)_2$ may not be stable.

Temperature and Pressure Measurements. Temperatures were measured with Type K inconel-sheathed thermocouples calibrated against the melting point of NaCl (801°C) in evacuated SiO_2 tubes placed within the pressure vessels with the same configuration as the Au capsules. Temperature gradients were minimized to less than 1°C by careful positioning of the vessel within the furnace and were measured within the vessel at pressure of 2000 bars. A variety of temperature controllers were used. The narrow range of temperature

Table 2. Comparison of unit cell dimensions of starting materials in this study and other materials

Phase	a, Å	b, Å	c, Å	β	V, $(Å)^3$	Experimental conditions or reference
Diopside	9.750 (2)	8.928 (3)	5.245 (2)	105°56' (2)	439.0 (2)	800°C, 2 kb, 48 hr.
Diopside	9.746 (4)	8.899 (5)	5.251 (6)	105°38' (4)	438.6 (3)	Clark, et al.(1969)
Enstatite	18.218 (6)	8.818 (3)	5.176 (2)	90°	831.5 (4)	830°C, 2 kb, 237 hr.
Enstatite	18.225 (1)	8.815 (1)	5.175 (1)	90°	831.3 (1)	Stephenson, et al. (1966)
Tremolite	9.814 (6)	18.063 (12)	5.275 (3)	104°39' (3)	904.6 (7)	750°C, 2 kb, 166 hr.
Tremolite	9.818 (5)	18.047 (8)	5.273 (2)	104°43' (2)	904.5 (3)	Ross, et al.(1969)
Tremolite	9.840 (5)	18.052 (7)	5.275 (5)	104°42' (2)	907.5 (11)	Zussman (1959)
Sanidine	8.600 (1)	13.008 (3)	7.178 (1)	116° 1' (1)	721.6 (2)	800°C, 2 kb, 618 hr.
Sanidine	8.610 (2)	13.033 (2)	7.174 (2)	116° 3' (2)	723.5 (2)	Stewart (1975)
Phlogopite	5.318 (2)	9.205 (4)	10.315 (3)	99°56' (2)	497.3 (2)	800°C, 2 kb, 960 hr.
Phlogopite	5.316 (1)	9.208 (2)	10.314 (5)	99°54' (2)	497.9 (3)	Wones (1967)

variation indicated in the tables is real and was achieved in most cases with controllers of platinum resistance type. Pressures were measured every 24 hours with factory calibrated bourdon gauges. All pressures deviated less than 0.5% of the quoted value.

Evaluation of Phase Assemblages. Both optical and x-ray methods were used to identify the products of each experiment. Visual and optical methods were used for the detection of H_2O-rich gas, and glass, whereas x-ray methods were best for the identification of the solid phases. Unit cells determined in this study are compared with those of other investigators in Table 2. The reactions between phlogopite, quartz, diopside, sanidine, and tremolite required seeding and careful evaluation of the x-ray patterns to determine growth and loss of phases. Powder diffraction pattern peak intensities were measured both on chart paper and by computer evaluation of digitized output (Hellman, 1971). The two methods proved to be comparable. Peaks used were phlogopite, 001, $11\bar{3}$; diopside, $\bar{2}21$, $\bar{6}01$; sanidine, 130; tremolite, 110, 310, 151; and quartz, 112. Six sets of ratios were used: Ph001/Tr110; $Ph11\bar{3}$/Tr310; Di221/Tr310; $Di\bar{6}01$/Tr151; Sa130/$Ph11\bar{3}$; $Di\bar{6}01$/$Ph11\bar{3}$. Values for the measured ratios are given in Table 4. In all cases reported as growth of phases in Table 3, changes in intensity were always greater than those observed in repeated measurements of the same assemblage.

STABILITY OF PHLOGOPITE AND QUARTZ

At low H_2O pressures, phlogopite and quartz react to form sanidine, enstatite, and an H_2O-rich gas. This reaction terminates between 830°C and 850°C and 400 bars and 500 bars at an invariant point, (Tr'), involving silicate melt. Luth (1967) discussed the topology of this invariant point which is reproduced here schematically in Figure 1. The chemography of the solid phases and melt is projected from the H_2O apex onto the three components $KAlSiO_4$, Mg_2SiO_4 and SiO_2.

The stability of phlogopite and quartz is controlled by four reactions:

$$\underset{\text{phlogopite}}{KMg_3AlSi_3O_{10}(OH)_2} + \underset{\text{quartz}}{3SiO_2} \rightleftarrows \underset{\text{sanidine}}{KAlSi_3O_8} + \underset{\text{enstatite}}{3MgSiO_3} + \underset{\text{gas}}{H_2O} \quad \text{(L)}$$

$$\underset{\text{phlogopite}}{KMg_3AlSi_3O_{10}(OH)_2} + \underset{\text{quartz}}{nSiO_2} + \underset{\text{sanidine}}{mKAlSi_3O_8} + \underset{\text{gas}}{xH_2O} \rightleftarrows \underset{\text{liquid}}{\text{melt}} \quad \text{(En)}$$

$$\underset{\text{phlogopite}}{KMg_3AlSi_3O_{10}(OH)_2} + \underset{\text{quartz}}{(3+n)SiO_2} + \underset{\text{gas}}{xH_2O} \rightleftarrows \underset{\text{enstatite}}{3MgSiO_3} + \underset{\text{liquid}}{\text{melt}} \quad \text{(Sa)}$$

$$\underset{\text{phlogopite}}{KMg_3AlSi_3O_{10}(OH)_2} + \underset{\text{quartz}}{(3+n)SiO_2} \rightleftarrows \underset{\text{sanidine}}{xKAlSi_3O_8} + \underset{\text{enstatite}}{3MgSiO_3} + \underset{\text{liquid}}{\text{melt}} \quad \text{(V)}$$

(L) occurs at pressures below the invariant point, the others above

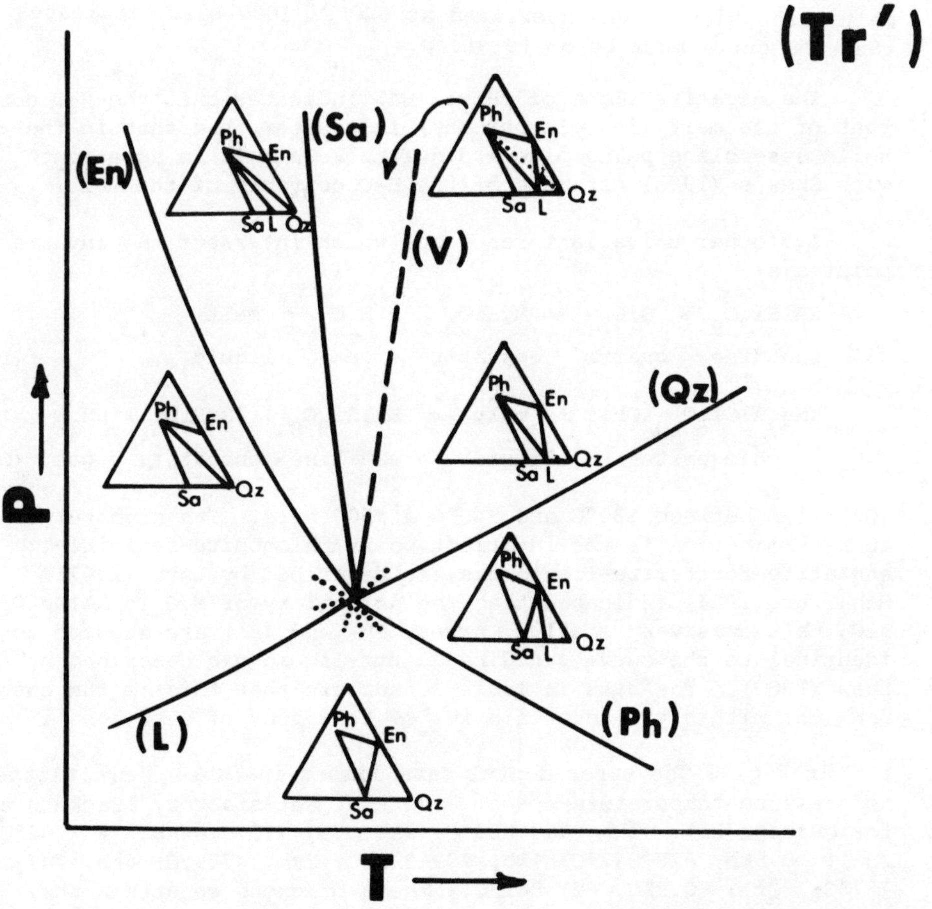

Figure 1. Schematic representation of invariant point (Tr'). Abbreviations as given in Table 3. Projection from H_2O onto $KAlSiO_4$-Mg_2SiO_4-SiO_2 plane. Reaction labels are given in text. Solid lines represent stable reactions; dotted lines are metastable extensions of the stable portions of reactions; the dashed line represents a condensed reaction in which no vapor participates. Reaction labels are given in text.

that point. Reactions (En) and (Sa) are the ones for which determinations were made in this study. Reaction (V) was not investigated directly, but the presence of sanidine and enstatite in a melt with which no gas coexisted at 840°C, 1000 bars, indicates that the curve must be quite steep.

The negative slope of curve (Sa) indicates that the H_2O content of the melt (in mole percent) is greater than that in the solid assemblage phlogopite + 3 quartz. This is in agreement with Shaw's (1963) estimate of the H_2O contents of the melt.

Two other univariant reactions which intersect the invariant point are:

$$\underset{\text{sanidine}}{KAlSi_3O_8} + \underset{\text{quartz}}{SiO_2} + \underset{\text{enstatite}}{MgSiO_3} + \underset{\text{gas}}{H_2O} \rightleftarrows \underset{\text{liquid}}{\text{melt}} \qquad \text{(Ph)}$$

$$\underset{\text{phlogopite}}{KMg_3AlSi_3O_{10}(OH)_2} + \underset{\text{liquid}}{\text{melt}} \rightleftarrows \underset{\text{sanidine}}{KAlSi_3O_8} + \underset{\text{enstatite}}{3MgSiO_3} + \underset{\text{gas}}{nH_2O} \qquad \text{(Qz)}$$

(Qz) lies between 830°C and 840°C at 500 bars. Its high temperature termination is the invariant point phlogopite-sanidine-enstatite-forsterite-liquid-gas, as described by Luth (1967). Schairer (1954) indicated that the solubility of MgO in $KAlSi_3O_8$-SiO_2 melts was very small. Curves (Ph) and (En) are assumed to be identical to the curve sanidine-quartz-liquid-gas described by Shaw (1963). The data in Table 3 indicate that this is the case for (En) within the uncertainties of this body of data.

In Fig. 2 the experimental data listed in Table 3 are plotted in pressure-temperature space. The most satisfactory brackets are for curves (En): 1kb, 789-803°C; 2kb, 757-775°C; 4kb, 743-752°C; (Sa): 0.5kb, 828-842°C; 1kb, 799-812°C; 2kb, <779°C; 4kb, 747-757°C; (Qz): 0.5kb, 827-842°C. These brackets constrain the invariant point to the values of 830-840°C and 400-500 bars. This becomes the constraint on reaction (L) which leads to an estimate of the Gibbs Free Energy of phlogopite at that temperature (1108°K).

The assemblage quartz + phlogopite has a maximum stability of about 835°C in the presence of H_2O-rich gas. At pressures above 450±50 bars this assemblage participates in melting reactions. An analogous curve for the iron analogue, quartz + annite, would be stable to about 4000 bars and 735°C, where it would intersect an invariant point containing the assemblage quartz-annite-sanidine-fayalite-melt-gas. The substitution of F for OH would raise the thermal stability of phlogopite to a temperature which could be considerably in excess of 835°C (Munoz and Ludington, 1974).

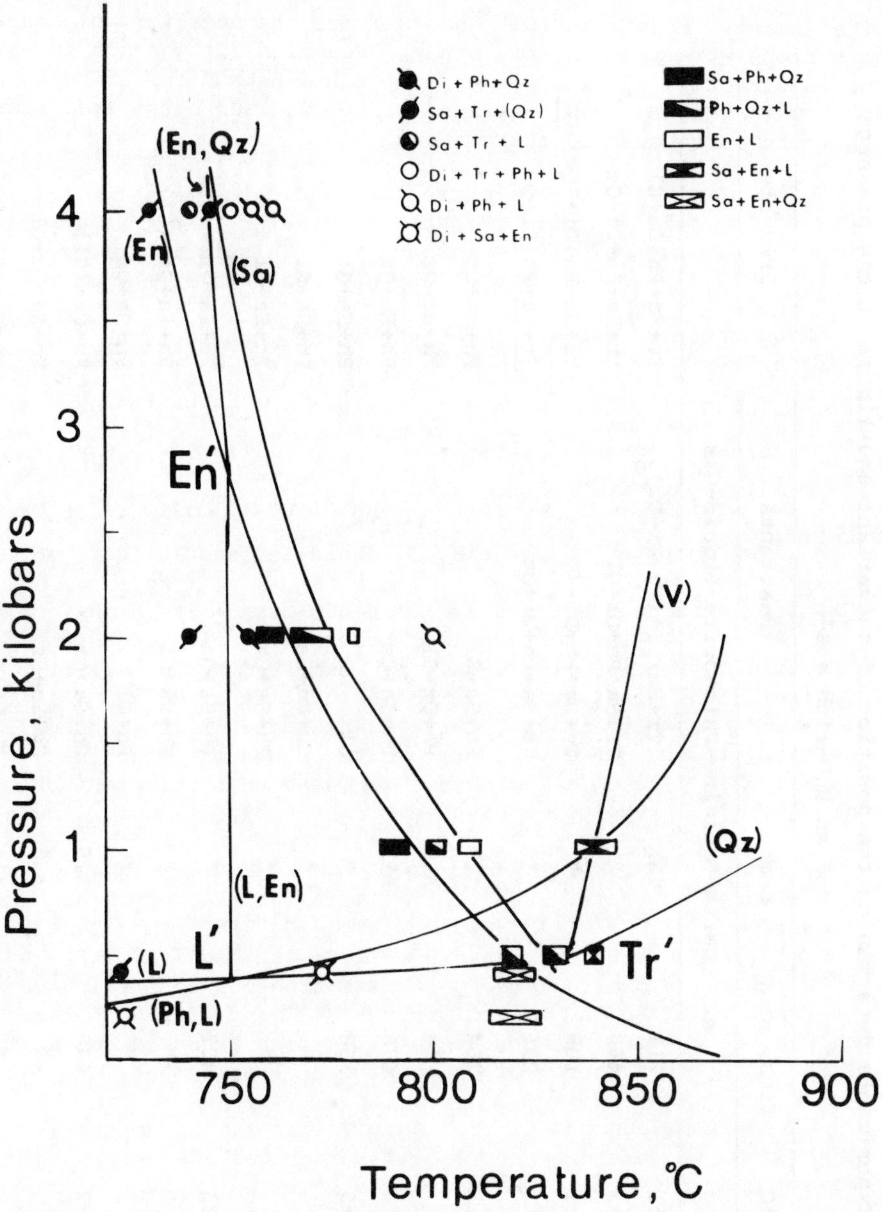

Figure 2. Experimental data for the system $KAlSiO_4-Mg_2SiO_4-SiO_2-H_2O$ (rectangular symbols) and the system $CaMgSi_2O_6-MgO-KAlSiO_4-SiO_2-H_2O$ (circular symbols) plotted on pressure versus temperature.

Table 3. Experiments determining the position of univariant equilibria in the presence of an H_2O-rich gas[1/]

P, bars	T, °C	Time, hrs.	Comp.	Reactants	Products
a. Quartz-Sanidine-Phlogopite-Liquid-Gas					
4000	751±1	211	TRK	$(Tr+Sa)_{50}+(Ph+Di+Qz)_{50}$	Tr+Sa+Ph+Di+Gl
"	745±2	455	"	Tr+Sa+Ph+Di+Gl	Tr+Sa+Ph+Di+Qz
"	740±1	188	"	Tr+Sa+Ph+Di+(Qz)	Tr+Sa+Ph+Di+(Qz)+<u>Gl</u>
"	730±1	236	"	Tr+Sa+Ph+Di+Qz	<u>Tr</u>+<u>Sa</u>+Ph+Di+Qz
2000	760±3	264	A11	Ph+Sa+Gl	Ph+Sa+Qz
"	"	"	A10	Ph+Qz+Gl	Ph+Sa+Qz
"	"	478	A11	Ph+Sa+Qz	Ph+Sa+Qz
"	770±5	358	"	Gl	Ph+Sa+Gl
"	"	"	"	Cr+En+Sa	Ph+Sa+Gl
"	"	"	"	Ph+Sa+Qz	Ph+Sa+Gl
1000	791±3	403	"	Ph+Sa+Qz	Ph+Sa+Qz
"	"	"	"	Ph+Sa+Gl	Ph+Sa+Qz+(Gl)
"	"	"	"	Ph+Gl	Ph+Sa+Qz
"	801±2	1448	A25	Ph+Sa+Qz	Ph+Qz+(Gl)
"	"	"	A36	Ph+Sa+Qz	Ph+Sa+Qz+Gl
"	"	"	A11	Ph+Sa+Qz	Ph+Sa+Gl

Table 3. Continued

P, bars	T, °C	Time, hrs.	Comp.	Reactants	Products
500	819±2	603	A36	Ph+Qz+En+Gl	Ph+Sa+Gl
			b. Quartz-Phlogopite-Enstatite-Liquid-Gas		
4000	756±1	286	TRK	Tr+Sa+Ph+Di+Qz	Ph+Di+Tr+Sa+Gl
"	"	"	"	Tr+Sa+Ph+Di+Qz	Ph+Di+Tr+Gl
"	750±3	212	"	$(Tr+Sa)_{50}+(Ph+Di+Qz)_{50}$	Ph+Di+Qz+Tr+Gl
"	745±2	455	"	Tr+Sa+Ph+Di+Gl	Tr+Sa+Ph+Di+Qz
2000	780±1	1858	PQ	Ph+En+Gl	Ph+En+Gl
"	"	"	A9	Gl	
"	"	"	A15	Ph+Qz	Ph+(En)+Gl
1000	801±2	1448	A25	Ph+Qz+(Sa)	Ph+Qz+Gl
"	809±3	1270	A38	En+Gl	Ph+En+Gl
"	"	"	A25	Ph+Qz+Sa	Qz+En+Gl
500	830±3	188	PQ	Ph+Qz	Ph+Qz
"	830±2	210	A36	Glass	Ph+Qz+En+Gl
"	840±4	168	A3	Ph+Qz+Gl	Ph+Qz+Sa+En+Gl
"	840±2	115	PQ	Ph+Qz	Sa+En+(Gl)
"	839±2	600	PQ	Sa+En	Sa+En

Table 3. Continued

P, bars	T, °C	Time, hrs.	Comp.	Reactants	Products
				c. Quartz-Phlogopite-Sanidine-Enstatite-Gas	
400	820±5	860	PQ	Ph+Qz	Sa+En+Ph+Qz
"	774±1	284	TRK	$(Tr+Sa)_{50}+(Ph+Di+Qz)_{50}$	Sa+Di+En
100	820±6	930	PQ	Ph+Qz	Sa+Px+Ph+Qz
				d. Phlogopite-Liquid-Sanidine-Enstatite-Gas	
500	830±3	240	A36	Gl	Ph+Oz+En+Gl
"	"	188	PQ	Ph+Qz	Ph+Qz
"	840±4	168	A3	Ph+Qz+Gl	Ph+Qz+Sa+En+Gl
"	840±2	237	PQ	Ph+En+Gl	Sa+En+(Gl)
"	839±2	600	"	Sa+En	Sa+En
400	830±3	339	"	Ph+Qz	Sa+En+Qz+Gl
				e. Phlogopite-Diopside-Liquid-Tremolite-Sanidine-Gas	
4000	761±1	236	TRK	Tr+Sa+Ph+Di+Qz	Ph+Di+(Qz)+Tr+(Sa)+Gl
"	756±1	286	"	Tr+Sa+Ph+Di+Qz	Ph+Di+Tr+Gl
"	751±1	211	"	$(Tr+Sa)_{50}+(Ph+Di+Qz)_{50}$	Ph+Di+Tr+Sa+Gl
"	745±2	455	"	Ph+Di+Tr+Sa+Gl	Tr+Sa+Ph+Di+Qz

Table 3. Continued

P, bars	T, °C	Time, hrs.	Comp.	Reactants	Products
				f. Phlogopite-Diopside-Quartz-Tremolite-Sanidine-(Gas)	
2000	800±5	214	TRK	Tr+Sa	Ph+Di+(Tr)+Gl
"	770±1	407	"	$(Tr+Sa)_{50}+(Ph+Di+Qz)_{50}$	Ph+Di+Qz+(Tr)+Gl
"	755±1	671	"	$(Tr+Sa)_{50}+(Ph+Di+Qz)_{50}$	Di+Ph+Qz+(Tr)
"	740±5	307	"	$(Tr+Sa)_{50}+(Ph+Di+Qz)_{50}$	Tr+Sa+(Ph)
400	774±1	284	"	$(Tr+Sa)_{50}+(Ph+Di+Qz)_{50}$	Sa+Di+En
"	726±1	235	"	Sa+Di+En	Tr+Sa
200	723±1	207	"	Tr+Sa+Ph+Di+Qz	Sa+Di+En

1/ Abbreviations of phases: En, enstatite; Gl, glass, Ph, phlogopite; Qz, quartz; Sa, sanidine; Tr, tremolite; Di, diopside. (Sa) minor amount of phase. Sa phase increased during experiment. $(Tr+Sa)_{50}$ + $(PH+Di+Qz)_{50}$, 50% (wt.) Tr + Sa mixed with 50% Ph+Di+Qz. Cr, cristobalite.

Stability of Phlogopite, Quartz and Diopside

Superposing Boyd's (1959) curve for the reaction

$$\underset{\text{tremolite}}{Ca_2Mg_5Si_7O_8O_{22}(OH)_2} \quad \underset{\text{diopside}}{2CaMgSi_2O_6} + \underset{\text{enstatite}}{3MgSiO_3} + \underset{\text{quartz}}{SiO_2} + \underset{\text{gas}}{H_2O} \quad (Ph,Sa,L)$$

on the data for phlogopite and quartz produces two intersections which are the invariant points:

sanidine-tremolite-phlogopite-diopside-quartz-enstatite-gas (L')

liquid-tremolite-phlogopite-diopside-enstatite-quartz-gas (Sa')

This generates a network of equilibria schematically shown in Figure 3.

This diagram, as drawn, makes the following assumptions. There is a constant ratio of 1:1 between K_2O and Al_2O_3 in the phases sanidine, phlogopite, and liquid. Quartz and H_2O-rich gas are always present. There is no significant CaO content in any phase except diopside and tremolite (see Yoder and Upton, 1971).

The reaction

$$\underset{\text{tremolite}}{Ca_2Mg_5Si_8O_{22}(OH)_2} + \underset{\text{sanidine}}{KAlSi_3O_8} \rightleftarrows$$

$$\underset{\text{phlogopite}}{KMg_3AlSi_3O_{10}(OH)_2} + \underset{\text{diopside}}{2CaMgSi_2O_6} + \underset{\text{quartz}}{4SiO_2} \quad (En,L,V)$$

is generated by the intersection of the two reactions (Ph,Sa,L) and (L) and requires that tremolite + sanidine be the low temperature assemblage. At higher pressures in the presence of an H_2O-rich gas, this curve will terminate at an invariant point involving reaction (En): sanidine-tremolite-phlogopite-diopside-quartz-liquid-gas, (En'). Two new reactions are generated:

$$\underset{\text{phlogopite}}{KMg_3AlSi_3O_{10}(OH)_2} + \underset{\text{diopside}}{2CaMgSi_2O_6} + \underset{\text{quartz}}{(4+n)SiO_2} \rightleftarrows$$

$$\underset{\text{tremolite}}{Ca_2Mg_3Si_8O_{22}(OH)_2} + \underset{\text{liquid}}{\text{melt}} \quad (Sa,En)$$

$$\underset{\text{tremolite}}{Ca_2Mg_5Si_8O_{22}(OH)_2} + \underset{\text{sanidine}}{mKAlSi_3O_8} + \underset{\text{gas}}{xH_2O} \rightleftarrows$$

$$\underset{\text{phlogopite}}{KMg_8AlSi_3O_{10}(OH)_2} + \underset{\text{diopside}}{2CaMgSi_2O_6} + \underset{\text{liquid}}{\text{melt}} (Qz,En)$$

Experiments defining reaction (Qz,En) are included in Table 3 and Figure 2.

At 2000 bars, the reaction (En,L,V) is located between 735° and 756°C. At 4000 bars, the invariant point (En') has been exceeded and the reaction (Qz, En) is located between 743° and 752°C.

THERMODYNAMIC CONSIDERATIONS

Using the data of Hemingway and Robie (1977) and Robie, Hemingway and Fisher (1977) and the properties of H_2O (Fisher and Zen, 1971), we can estimate values for the free energy of hydrous

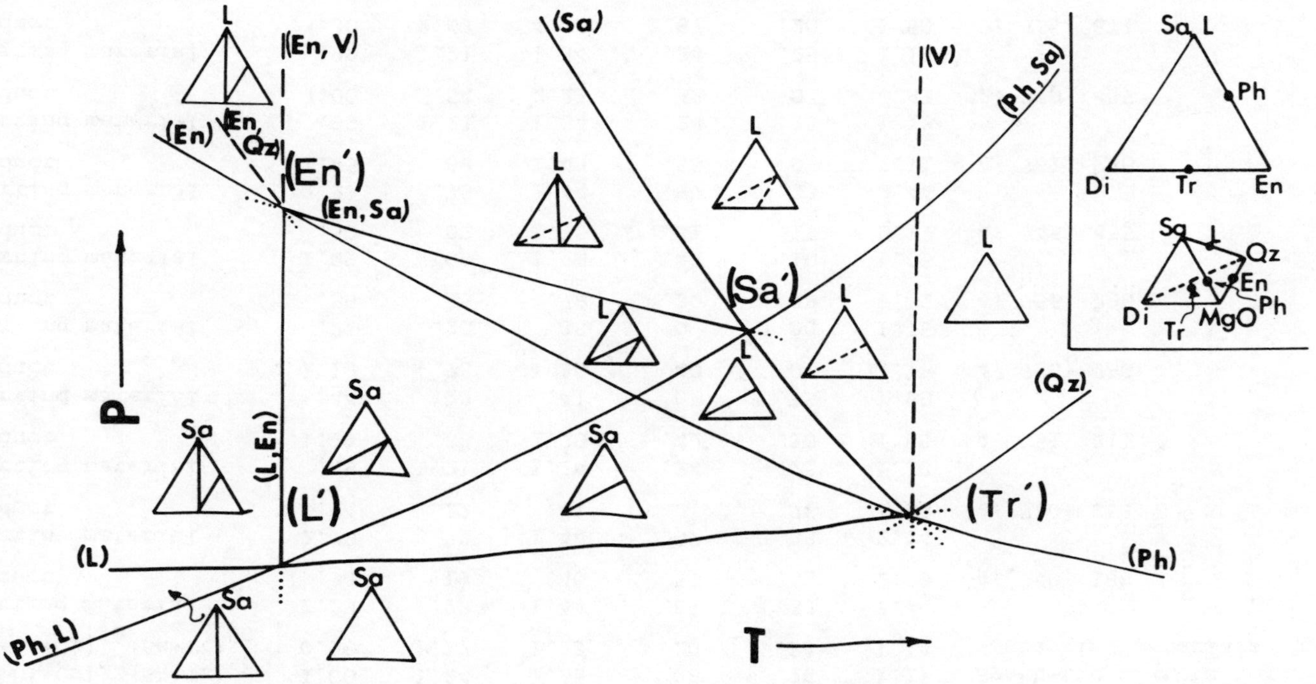

Figure 3. Schematic representation of portions of system $CaMgSi_2O_6$-MgO-$KAlSiO_4$-SiO_2-H_2O where H_2O and SiO_2 are always present. It is assumed that CaO is only present in the phases Di (diopside) and Tr (tremolite). Solid, dotted, and dashed reactions have the same meanings as in Fig. 1.

Table 4. Comparison of X-ray diffraction peak ratios for starting materials and products of experiments on the composition TRK

	Ph 001 / Tr 110	Ph 11$\bar{3}$ / Tr 310	Di 221 / Tr 310	Di $\bar{6}$01 / Tr 151	Sa 130 / Ph 11$\bar{3}$	Di $\bar{6}$01 / Ph 11$\bar{3}$	Run conditions kilobars, °C, hrs.
$(Ph+Di+Qz)_{50}(Sa+Tr)_{50}$	1.00	1.24	1.35	.28	.26	1.11	Synthetic mixture for
$(Ph+Di+Qz)_{50}(Sa+Tr)_{50}$	0.98	1.17	1.33	.20	.24	1.14	starting materials
Starting material	2.28	.73	1.44	.67	.33	2.44	
Product	.90	.19	.40	.57	.33	2.08	4; 740; 188
Starting material	2.28	.73	1.44	.67	.33	2.44	
Product	.44	.22	.41	0	.75	1.88	4; 730; 236
Starting material	.99	1.21	1.34	.24	.25	1.13	
Product	1.29	.69	1.28	.42	.10	1.85	4; 751; 211
Starting material	.44	.22	.41	0	.75	1.88	
Product	1.12	.73	1.30	.40	0	1.78	4; 756; 286
Starting material	.21	.20	.35	0	1.00	1.75	
Product	.86	.45	.78	.30	.28	1.72	4; 756; 286
Starting material	1.29	.69	1.28	.42	.10	1.85	
Product	.50	.62	1.35	.31	.17	2.17	4; 745; 455
Starting material	2.28	.73	1.44	.67	.33	2.44	
Product	1.45	.68	1.43	.55	0	2.11	4; 761; 236
Starting material	.99	1.21	1.34	.24	.25	1.13	
Product	3.00	2.54	2.21	.43	0	.87	2; 770; 407
Starting material	.99	1.21	1.34	.24	.25	1.13	
Product	3.30	2.67	4.00	1.67	.20	1.50	2; 755; 671
Starting material	.99	1.21	1.34	.24	.25	1.13	
Product	.42	.81	.56	0	.55	.68	2; 740; 307

phlogopite at 750°C, 2000 bars, and at 835°C, 450 bars. Without information on the heat capacity and entropy of hydrous phlogopite, this data is not directly comparable to recent estimates for phlogopite made by Bird and Anderson (1973), but can be compared with data of Ludington and Munoz (1975).

The equilibrium constant for the exchange reaction,

$$1/2KMg_3AlSi_3O_{10}(OH)_2 + HF = 1/2KMg_3AlSi_3O_{10}F_2 + H_2O$$

is given by Ludington and Munoz (1975) as $\log K = \frac{2100}{T} + 1.523$.

Knowing free energy values for fluorphlogopite, HF and H_2O permits the derivation of free energy for phlogopite at any temperature, T. Robie (in preparation; personal communication) has recompiled the thermodynamic data for fluorphlogopite by incorporating the disorder term for the mixing of Al and Si in the tetrahedral layer (Ulbrich and Waldbaum, 1976) and new values for the enthalpy of $Al(OH)_3$ (Hemingway and Robie, 1977). Values used in this study for fluorphlogopite and other minerals are given in Table 5 along with the values derived for phlogopite. The discrepancy between this study and Ludington and Munoz (1975) is 24 kJ at 750°C and 14 kJ at 835°C. The problem of the composition of tremolite makes the calculations at 750°C somewhat questionable until a rigorous stoichiometry can be proven.

The low value of (dP/dT) for the reaction (L) is due to the very large volume term involving H_2O for this reaction. The activity of SiO_2 has a profound effect on the stability of phlogopite as has been pointed out by Luth (1967) and Yoder and Kushiro (1969). A whole series of stability curves exist for phlogopite in which the activity of SiO_2 is successively buffered by quartz, enstatite-forsterite, leucite-sanidine, and leucite-kalsilite. In contrast, the maximum stability of tremolite is in the presence of quartz. Thus we see that the stability of phlogopite relative to tremolite is a function of the activities of SiO_2, $KAlSi_3O_8$, H_2O, and F.

GEOLOGICAL APPLICATIONS

The network demonstrated in Figure 3 has some application to the crystallization sequence of mafic silicates formed in igneous rocks. In K_2O-rich magmas, the early ferromagnesian phase would be pyroxene. However, the next phase to appear would be dependent on H_2O activity. At a low activity of H_2O, phlogopite (biotite) would appear, followed by amphibole. At higher H_2O activities, amphibole would appear early. Ewart et al. (1975) have described such sequences from ash flow tuffs in New Zealand, where saturation of the magma with H_2O is presumed to be the case. However, in quartz latites described by Ransome (1898) and in granodiorites of the Sierra Nevada (Bateman et al., 1965), biotite-augite assemblages have been described. Reaction (Sa, En)(Figures 2 and 3) represents the maximum H_2O pressure at which phlogopite and

Table 5. Estimated thermodynamic properties of phlogopite and other values

Phase	ΔG_{1023}, kj	ΔG_{1108}, kj	Reference
Enstatite, $MgSiO_3$	---	-1226.2±5.0	Robie and Waldbaum, 1968
Sanidine, $KAlSi_3O_8$	-3200.7±4.2	-3134.1±4.2	Hemingway and Robie, 1977
Quartz, SiO_2	-726.4±1.7	-711.6±1.7	Robie and Waldbaum, 1968
Gas, H_2O at 450 bars	---	-130.6±0.4	Fisher and Zen, 1971
Diopside, $CaMg(SiO_3)_2$	-2616.5±9.2	---	Robie and Waldbaum, 1968
Tremolite, $Ca_2Mg_5Si_8O_{22}(OH)_2$	-9874.2±17.2	---	Robie and Waldbaum, 1968
Gas, H_2O, at 1 bar	-191.4±0.2	-186.7±0.2	Robie and Waldbaum, 1968
Gas, HF, at 1 bar	-227.5±0.4	-227.2±0.4	JANAF tables
Fluorphlogopite	-5224.3±5	-5119.8±5	Robie, personal comm., in prep.
Phlogopite, $KMg_3AlSi_3O_{10}(OH)_2$	-4936.3±22	-4808.5 10	This study
	-4912.1±13	-4794.8 13	Ludington and Munoz (1975)

diopside crystallize before tremolite in a crystallizing magma.

Biotite-diopside-quartz assemblages occur in high grade regional metamorphic rocks (Hewitt, 1973; Thompson, 1975). The higher the Fe content, the lower the temperature at which this assemblage takes the place of actinolite-K-feldspar assemblages. Melson (1966) described such a change of assemblage in the contact aureole of a stock in Montana. The effect of Fe is further emphasized by the common occurrence of hedenbergite and Fe-rich biotite in granitic rocks, whereas ferrotremolite is virtually unknown in such rocks. This is not the case for Na-rich amphiboles, where arfvedsonite-alkali feldspar assemblages are common. Regional metamorphic rocks containing orthopyroxene and K-feldspar i.e., charnockites, must have crystallized in H_2O-poor environment, or have a significant Fe content. If sufficient H_2O were present, such rocks would become either quartz-biotite schists or silicate melts. The low H_2O activity can be accounted for either by the intrinsic composition of the rocks (metavolcanics) or by the introduction of CO_2 into the system by virtue of decarbonation reactions. The fact that many charnockites are found in marble-bearing terrains is in accord with the requirement for low H_2O activity.

REFERENCES

Bateman, P.C., Lorin, D., Clar, N., Huber, N.K., Moore, J.G. and Rinehart, C.D. U.S. Geol. Survey Prof. Paper 414-D (1963).

Bird, G.W. and Anderson, G.M. Am. Jour. Sci., 273, 84-91 (1973).

Boyd, F.R. in Researches in Geochemistry (P.H. Abelson, Ed.), Wiley & Sons, New York, 377 (1959).

Burnham, C.W., Holloway, J.R. and Davis, N.F. Geol. Soc. Am. Special Paper, 132 (1969).

Clark, J.R., Appleman, D.E. and Papike, J.J. Mineral. Soc. Amer., Special Paper No.2, 31 (1969).

Deer, W.A., Howie, R.A. and Zussman, J. Rock Forming Minerals v.2, Chain Silicates, Longman, Green & Co., London, 379 p. (1963).

Ewart, A., Hildreth, W. and Carmichael, I.S.E. Contr. Min. Pet., 51, 1 (1975).

Fisher, J.R., Zen, E-an. Am. Jour. Sci., v.270, 297 (1971).

Hellman, Marshall, S. U.S.G.S. Computer Cont. No.9 (1971).

Leake, B.E. A Catalog of Analyzed calciferous and subcalciferous amphiboles together with their nomenclature and associated minerals. Geol. Soc. Am. Special Paper 98 (1968).

Ludlington, S.D. and Munoz, J.L. Geol. Soc. Am. Abs. with Prog., v.7, 1179 (1975).

Luth, W.C. Jour. Petrology, 8, 372 (1967).

Melson, W. Am. Min., 51, 402 p. (1966).

Munoz, J.L. and Ludlington, S.D. Am. Jour. Sci., 274, 396 (1974).

Ransome, F.L. U.S. Geol. Survey Bull. No.89 (1898).

Robie, R.A. and Waldbaum, D.R. U.S. Geol. Survey Bull., 1259 (1968).
Ross, M., Papike, J.J. and Shaw, K.W. Mineral. Soc. Amer. Special Paper No.2, 275 (1969).
Schairer, J.F. Am. Ceram. Soc. Jour., 37, 501 (1954).
Shaw, H.R. Am. Mineral., 48, 883 (1963).
Stephenson, D.A., Sclar, C.B. and Smith, J.V. Mineral. Mag., v.35, 838 (1966).
Stewart, D.B. Feldspar Mineralogy, Mineral. Soc. Am. Short Course, vol.2, St-1-22, (1975).
Thompson, A.B. Contr. Mineral. Pet., 53, 105 (1975).
Troll, Georg, and Gilbert, M.C. Am. Mineral., 57, 1386 (1972).
Ulbrich, H.H. and Waldbaum, D.R. Geochimica et Cosmochim. Acta, v.40, no.1, 1 (1976).
Wones, D.R. Geochimica et Cosmochim. Acta, v.3, 2248 (1967).
Wones, D.R. and Dodge, F.C.W. Geol. Soc. Am. Sp. Paper, 203 (1966)
Yoder, H.S. and Kushiro, I. Am. Jour. Sci., 267A, 558 (1969).
Yoder, H.S. and Upton, B.J.G. C.I.W. Yearbook 70, 108 (1971).
Zussman, J. Acta Crystallogr., vol.12, 309 (1959).

STUDY PROBLEMS

1. Given the enclosed chemical analyses from the same rock in Portugal (Neiva, 1975, 1976), calculate structural formulae for the muscovite and biotite.

   a) Using all elements (including trace elements) (Analysis A).

      1. Assuming cationic charge of + twenty-two (22).
      2. Assuming twelve (12) anions.
      3. a. Six cations in octahedral and tetrahedral sites of muscovite.
         b. Seven cations in octahedral and tetrahedral sites of biotite.

   b) Using only major elements (Analysis B) and obtaining H_2O by difference.

      1. Assuming cationic charge of + twenty-two (22).
      2. Assuming twelve (12) anions.
      3. a. Six cations in octahedral and tetrahedral sites of muscovite.
         b. Seven cations in octahedral and tetrahedral sites of biotite.

   This problem considers the evaluation of chemical analyses of micas and suggests the pitfalls of microprobe analyses.

2. Are the biotite and muscovite in equilibrium?

3. Given the compositions of the feldspars, what can you say about the activities of volatile components in the melts?

	Muscovite		Biotite	
	A	B	A	B
SiO_2	45.72	45.72	35.17	35.17
TiO_2	0.26	0.26	1.70	1.70
Al_2O_3	33.73	33.73	21.49	21.49
Fe_2O_3	n.d	n.d	2.54	n.d
FeO	3.45	3.45	19.58	21.87
MnO	0.10	0.10	0.95	0.95
MgO	1.49	1.49	4.95	4.95
CaO	n.d	n.d	n.d	n.d
Na_2O	0.54	0.54	0.19	0.19
K_2O	10.27	10.27	8.97	8.97
Cl	0.11	0.11	0.11	0.11
F	0.27	0.27	0.60	0.60
H_2O^+	4.51	4.09	3.95	4.27
Ga	103		92	
Nb			362	
Zn	380		543	
Sn	205		140	
Li	2100		6000	
Ni			8	
Zr			84	
Cu	39		55	
Ba	60		55	
Rb	1300		1800	
Cs	173		1000	

OPAQUE MINERALS AS SENSITIVE OXYGEN BAROMETERS AND GEOTHERMOMETERS IN LUNAR BASALTS

A. El Goresy, Max-Planck-Institut für Kernphysik, 6900 Heidelberg, Germany

and

E. Woermann, Institut für Kristallographie,Rheinisch-Westfälische Technische Hochschule, 5100 Aachen, Germany

I. INTRODUCTION: Lunar basalts can be classified according to their chemistry in two distinct categories: 1. TiO_2-rich basalts with a TiO_2 content ranging from 7 to 14 wt. %. These basalts were recovered from the Apollo 11 (Mare Tranquillitatis) and Apollo 17 (Taurus-Littrow site at the southeast rim of Mare Serenitatis) landing sites, both on the side of the moon facing the east side of the earth. 2. TiO_2-poor basalts with TiO_2-content ranging from 0.6 to 4 wt.%. Samples of this category were collected from the Apollo 12(Oceanus Procellarum), Apollo 15 (at the east rim of Mare Imbrium), and Luna 16(Mare Fecunditatis). All these basalts without exception differ from their terrestrial counterparts mainly in their extremely low oxygen fugacities. This feature is manifested by the ubiquitous occurrence of metallic iron as a primary phase, and the presence of textures indicative of subsolidus reduction with metallic Fe as one of the reaction products.

Opaque oxide assemblages coexisting with metallic iron in lunar basalts belong to three distinct solid solution series: a) chromite-ulvöspinel series; b) ilmenite-geikielite series, c) armalcolite-anosovite (Ti_3O_5) series in the system MgO-FeO-Al_2O_3-Cr_2O_3-TiO_2. Experimental investigations of this system will thus be helpful in gaining information on the petrology of lunar basalts. The general condition requiring equilibration of the oxide phases with metal and correspondingly a close control of oxygen fugacities offers a favourable basis for thermodynamic treatment of the experimental data.

Emphasis will be given to the spinel solid solution series

D. G. Fraser (ed.), Thermodynamics in Geology, 249-277.

due to its potential as a sensitive indicator of the crystallization sequence of the basalts, and to the armalcolite-anosovite series due to its importance as an indicator for oxygen fugacity during crystallization. Subsolidus reduction reactions involve opaque oxides of all three solid solution series and hence textures and assemblages of members of all series in addition to the reaction fayalite $\longrightarrow$ Fe + SiO_2 + O_2 will be discussed in comparison with laboratory data.

II. SPINEL SOLID SOLUTION SERIES

Lunar spinels are one of the dominating opaque oxides in TiO_2-poor basalts. They are members of a complex solid solution series with six endmembers. 1. Magnesiochromite, $MgCr_2O_4$, 2. Chromite $FeCr_2O_4$, 3. Spinel $MgAl_2O_4$, 4. Hercynite $FeAl_2O_4$, 5. Mg-ulvöspinel $MgTi_2O_4$ and 6. Ulvöspinel Fe_2TiO_4. The dominating endmembers are chromite, ulvöspinel, and to a lesser extent hercynite. The paragenetic sequence during crystallization from a lunar basaltic melt indicates initial precipitation of spinels enriched in $MgCr_2O_4$ molecule followed by members enriched in $FeCr_2O_4$ molecule and finally members enriched in Fe_2TiO_4 molecule. In many mare basalts numerous spinel grains were found to display this sequence in the same grain with the $MgCr_2O_4$-rich spinel in the core and the Fe_2TiO_4-rich spinel comprising the outermost mantle with the $FeCr_2O_4$ rich spinel in between. The boundaries between the chromite-rich core or zones and the ulvöspinel-rich mantle can be gradational or abrupt. Gradational zoning from chromite-rich to ulvöspinel-rich zones indicate continuous precipitation at a temperature at which the solid solution between chromite and ulvöspinel was complete (Nehru *et al.*, 1974, El Goresy *et al.*, 1976). On the other hand, abrupt zoning between chromite and ulvöspinel is suggestive of crystallization of ulvöspinel at lower temperatures at which the solid solution between the two endmembers was interrupted by a solvus (Nehru *et al.*, 1974, El Goresy *et al.*, 1976). Both gradational and abrupt zoning in lunar spinels is a direct result of the chemical changes which took place in the cooling basaltic liquid during growth of these spinels. Growth of spinel grains may have spanned a great deal of the cooling history of mare basalts and hence precipitation of major silicate phases causing changes in abundances of major and minor elements e.g. build-up of or depletion in oxides should be reflected in the type of zoning and the chemical variation of a zoned spinel grain. Tracing of the change in spinel chemistry in individual zoned grains with the electron microprobe should help in reconstructing the crystallization sequence of mare basalts.

A distinct textural feature characterizes the sharp chromite-ulvöspinel zoning in fine grained pigeonite basalts. The chromite cores display throughout idiomorphic sharp boundaries to ulvöspinel without any sign of reaction prior to ulvöspinel precipitation. In contrast coarse grained olivine and pigeonite basalts contain corroded and rounded chromite cores in sharp contact with an ulvöspinel rim. In coarse grained rocks, considerable reaction must have taken place between basaltic liquid and chromite thus removing the outer rims of the chromite prior to precipitation of the late chromian ulvöspinel. Ti-chromites with gradational zoning from all sides to chromian ulvöspinel as well as grains with gradational zoning on one side and sharp zoning on the other side of the grain also occur together with spinels with corroded chromite cores in the same basalt. Only the unprotected grains or sides of the grains were subjected to this reaction. It is important to determine the compositions of removed outer rims (before ulvöspinel crystallization) and at which stage in the crystallization sequence this reaction took place. In addition to the spinel mentioned above in coarse grained olivine and pigeonite basalts idiomorphic inclusions of spinels in olivine without any sign of optical zoning also occur. Chemical zoning of these spinel inclusions reflects the chemical changes of the basaltic liquid before and during crystallization of olivine.

Study of chemical zoning in individual spinel grains of all generations in coarse grained rocks and idiomorphic sharp zoning in fine grained basalts revealed the presence of three major, but distinct variation trends characterized by certain coupled oxide variations from spinel core to rim (El Goresy et al., 1976):
1. Slight increase in $TiO_2/TiO_2+Cr_2O_3 + Al_2O_3$, decrease in $Cr_2O_3/Cr_2O_3 + Al_2O_3$, increase in V_2O_3, decrease in FeO/FeO + MgO; FeO/FeO + MgO (FFM) initial ratio of the core $\leqslant 0.9$.
2. Sharp increase in $TiO_2/TiO_2 + Cr_2O_3 + Al_2O_3$, increase in $Cr_2O_3/Cr_2O_3 + Al_2O_3$, decrease in V_2O_3, increase in FeO/FeO + MgO; initial FFM ratio of the core $\geqslant 0.9$.
3. Constant or slightly increasing $TiO_2/TiO_2 + Cr_2O_3 + Al_2O_3$, increase in $Cr_2O_3/Cr_2O_3 + Al_2O_3$, decrease in V_2O_3, sharp increase in FeO/FeO + MgO; initial FFM ratio of the core $\leqslant 0.85$.

The first two chemical trends were encountered mostly together in two successive chromite generations in many coarse grained olivine and pigeonite basalts. The first zoning trend is restricted to early spinels enclosed in olivine in coarse grained pigeonite and olivine basalts (Fig. 1a). This trend is unique as it indicates an increase in the concentrations of Al_2O_3, V_2O_3, and MgO during the growth of the chromite. This chemical zoning thus indicates precipitation of chromite from a silicate liquid with continuous build-up of Al_2O_3, V_2O_3, and MgO contents. Crystallization of this chromite very probably took place before crystallization of plagioclases, pyroxene, and either before or during olivine precipitation. Spinels displaying both gradational and sharp zoning on opposite sides of the same grain are charac-

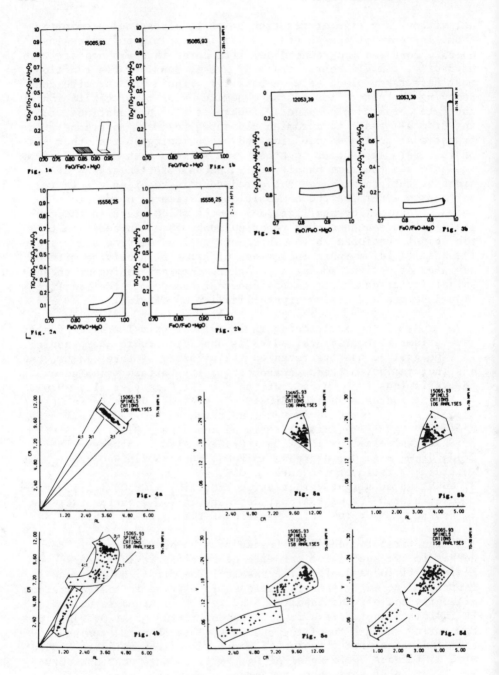

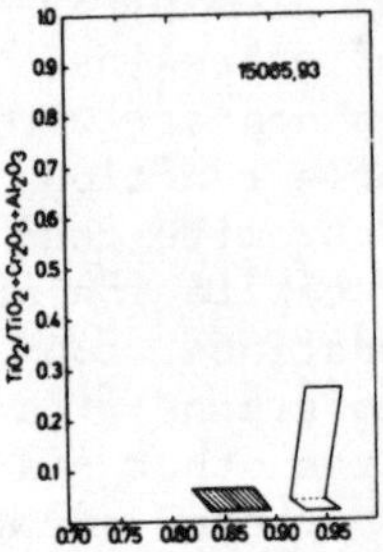

Fig. 1a

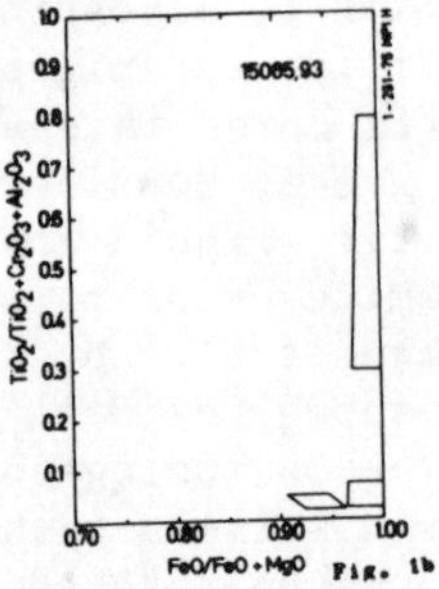

Fig. 1b

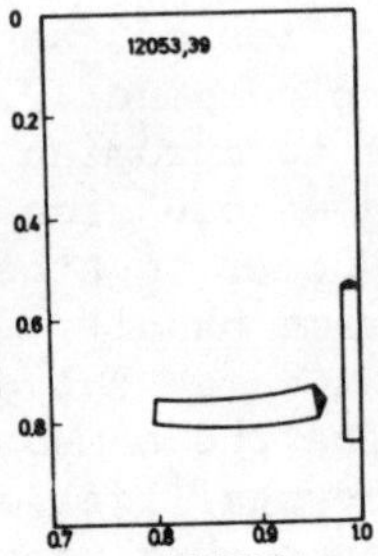

Fig. 3a

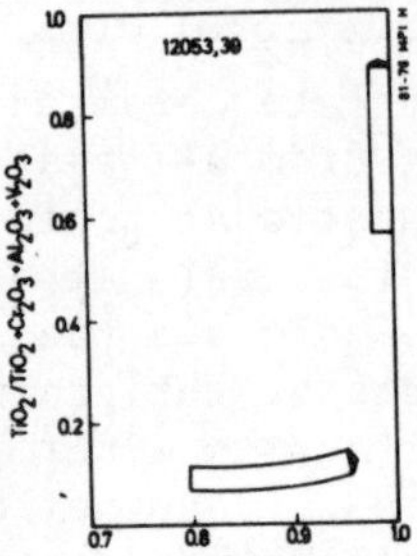

Fig. 3b

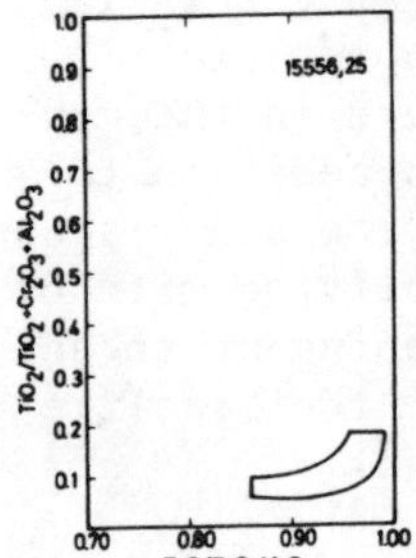

Fig. 2a

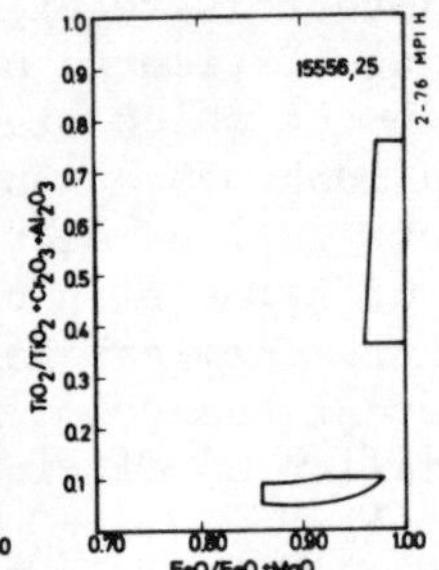

Fig. 2b

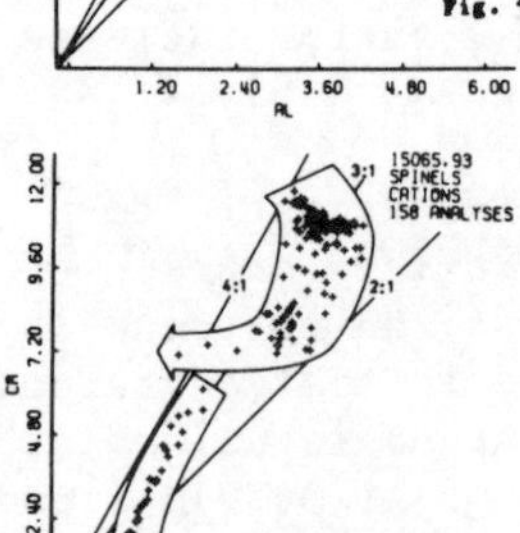

Fig. 4a

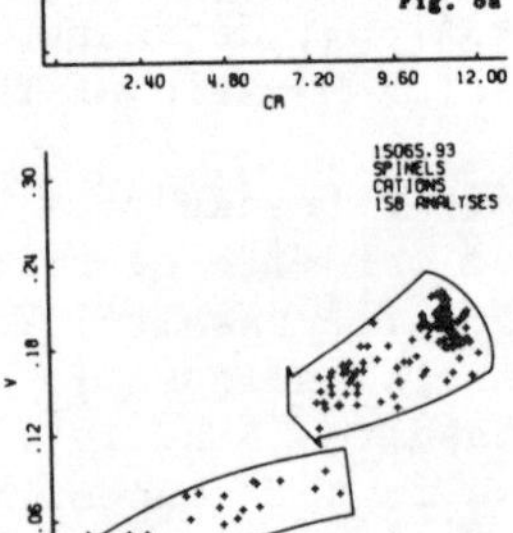

Fig. 5a

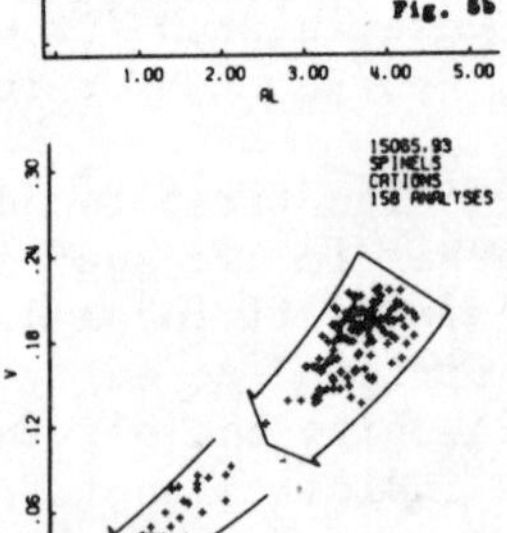

Fig. 5b

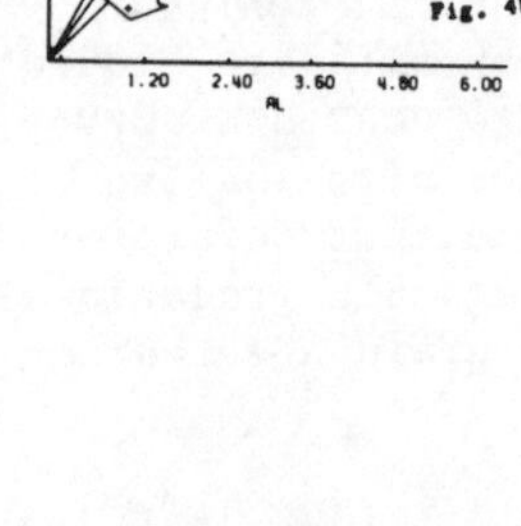

Fig. 4b

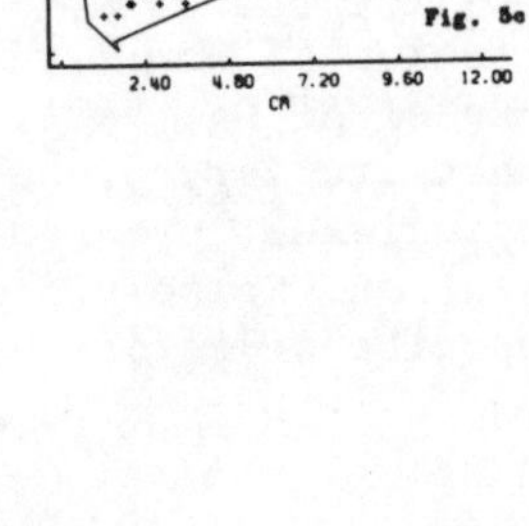

Fig. 5c

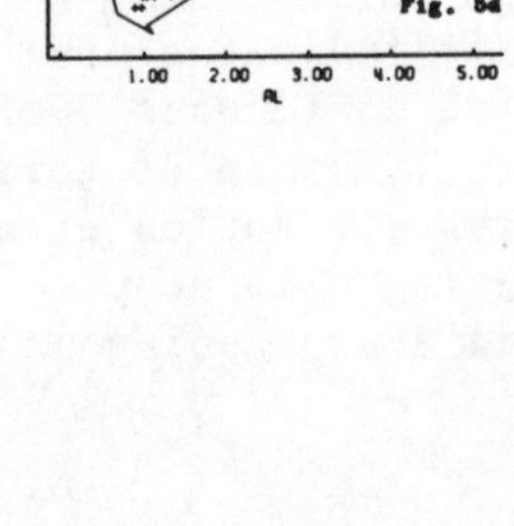

Fig. 5d

FIGURE CAPTIONS

Figure 1: Projection of the rectangular face of the spinel prism with the first and second zoning trends in a pigeonite basalt. a. Compositional variation in early spinels (hachured area) and gradational zoning on the shielded sides of spinels of the second trend. b. Compositional variation on the corroded side and the late ulvöspinel demonstrating the gap after removal of the gradationally zoned layers.

Figure 2: Projections of the rectangular face of the spinel prism displaying the third zoning trend in an olivine basalt a. sharp increase in the FFM ratio and sharp rise in the $TiO_2/TiO_2+Cr_2O_3+Al_2O_3$ only in the terminating stage. b. Corrosion of the TiO_2-rich zones and compositional variation of the late stage ulvöspinel.

Figure 3: a. Base of the spinel prism showing the third compositional variation trend of idiomorphic chromite core and late ulvöspinel in an Apollo 12 pigeonite basalt.b.Rectangular face of the prism displaying the third zoning with the sharp increase in FFM ratio but without any significant increase in $TiO_2/TiO_2+Cr_2O_3+Al_2O_3$.

Figure 4: Cr-Al substitutional trends in spinels in an Apollo 15 pigeonite basalt. a. Cr-Al substitutions for early spinels. b. Cr-Al substitutional trends for the second chromite type and late ulvöspinel.

Figure 5: V-Al and V-Cr substitutional relationships in a pigeonite basalt. a. and b. substitutional trends of early spinels indicationg increase in V and Al from core to rim. c. and d. substitutional trends for later chromite and ulvöspinel.

Figure 6: Fe-Mg substitutional trends for various zoned spinels in a pigeonite basalt. Antipathetic negative trends for early spinels is pronounced. All chromite generations with various FFM ratios emerge.

Figure 7: Melting relations of Apollo 17 samples 74275. Triangular points are those of O'Hara and Humphries (1975)at their stated oxygen fugacities. The iron-wüstite (Fe-FeO) curve is shown as reference.

Figure 8: Mg-Fe cationic substitutional relationship (based on 5 oxygens) for tan and gray armalcolite.

Figure 9: MgO Arm./MgO Ilm. versus Cr_2O_3 Arm./Cr_2O_3 Ilm. relationship for gray armalcolites and mantling ilmenites.

Figure 10: Schematic plot of possible phase equilibria in the system $FeO-TiO_2-SiO_2$ in equilibrium with metallic iron.

Figure 11: Stability of the iron-titanium oxide phases in the system Fe-Ti-O.

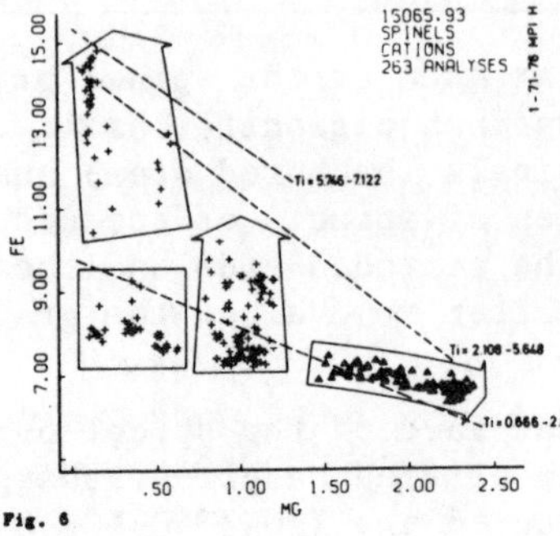

Fig. 6

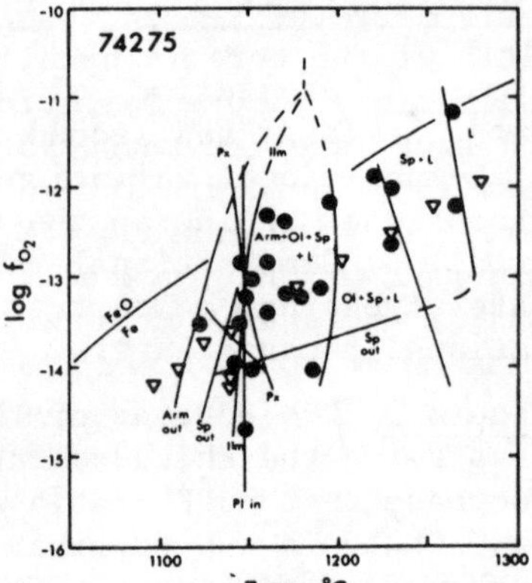

Fig. 7

TABLE 1

14053	70017; 70035; 70135
1. Ulvöspinel — ilmenite+Fe	1. Ulvöspinel — ilmenite+Fe
2. Fayalite — Fe+silica	2. Ilmenite — spinel+rutile+Fe
3. Incipient breakdown of ilmenite (only very few grains)	3. No fayalite breakdown

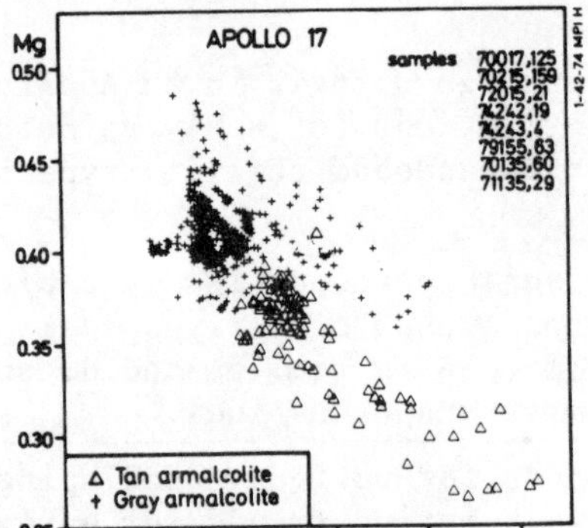

Fig. 8

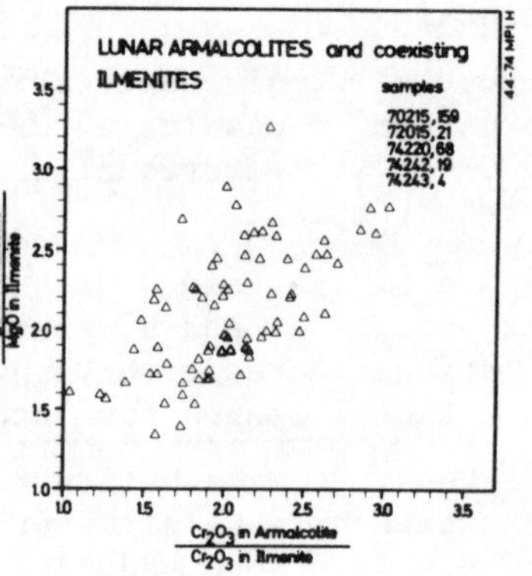

Fig. 9

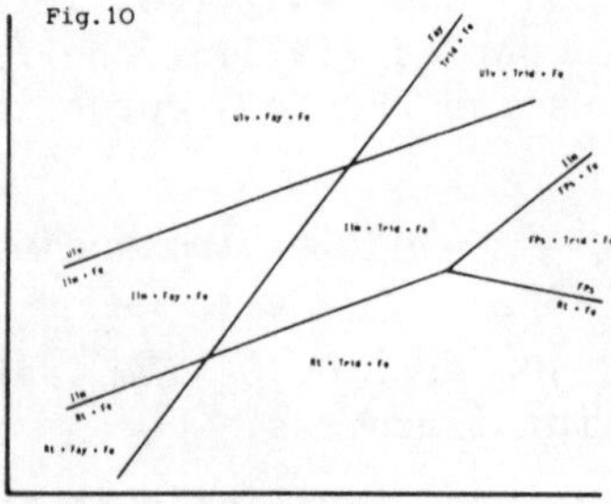

Fig.10

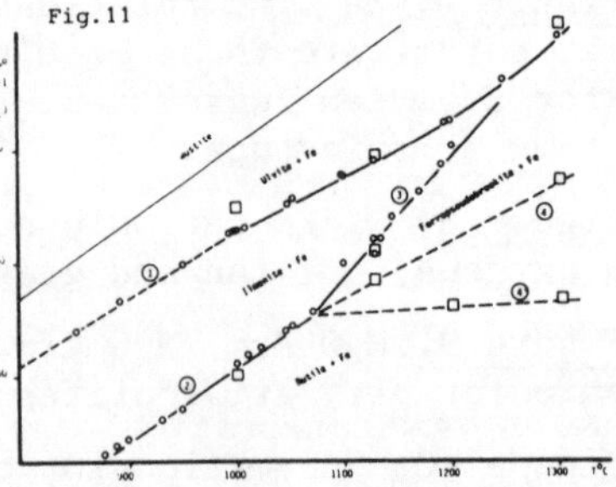

Fig.11

terized by the second chemical trend (Fig. 1a). The initial FFM ratio of the core is usually higher than that of early chromites indicating: a. precipitation from a liquid with higher FFM ratio and thus later in the crystallization sequence, and b. cooling basaltic liquid did indeed subsequently precipitate various generations with various FFM ratios due to changes in its composition and at changing cooling rates. On the sides of the grains with gradational zoning increase in the $TiO_2/TiO_2 + Cr_2O_3 + Al_2O_3$ ratio from core to rim from 0.02 to 0.48, thus filling the Apollo 12 gap, were recorded. On the opposite unprotected side the basaltic liquid reacted with the spinel thus removing the TiO_2-rich zones and corroding the spinel down to its core before it started precipitation of the late Cr-ulvöspinel on the corroded core (Fig. 1b). There is indication that this process took place repeatedly in the course of crystallization and as many as three generations of Ti-chromites with various initial FFM ratios could be encountered in the same basalt. This chemical trend is indicative of co-precipitation with pyroxene but before massive crystallization of plagioclase took place. The third trend, though characteristic for the fine grained pigeonite basalts, was also encountered in some coarse grained olivine and pigeonite basalts. The Ti-chromite cores have usually a low initial FFM ratio (0.80-0.85)- The ratio increases sharply from core to rim (Fig. 2a & 3b) with slight increase in $TiO_2/TiO_2 + Cr_2O_3 + Al_2O_3$ and decrease in V_2O_3 content thus demonstrating crystallization from a liquid with continuous build-up of FeO content. In the coarse grained rocks resorption of TiO_2-rich outer rims also took place and studies indicate that the trend did show a sharp increase in Fe_2TiO_4 molecule before corrosion took place (Fig. 2a & b). The initial zoning path is, however, identical in both fine and coarse grained rocks (Fig 2a & 3b). At the terminating stage a sharp rise of the $TiO_2/TiO_2 + Cr_2O_3 + Al_2O_3$ is observed only in coarse grained rocks to values up to 0.2. This trend is indicative of coprecipitation of spinel and pyroxene but before plagioclase from a liquid continuously enriched in FeO.

Cationic Relationships and Substitutions

Cationic substitutions in a zoned spinel in the B site involving the trivalent cations Cr, Al, and V are extremely sensitive to changes in the concentrations of these elements in the liquid Any drastic changes may indicate crystallization of a major silicate phase like pyroxene or plagioclase. Changes in the substitutional trends for the divalent cations Fe and Mg are also dependent on changes in their relative concentrations in the cooling liquid due to crystallization of olivine and/or pyroxene. Substitutional trends for the pairs Cr-Al, Cr-V, Al-V, and Fe-Mg offer an accurate way to trace the crystallization sequence. Precipitation of major silicate phases usually involve several elements and hence should be documented in the majority of the trends involving the different pairs.

a. Cr-Al Substitutional Trends

In coarse grained olivine and pigeonite basalts with spinel generations displaying the first and second zoning trends, three distinct Cr-Al substitutional trends were encountered (Fig. 4). Early chromites show a unique trend that varies from core to rim from Cr/Al ratios 4:1 to 2:1 (Fig. 4a). The negative slopes of the Cr/Al substitutional trends approaches a 1:1 ratio. This trend again demonstrates that the early chromites grew from a liquid with continuous build-up of Al and thus before crystallization of plagioclase. This trend is entirely different from the trends of other chromite generations and ulvöspinel. Spinels with the second chemical zoning trend display a curved substitutional trend starting for cores at a Cr/Al ratio of 4:1 with a steep positive slope to 2:1 and with a sharp turn back to ratios higher than 4:1 (Fig. 4b). This curvature is indicative of change in the activities of Cr_2O_3 and Al_2O_3 due to subsequent crystallization of pyroxene and plagioclase. The steep drop in Cr in the upper part of the first branch is indicative that at that stage pyroxene along with Ti-chromite was the main crystallizing phase. The sharp turn back to much lower Al concentration signals the entry of plagioclase as a major crystallizing phase. The Cr/Al substitutional trend for Cr-ulvöspinel (lower branch in Fig. 4b) is indicative of crystallization from a liquid continuously depleted in Cr and Al due to simultaneous precipitation of spinel, pyroxene, and plagioclase.

b. V-Cr and V-Al Substitutional Trends

Though a trace element in spinels , vanadium is a valuable element because of its partitioning between chromite and ulvöspinel and its preference to pyroxene (Laul and Schmitt, 1973). It provides a sensitive check for pyroxene entry as a crystallizing phase. V-Cr and V-Al substitutional trends provide a good control for the positions of both pyroxene and plagioclase in the crystallization sequence. The positive V-Al sympathetic trend and the negative V-Cr relationship of early spinels (Fig. 5a & b) demonstrate the unique chemistry of those early chromites and indicate crystallization before pyroxene and plagioclase. The trends for chromites with the second chemical zoning trend and late ulvöspinel are antipathetic both for V-Cr and V-Al (Fig. 5c & d). The late stage Cr-ulvöspinel crystallized along with both pyroxene and plagioclase as seen from the lower trend in Figs. 5c and 5d.

c. Fe-Mg Substitutional Trends

Several features can be recognized in the Fe-Mg substitutional trends shown in Figure 6. All the substitutional trends observed in the Cr-Al, V-Cr, V-Al diagrams are encountered in Fig. 6. All chromite and ulvöspinel generations with various initial FFM ratios emerge. Early spinels display a unique antipathetic trend with negative slope with increasing Mg substitutions for Fe from core to rim. Haggerty argued (1972 a, 1972 b, 1972 c) that the divalent Fe versus Mg relationship is poor to totally incoherent and that distinct linear slopes do emerge if the Ti content of the spinels is considered. Each of the slopes proposed by Haggerty would correlate spinel compositions with similar $TiO_2/TiO_2 + Cr_2O_3 + Al_2O_3$ ratios. This attempt does not reflect the real substitutional trend and even obscures the Fe-Mg relationship in the zoned spinels. This conclusion is demonstrated in Fig. 6 where vertical trends with sharply increasing Fe substitutions for Mg from core to rim for the later chromite and Cr-ulvöspinel generations are evident. In fact, these steep trends cross the several slope lines for compositions with various Ti-contents proposed by Haggerty. Furthermore, slopes constructed for compositions (Haggerty, 1972 a, 1972 b, 1972 c) with similar Ti-contents connect spinels of various generations crystallized at different times with different initial FFM ratios and the relationship observed for early spinels would then completely disappear. The Fe-Mg substitutional trends for spinels with the second chemical zoning demonstrate a continuous increase in the FFM ratio of the liquid and after precipitation of olivine documenting coprecipitation of pyroxene.

III. ARMALCOLITE SERIES

Armalcolite $(Fe,Mg)Ti_2O_5$ was first discovered in TiO_2-rich basalts collected from the Apollo 11 site (Anderson _et al._,1970). The mineral also occurs in numerous TiO_2-rich basalts from the Taurus-Littrow site (Apollo 17). In many TiO_2-rich basalts of the Apollo 11 and Apollo 17 sites armalcolite is one of the major opaque oxides. Textural variations and compositional similarity between TiO_2-rich basalts of both landing sites is suggestive of a similar source of the rocks. The TiO_2-rich basalts can be classified into two major types: 1. plagioclase poikilitic ilmenite basalts and 2. olivine porphyritic ilmenite basalts. So far, the first basalts type was not encountered in the Apollo 11 site and is restricted to the Apollo 17 samples. Recent investigations indicate that armalcolite is a member of the solid solution series karooite ($Mg Ti_2O_5$)-ferropseudobrookite ($FeTi_2O_5$)-anosovite (Ti_3O_5) (Papike _et al._, 1975, Wechsler _et al._, 1975, El Goresy _et al._, 1974). Of special interest are textural relationships between armalcolite and the coexisting silicates and opaque oxides in different rocks as well as variations of armalcolite chemistry. Two optically different armalcolite types were reported in several Apollo 17 basalts (Haggerty ,1973, El Goresy _et al._, 1974).

a. a grey variety usually mantled by Mg-rich ilmenite present in olivine porphyritic ilmenite basalts, b. a tan variety encountered in plagioclase poikilitic basalts. Although in both rock types armalcolite was early in the crystallization sequence, the ilmenite-pyroxene crystallization sequence after armalcolite precipitation is different in the two rock types.

1. Plagioclase poikilitic ilmenite basalts: In this rock type olivine and Cr-ulvöspinel were the first minerals to crystallize followed by tan armalcolite followed by titanaugite and pigeonite. Ilmenite, plagioclase and then crystobalite were the last minerals in subsequent order. In these rocks massive ilmenite precipitation started after the majority of the pyroxenes crystallized. Armalcolite occuring in these rocks is only of the tan variety and is exclusively present as inclusions in the pyroxenes. This crystallization sequence is not dependent on the cooling rate.

2. Olivine porphyritic ilmenite basalts: These rocks usually contain olivine phenocrysts (as a quensh phase) with ulvöspinel inclusions. The two minerals also here were the first phases to crystallize (El Goresy et al., 1974). Both were then followed by grey armalcolite, then ilmenite and at last augite, plagioclase and then tridymite. Textures in Apollo 11 and 17 rocks are identical. Characteristic for these rocks is that the majority of the armalcolite grains is surrounded by ilmenite rims. Shape and width of the ilmenite rim vary from grain to grain. Usually, the armalcolite grains are surrounded by continuous ilmenite mantles and the composite grain still displays armalcolite morphology.

Origin of ilmenite rims around armalcolite: Ilmenite mantles around grey armalcolite are formed according to one or a combination of the following processes: a. Reaction between the cooling basaltic liquid and early crystallized armalcolite

$$FeTi_2O_5 + FeO \text{ (from melt)} \longrightarrow 2\ FeTiO_3$$

The reaction between the cooling basaltic liquid and armalcolite is evidently the major process responsible for the formation of ilmenite rims around armalcolite (Lindsley et al., 1974, El Goresy et al., 1974). Many armalcolites show reactions only on certain sides, namely where the basaltic liquid had free path to the armalcolite crystals (El Goresy et al., 1974, Papike et al., 1974). Ilmenite formed due to this reaction was apparently enriched in TiO_2 in contrast to primary ilmenites precipitated directly from the basaltic liquid, since the ilmenite mantles usually show numerous rutile inclusions which probably exsolved on cooling. b. Reaction between chromian ulvöspinel and armalcolite. In an ideal case, pure ulvöspinel would react with armalcolite to form ilmenite according to the reaction

$$Fe_2TiO_4 + FeTi_2O_5 \longrightarrow 3\ FeTiO_3$$

since ulvöspinel in TiO_2-rich basalts in a broad sense is a member of ulvöspinel-chromite solid solution series, secondary Ti-chromite will precipitate in addition to ilmenite as a result of this reaction. Normally, the boundaries of the original ulvöspinel grain are still visible whereby the newly formed chromite deposited between ulvöspinel and ilmenite. In advanced stages the reaction is also accompanied by exsolution of ilmenite from the host ulvöspinel. c. Harzman and Lindsley (1973) and Lindsley et al. (1974) report that armalcolite heated within its stability field with metallic iron yields $ilmenite_{ss}$ + an armalcolite enriched in the anosovite (Ti_3O_5) molecule according to the reaction $4\ FeTi_2O_5 + Fe^o \longrightarrow 5\ FeTiO_3 + "Ti_3O_5"$ (in armalcolite host). Armalcolites with textures and chemistry indicative of this reaction were reported from Taurus-Littrow basalts (El Goresy et al. , 1974). d. Pure ferropseudobrookite ($FeTi_2O_5$) decomposes to ilmenite + rutile at and below 1140 ± 10°C (Harzman and Lindsley, 1973). This breakdown requires the presence of ilmenite and rutile in almost 1:1 ratio: only a few grains of armalcolite in both landing sites were found mantled by ilmenite and rutile satisfying this reaction. e. In olivine porphyritic ilmenite basalts ilmenite precipitates after armalcolite and before pyroxene. Simple growth of ilmenite is to be expected and is in fact a frequent phenomenon.

Phase equilibrium studies of synthetic TiO_2-rich basalts as a function of fo_2 indicate that there is a direct relationship between the crystallization sequence and fo_2 for basaltic liquids with the same composition (Usselman et al., 1975). The observed difference in the crystallization sequences in both rock types was found to be a function of oxygen fugacity (Usselman et al., 1975) (Fig. 7).

Chemistry of armalcolite: Numerous electron microprobe analyses on tan and grey armalcolites suggested that both armalcolite varieties are cation deficient (Ti is calculated as tetravalent and the formula based on 5 oxygens) since the total number of cations never totalled 3. Total number of cations in more than 400 electron microprobe analyses range from 2.91 to 2.97. El Goresy et al. (1974) suggested that this is an indirect evidence for the presence of Ti^{3+} in armalcolite. Wechsler et al. (1975) calculated 4 to 10% $Ti_2^{3+}Ti^{4+}O_5$ component for many lunar armalcolites. Haggerty (1973) reports that tan and grey armalcolites are compositionally indistinguishable in terms of major element abundances. However, the grey armalcolite variety is characterized by relatively higher Cr_2O_3 and MgO contents than the tan variety (El Goresy et al., 1974). The Mg versus Fe cationic distribution for tan and grey armalcolites are shown in Fig. 8. Two important features are recognized: a) There is, indeed, a compositional bimodal distribution of tan and grey armalcolites with slight overlap of the fields. b) The Mg-Fe substitutional relationship for tan armalcolite variety is almost coherent with a negative slope indicating Fe substituting for Mg. The Mg-Fe substitutional relationship of grey armalcolite is not coherent. This may be due to enrichment of armalcolite in Mg due to reactions a,b, c, and d described above. Electron microprobe analyses indicate such a preference of Mg to armalcolite than to the mantling ilmenite (El Goresy et al., 1974). A similar compositional bimodal distribution was also found for the Mg-Cr relationship whereby grey armalcolite is higher in both Mg and Cr contents. The Mg-Cr substitutional relationship is also strongly suggestive of partitioning of Cr between armalcolite and mantling ilmenite with strong preference of Cr to armalcolite. The partitioning of Mg and Cr between armalcolite and ilmenite with preference to armalcolite is demonstrated in the positive slope of the data points in Figure 9. This slope, however, may not represent the actual partitioning relationship between armalcolite and the mantling ilmenite, since the majority of ilmenite mantles exsolved rutile after their formation which may have caused an additional redistribution of both Mg and Cr between ilmenite and rutile.

IV. SUBSOLIDUS REDUCTION REACTIONS

a. Natural Assemblages

Very probably many of the lunar magmas were already in a reduced state at the time of extrusion. Considerable variation in the intensity of reduction was reported among rocks of different landing sites and even of the same landing site. The reduction processes could either be of exogenic or endogenic nature. An attempt to understand the processes responsible for reduction reactions were made by Mao et al. (1974) who demonstrated that trapped solar wind hydrogen is probably an important reducing agent on the surface of the moon, especially during the flow of basaltic liquid on hydrogen enriched regolith. Evidence for five subsolidus reactions for which experimentally determined buffer curves exist, was reported from the lunar samples of all landing sites. The experimentally determined buffer curves involve the following reactions (Sato et al., 1973, Williams, 1971, Taylor et al., 1972; Lindsley et al., 1974):

$$2Fe_2TiO_4 \longrightarrow 2FeTiO_3 + 2Fe + O_2 \qquad (1)$$

$$2FeTiO_3 \longrightarrow 2TiO_2 + 2Fe + O_2 \qquad (2)$$

$$4FeTiO_3 \longrightarrow 2FeTi_2O_5 + 2Fe + O_2 \quad (3)$$

$$2FeTi_2O_5 \longrightarrow 4TiO_2 + 2Fe + O_2 \quad (4)$$

$$Fe_2SiO_4 \longrightarrow SiO_2 + 2Fe + O_2 \quad (5)$$

In applying these reactions only estimations of the degree of reduction and temperature can be made, since the reduced lunar minerals are by no means pure compounds as compared to the experimentally investigated phases. Furthermore, the host minerals subjected to reduction change their composition continuously as the reaction proceeds to the right hand side of the above equations. These compositional variations were reported by El Goresy et al. (1972) and Haggerty (1972) who demonstrated that chromian ulvöspinel changes its composition continuously upon subsolidus reduction to become enriched in the chromite molecule. In comparison to equation 2 lunar magnesian ilmenites breakdown to rutile + spinel + Fe (Haggerty, 1973). Host ilmenite also changes its composition to become enriched in the geikielite molecule. The activity of Fe_2SiO_4 in lunar olivine decreased continuously upon reduction to Fe + SiO_2. Reactions 3 and 4 were found to be induced upon flow of basaltic liquid on regolith loaded with solar wind hydrogen (Mao et al., 1974) and hence are regarded as exogenic reactions. The reduction described by Mao et al. (1974) indicates that extensive reduction of lunar basalts probably first took place on the surface of the moon during the eruption process.

Figure 10 displays schematically the oxygen fugacity curves of the univariant reactions listed above after Sato et al. (1973) and Taylor et al. (1972) and extrapolated down to below 700°C. Among the univariant reactions described above, the I-Q-F curve holds a key position. The quartz-fayalite-iron (I-Q-F) univariant curve intersects the ulvöspinel-ilmenite-iron (Fe-il-uv) curve at 950°C and the ilmenite-rutile-iron (Fe-ru-il) curve at roughly 830°C (Taylor et al., 1972). As will be shown later, the uncertainty of these two intersections is quite large towards higher temperatures due to uncertainties in the slope of the I-Q-F univariant curve. Apollo 14 samples 14053 and 14072 were previously considered to be the most reduced basalts then, due to the presence of the breakdown of fayalite to silica + Fe in addition to the extensive reduction of chromian ulvöspinel to Ti-chromite + ilmenite + Fe (El Goresy et al., 1972, Haggerty, 1972). Taylor et al. (1972) conclude from their experimental studies that the presence of fayalite reduction in these two samples is direct evidence that these two rocks have undergone more reducing conditions than any other rocks, terrestrial or lunar reported from Apollo 11 through Apollo 14. A comparison between the breakdown assemblages on these two rocks and in several TiO_2-rich basalts as well as the application of the first three reactions negates a severe reduction as proposed by Taylor et al. (1972). Table 1 displays the assemblages involved in the univariant reactions 1 to 3 in sample 14053 as compared to assemblages in several Apollo 17 samples. Comparison between the assemblages in 14053 and in the three Apollo 17 basalts strongly suggests that the rocks of the two landing sites had different thermal histories. Very probably, samples 14053 and 14072 were subjected to reduction during a heating event to a temperature close or in excess of 950°C. A mechanism of reduction upon heating above this temperature at the intersection of the two univariant curves at the given fo_2 explains best the simultaneous breakdown of fayalite to Fe + silica and ulvöspinel to ilmenite + Fe. Textures and mineralogy of samples 14053 and 14072 do not support reduction during an initial cooling process. Apollo 17 samples 70017, 70035, and 70135 display extensive subsolidus reduction of Cr-ulvöspinel to Ti-chromite + ilmenite + Fe and of Mg ilmenite to Al-Ti chromite + rutile + Fe. Accessory fayalite in these samples does not show any sign of reduction to Fe + silica. Alone on the basis of the pervasive reduction of ilmenite and neglecting fayalite (Taylor et al., 1972, El Goresy et al., 1972), one may conclude that these samples were severely reduced. Such conclusion is indeed unrealistic. The three assemblages shown in Table 1 suggest that the reduction of these rocks took place below 830°C probably during cooling. These assemblages have been reduced to a fo_2 below the Fe-ru-il curve but above the I-Q-F curve. Haselton and Nash (1975) discussed a model of isochemical reaction formerly proposed by Lindsley et al. (1974) of opaque oxides with part of Ti present as Ti^{3+} in the initial phases to explain the subsolidus reactions observed. An important aspect of this model is the fact that the components of the assemblages are indeed not in their standard states. Such a system would display deviation from ideality. This deviation from stoichiometry indicates that the fo_2-curve will not be straight line. Experimental studies in the system Fe-Ti-O with assemblages coexisting with metallic Fe^o (Simons, 1974) indeed indicate a compositional variation as a function of temperature.

b. Experimental Results

Oxidation-reduction reactions in the system Fe-Ti-Si-O in equilibrium with metal

The oxidation-reduction reactions in the system Fe-Ti-O have been attracting attention by several groups of investigators, e.g. Taylor et al. (1972), Sato et al. (1973), Lindsley et al. (1974), Saha et al. (1974), Merritt et al. (1974), and Grey et al. (1974). Here new experimental data are presented. For detailed discussion of experimental procedures and for comparison with literature data reference is made to Simons et al. (1976).

The reaction $2\ Fe_2TiO_4 = 2\ FeTiO_3 + 2\ Fe + O_2$ (1)

Under constant pressure this reaction is monovariant. The free enthalpy-temperature curve as determined by CO_2/H_2 gas equilibration runs as well as from independent emf-measurements using the iron/wüstite assemblage as a reference electrode[x] are reproduced in Figure 11. It appears that no straight line can be drawn through the data points in the $\Delta G^o/T$-diagram, but a distinct curvature is observed. Reproducibility of the experimental data, however, has only been attained with very long reaction times, i.e. during equilibration with defined gas mixtures for 24 hours at 1300°C to 120 hours at 1000°C, or by emf-measurements by keeping the temperature of the emf-cell constant for several hours before the cell voltage was registered.

The deviation of the equilibrium curve from a straight line may be attributed to disordering effects in the spinel phase, comparable to the effects observed by Jacob *et al.* (1975) and Navrotsky (1976), although this reaction is expected to require faster equilibration rates.

Because of its deviation from a straight line the equilibrium curve $Fe_2TiO_4 + FeTiO_3 + Fe$ cannot yet be presented by an analytical expression.

The reduction of ilmenite: Under reducing conditions ilmenite dissociates to form rutile and metallic iron according to the equation

$$2\ FeTiO_3 = 2\ TiO_2 + 2\ Fe + O_2 \qquad (2)$$

at temperatures below 1068°C. The data points for the monovariant equilibrium curve $FeTiO_3+TiO_2+Fe$ as obtained from CO_2/H_2 gas equilibration as well as from emf-measurements are reproduced in Fig. 11. Here a straight line can be constructed and the equation

$$\Delta G^o_{(2)} = -\ 564.8 + 0.131\ T\ KJ.mol^{-1}\ (O_2)$$

---

[x] $G^o_{(wüstite)} = -\ 528.8 + 0.130\ T\ KJ.mol^{-1}\ (O_2)$

is derived by least squares approximation. Above 1068°C the product of the ilmenite reduction is ferropseudobrookite and metallic iron, according to the reaction equation

$$4\ FeTiO_3 = 2\ FeTi_2O_5 + 2\ Fe + O_2 \qquad (3)$$

Although the experimental error in the data points is greater here than for reactions (1) and (2), it is possible to construct a straight line by applying a least squares approximation, leading to

$$\Delta G^o_{(3)} = -\ 672.5 + 0.211\ T\ KJ.mol^{-}(O_2)$$

The ilmenite reduction curves intersect at one atmosphere total pressure in a point located at 1068 ± 1°C and $fo_2 = 10^{-15,16}$ (Woermann et al., 1976). In this isobarically invariant point the phases ilmenite + ferropseudobrookite + rutile + metallic iron coexist. The ferropseudobrookite phase contains 13.5 % of the Ti_3O_5-molecule. Since, according to Haggerty et al. (1969), pure ferropseudobrookite decomposes to TiO_2 + $FeTiO_3$ at 1140°C, it appears that the Ti_3O_5-molecule stabilizes the pseudobrookite structure. At the same time, a finite activity of Ti^{3+} in the coexisting rutile and ilmenite phases is required. The mutual agreement among literature data is rather disappointing. Obviously, many investigators did not realize that the limit of stability of ilmenite shows a distinct discontinuity. Others, who were well aware of this fact, accepted the temperature of decomposition of pure $FeTi_2O_5$ at 1140°C as the point of intersection of the reaction lines. As a consequence, straight lines have repeatedly been forced through data points belonging to different reaction curves. Under pressure the ferropseudobrookite + ilmenite + rutile + iron equilibrium is shifted towards higher temperatures (Figure 12a) i.e. the stability of the ferropseudobrookite phase is reduced. This is in agreement with the observations by Lindsley et al. (1974) on the stability of the isomorphous karrooite $MgTi_2O_5$ (Figure 12b).

The reduction of ferropseudobrookite: According to the simplified equation

$$2\ FeTi_2O_5 = 4\ TiO_2 + 2\ Fe + O_2 \qquad (4)$$

ferropseudobrookite is reduced to rutile and metallic iron. The reaction is more complicated, however, due to the fact that titanium is progressively reduced to the trivalent state with rising temperature, resulting in a continuous change of composition of the coexisting oxide phases. In spite of this complication an isobarically monovariant equilibrium curve ferropseudobrookite + rutile + metal in the ternary system Fe - Ti - O is required by phase rule arguments. Attempts to bracket this curve have so far been unsuccessful. A metastable range of coexistence of three phases has been observed instead within the two limiting curves (4) and (4') of Figure 11.

The system Fe - Ti - O in equilibrium with metallic iron: The characteristic features of this system are disclosed in Fig. 11. The ferropseudobrookite stability field has a triangular shape between the curves (3) and (4'). The point of intersection of these two lines represents the lower limit of stability of ferropseudobrookite with respect to ilmenite and rutile at 1068°C. Similarly, the stability field of ilmenite is reduced towards higher temperatures. An intersection of the curves (1) and (2) would be indicating that ilmenite has an upper limit of stability with respect to an ulvite + ferropseudobrookite assemblage (Lindsley et al., 1974). The stability of ilmenite (+ iron) up to its melting point at 1395°C (Mac Chesney et al., 1961) is verified experimentally, however. Thus the curves (1) and (2) do not intersect below this temperature. This can only be achieved by a still further deflection of curve (1) towards higher $\Delta G^{o}_{(1)}$ - i.e. towards higher entropy with higher temperatures. Since the stability field of ferropseudobrookite was shifted to higher temperature with pressure (see Fig. 12), the same tendency must hold for the curve (3). Thus, the ilmenite stability range will be extended by pressure.

Phase equilibria in the system Fe - Si - Ti - O: Since the $Fe_2SiO_4 + SiO_2 + Fe$ phase assemblage was first proposed by Eugster et al. (1962) as an oxygen buffer, its equilibrium relations are well established. They have been reinvestigated by several authors. In this work again Eugster & Wones' equilibrium equation is confirmed. According to the data in Figure 13 a linear equation for the reaction

$$Fe_2SiO_4 = SiO_2 + 2\ Fe + O_2 \qquad (5)$$

with $\Delta G^{o}_{(5)} = -571.3 + 0.152\ T\ KJ.mol^{-1}(O_2)$ is derived.

A first attempt to combine thermodynamic information on the Fe - Ti - O and the Fe - Si - O systems has been made by Taylor et al. (1972). According to a schematic plot (see Figure 10) the equilibrium curves for various monovariant reactions may be superimposed upon each other. As long as the equilibria are mutually independent - i.e. no titanium dissolves in the silicates and no silicon in the oxide phases - equilibrium relationships may be directly derived from this diagram. Based on the new data provided above a quantitative determination of equilibrium states among silicates and oxides coexisting with metallic iron in the system Fe - Si - Ti - O is presented in Figure 14. From analyti-

Fig. 12a

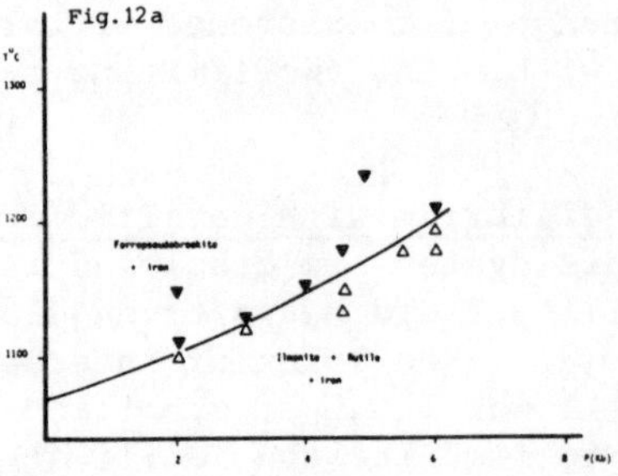

Fig. 12b

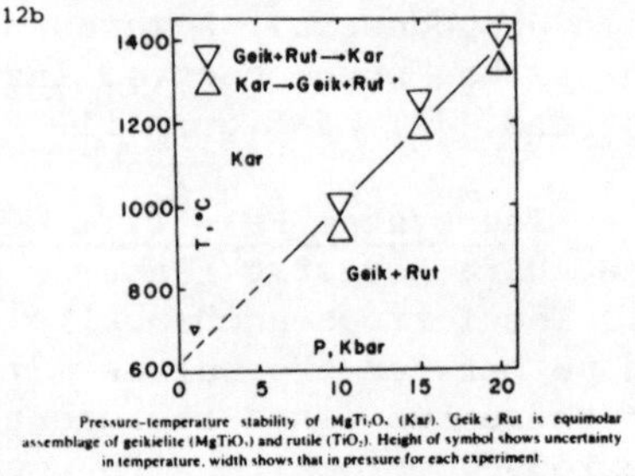

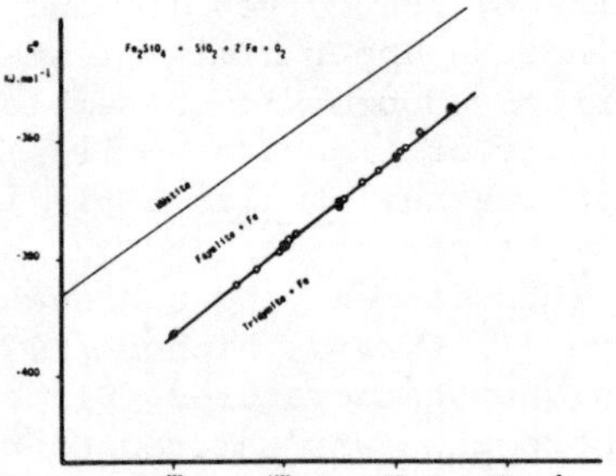

Fig. 13

Fig. 14

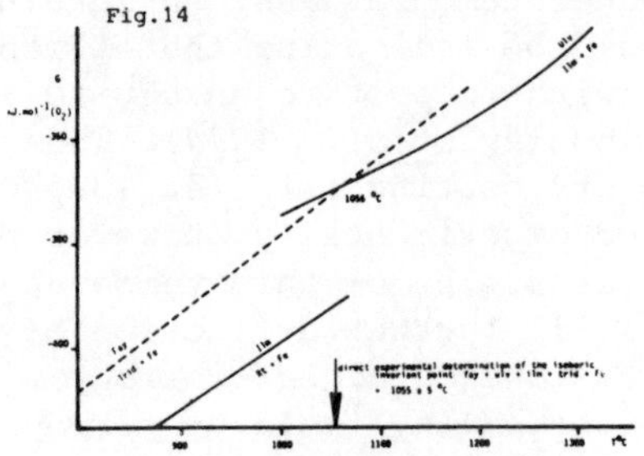

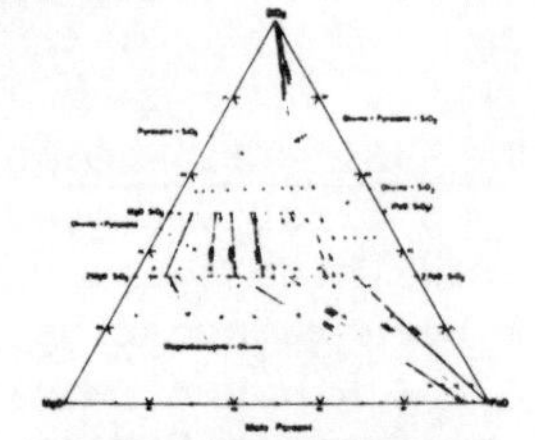

Fig. 15

Fig. 16

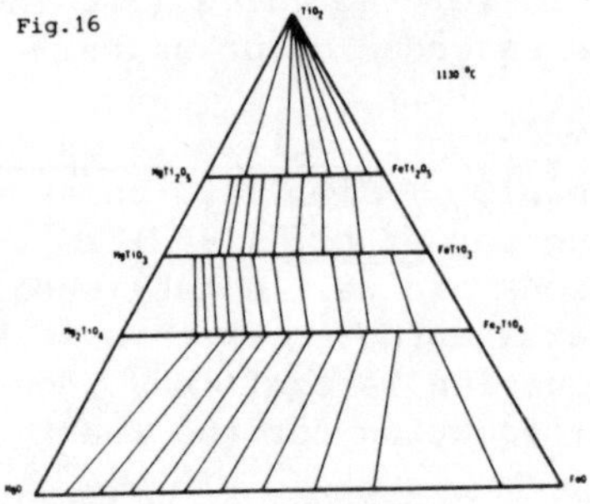

Figure 12: Pressure-temperature stability of the pseudobrookite phase. (a) in the system $FeO-TiO_2$ in equilibrium with metallic Fe. (b) in the system $MgO-TiO_2$ after D.H. Lindsley et al. (1974).

Figure 13: Equilibrium curve fayalite = tridymite + iron as determined by emf-measurements.

Figure 14: Phase equilibria in the system $FeO-TiO_2-SiO_2$ constructed from the monovariant reaction curves of Figs. 11 and 13 and as determined from direct experiment.

Figure 15: Conjugation lines and solidus isotherms in the system $MgO-FeO-SiO_2$ after Nafziger et al. (1967).

Figure 16: System $MgO-FeO-TiO_2$. Subsolidus phase relationships at 1130°C.

Fig. 17

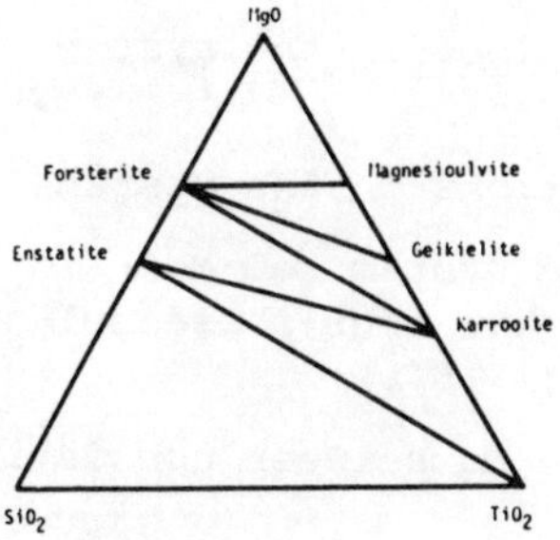

Fig. 19

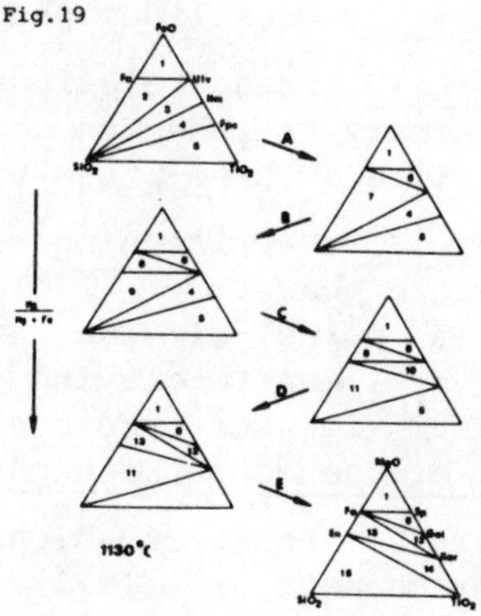

Fig. 18

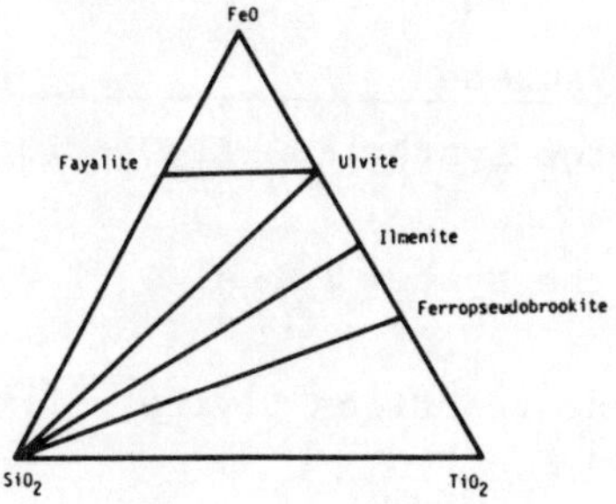

Fig. 20 a

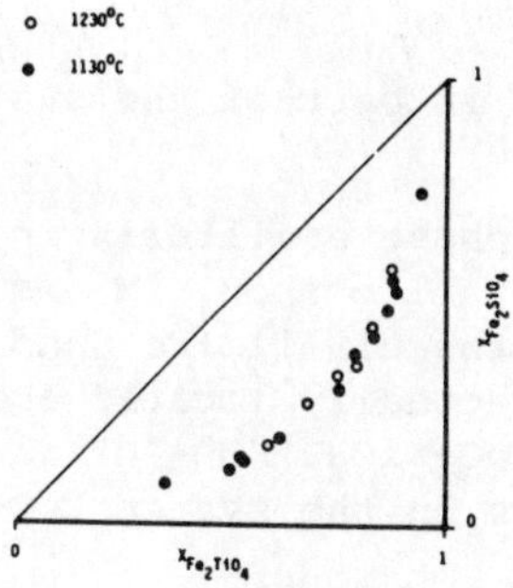

Fig. 20b

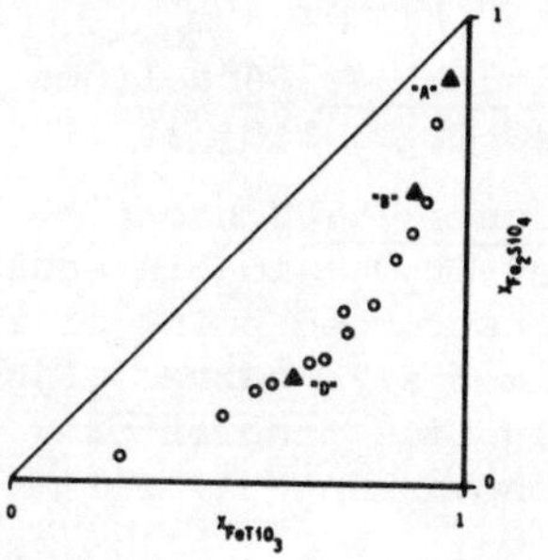

Fig. 20c

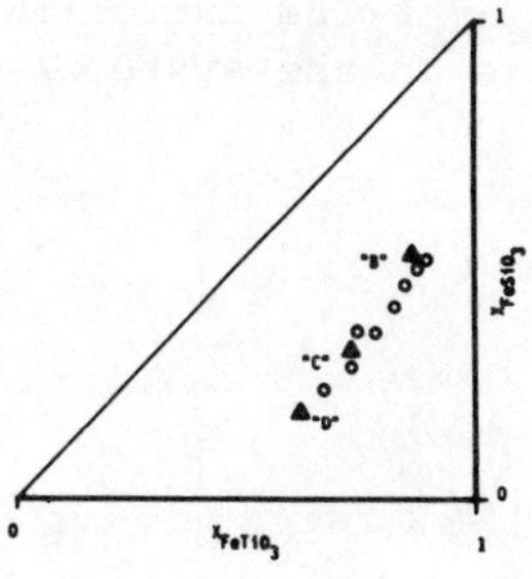

Fig. 20 d

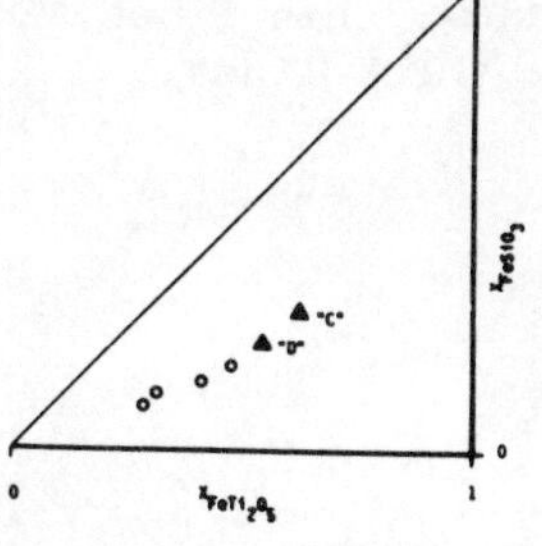

Figure 17: Subsolidus phase relations in the system $MgO-SiO_2-TiO_2$ after Massazza et al. (1958).

Figure 18: Subsolidus phase relations in the system $FeO-SiO_2-TiO_2$.

Figure 19: Phase relations in the system $FeO-MgO-SiO_2-TiO_2$. The quaternary tetrahedron contains five four-phase polyhedra (A-E) and fifteen three-phase volumes (1-15).

Figure 20: Partitioning of Fe and Mg between coexisting oxide and silicate phases:

a: spinel-olivine
b: ilmenite-olivine
c: ilmenite-pyroxene
d: pseudobrookite phase-pyroxene

Figure 21: Phase relations in the system $MgO-Al_2O_3-TiO_2$ at 1400°C after Muan et al. (1971).

Figure 22: Phase relations in the system $FeO-Al_2O_3-TiO_2$ at 1300°C after Muan et al. (1971).

Figure 23: $\Delta G^o/T$ curves for the reactions ulvite = ilmenite + iron and chromite = eskolaite + iron.
N-N = monovariant curve
ulvite + ilmenite + ferropseudobrookite in the system $FeO-Cr_2O_3-TiO$ (see text).

Figure 24: Subsolidus phase relations in part of the system $FeO-Cr_2O_3-TiO_2$ at 1150°C.

Figure 25: Schematic presentation of phase equilibria in the system $FeO-Cr_2O_3-TiO_2$ in equilibrium with metallic iron. "M" is the constructed point of intersection of the ulv+ilm+Fe- and the ilm+fps+Fe-curves (Fig. 10) which is actually located above the liquidus temperature, in the system $FeO-TiO_2$. "M"-"N" is the monovariant ulv + ilm + fps + Fe curve in the system $FeO-Cr_2O_3-TiO_2$.

Figure 26: Schematic phase relations in the system $FeO-Cr_2O_3-TiO_2$ in equilibrium with metallic iron. Equilibrium curves in the system $FeO-TiO_2$: thin lines. Equilibrium curve in the system $FeO-Cr_2O_3-TiO_2$: thick lines.

Fig.21

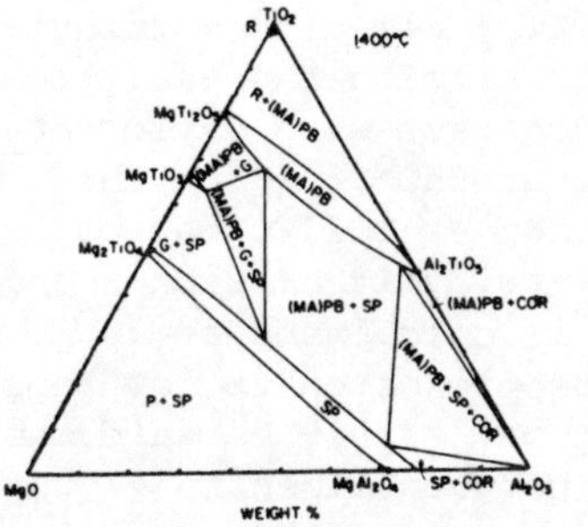

Fig.22

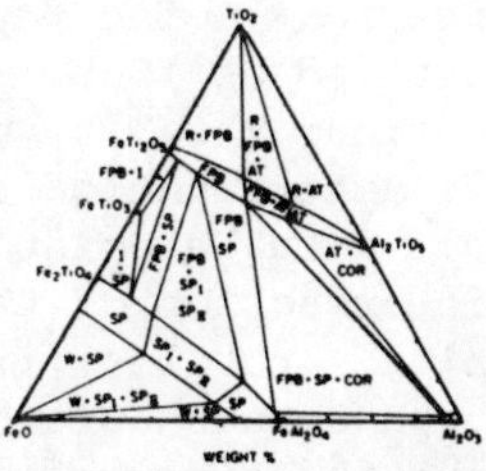

Phase relations in the system FeO-Al_2O_3-TiO_2 in contact with metallic iron at 1300 C. Abbreviations used have the following meanings: R = rutile, FPB = ferropseudobrookite, AT = Al_2TiO_5, I = ilmenite, SP = spinel, COR = corundum, W = wustite

Fig.23

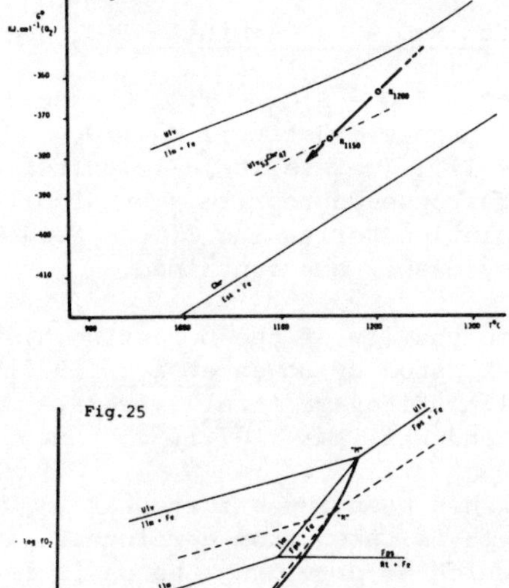

Fig.24

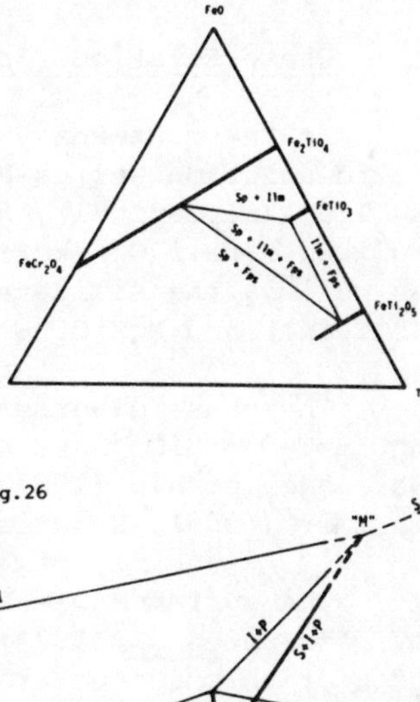

Fig.25

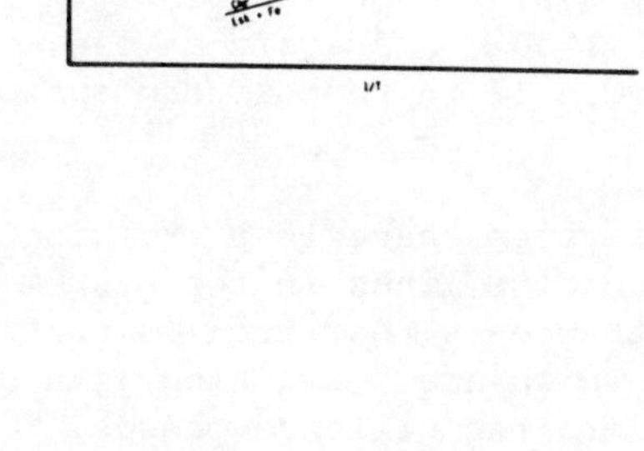

Fig.26

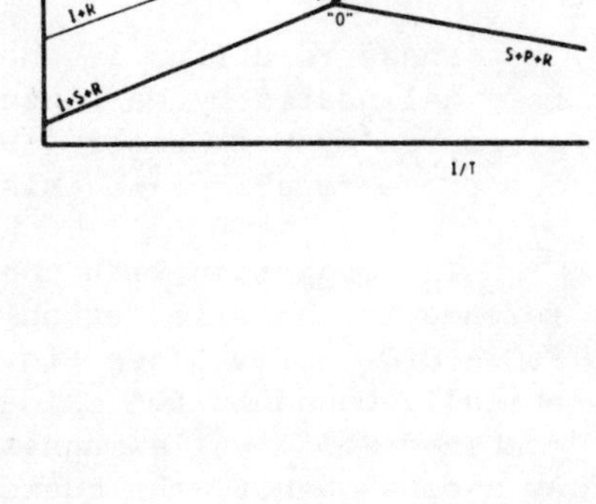

cal and graphical evaluations the point of intersection of the fayalite + SiO_2 + iron- and the ulvite + ilmenite + iron - curves is located at a temperature of 1056° ± 10°C. At lower temperatures the phase assemblage fayalite + ilmenite + iron will be stable, at temperatures above 1056°C, however, ulvite, tridymite, and iron will coexist. By direct experimental determination the isobarically invariant point fayalite + ulvite + ilmenite + tridymite + iron was located at 1055° ± 5°C, thus confirming the analytical evaluation. This is, however, in contrast with Taylor et al.'s (1972) determination of this point at 950°C. (From their analytical data, a point of intersection at 1027°C may be calculated) The intersection of the fayalite + tridymite + iron - and the ilmenite + rutile + iron - curves is calculated at 36°C. Certainly, high temperature data cannot be projected to low temperatures without inducing serious error. It may safely be said, however, that a possible equilibrium point between ilmenite + quartz + rutile + fayalite + iron must be at a temperature far too low to permit any equilibrium reaction to proceed for kinetical reasons

Phase Relations in the System MgO - FeO - SiO_2 - TiO_2

In the quaternary system MgO - FeO - SiO_2 - TiO_2 the oxide solid solution series MgO - FeO (magnesiowüstite), Mg_2TiO_4 - Fe_2TiO_4 (ulvöspinel), $MgTiO_3$ - $FeTiO_3$ (geikielite - ilmenite) and $MgTi_2O_5$ - $FeTi_2O_5$ (karrooite - ferropseudobrookite = armalcolite) as well as the silicate solid solution series (Mg_2SiO_4 - Fe_2SiO_4 (olivine) and $MgSiO_3$ - $FeSiO_3$ (pyroxene) are contained.

The phase diagram and thermodynamics of the bordering system MgO - FeO - SiO_2 have been investigated by Bowen et al. (1935), Nafzinger et al. (1967, Figure 15), Kitayama et al. (1968), Larimer (1968), Medaris (1969), and Williams (1971).

The system MgO - FeO - TiO_2 has been investigated at 1300°C by Johnson et al. (1971). Figure 16 is taken from a reinvestigation of this system at 1130°C, which is providing the basis for thermodynamic calculations in the quaternary system.

Phase relations in the ternary system MgO - SiO_2 - TiO_2 have been delineated by Massazza et al. (1958), and again at elevated pressures by Mac Gregor (1969). Figure 17 represents the subsolidus phase relations in this system.

In comparison with the latter system there is a striking difference in the orientation of conjugation lines in the system FeO - SiO_2 - TiO_2 (see Fig. 18). It appears that iron is preferentially bound in the oxide phases which are coexisting with uncombined SiO_2, while magnesium prefers the silicate phases. As a consequence, the quaternary system MgO - FeO - SiO_2 - TiO_2 shows rather complicated compatibility relations. The shift of several tie lines is required in moving from the FeO - SiO_2 -

TiO_2 side to the MgO - SiO_2 - TiO_2 side of the tetrahedron. The resulting phase relations are summarized in Fig. 19: Here the first diagram represents the ternary system FeO - SiO_2 - TiO_2 at 1 bar and 1130°C with five three-phase triangles, which are changing to corresponding three-phase polyhedra upon addition of magnesium, with variable compositions of all phases along the solid solution lines. The first tie line to be shifted is: spinel + tridymite = olivine + ilmenite. This four phase assemblage is represented by "A", with all four phases being defined at given temperature and pressure by phase rule requirements. Correspondingly the quaternary system MgO - FeO - SiO_2 - TiO_2 is subdivided into fifteen three phase polyhedra (numbers) and five four phase polyhedra (letters "A" - "E"), the compositions of the phases comprising the four phase polyhedra being:

"A" : Ulv-97 + Ilm-94 + Fay-86 + SiO_2

"B" : Ilm-88 + Fay-62 + Px -52 + SiO_2

"C" : Ilm-73 + Fps-63 + Px -27 + SiO_2

"D" : Ilm-60 + Fps-55 + Fay-27 + Px- 23

"E" : Fps-34 + Rt + Px -04 + SiO_2

where the number represents the molar percentage of the iron-bearing endmember of the respective solid solution series. Some of these four phase polyhedra are of special importance to lunar petrology.

The equilibrium phase assemblage ulvite + tridymite = ilmenite + fayalite has been discussed in detail for the ternary system FeO - SiO_2 - TiO_2 above. It appears that even by relatively minor additions of Mg^{2+} this four-phase equilibrium is shifted from 1056°C to 1130°C in "A". It follows that this phase assemblage can only be applied with great caution as lunar thermometer.

The phase volume "B" is defined by the reaction olivine + tridymite = pyroxene, while the ilmenite has no part in the reaction. In comparison with titanium free systems (Nafzinger et al., 1967, Kitayama et al. 1968) it appears that titanium does not have any influence on the state of equilibrium.

"C" is the first phase assemblage to contain the pseudobrookite phase armalcolite in equilibrium with pyroxene, while pyroxene + olivine coexist with armalcolite only in "D" and mixtures still richer in magnesium. This explains the observation that armalcolites from lunar basalts tend to have intermediate compositions around Fps55-Fps63. In bulk compositions with higher Fe/Fe + Mg ratios the armalcolite phase will only coexist with SiO_2-phases. No information is as yet available on the temperature dependence of the "C" and "D" polyhedra.

Partitioning of Fe and Mg between Coexisting Oxide and Silicate Phases

For selected pairs of solid solutions the partitioning of Fe and Mg between oxides and silicates is being determined. Tentative data are produced in Figure 20 (a-d). A pronounced feature is the concentration of iron in oxide and conversely of magnesium in silicate phases. A temperature dependence of the partitioning coefficients has been observed, although the effect is not pronounced enough to be applied as lunar thermometer.

Only the phase assemblage spinel + olivine can be investigated over the whole range of compositions, the majority of oxide - silicate pairs being cut off at certain compositional limits by corresponding four phase polyhedra in the quaternary system. In the range of high Fe/Fe + Mg-ratios, e.g., the phase assemblage olivine + ilmenite is replaced by the assemblage ulvite + silica, the olivine + ilmenite partitioning thereby being cut off by "A". It is noteworthy that the assemblage spinel + pyroxene is unstable in the whole range of compositions.

Phase Relations in the System FeO - Cr_2O_3 - TiO_2

The lunar spinels are members of the quinary system MgO - FeO - Al_2O_3 - Cr_2O_3 - TiO_2. This system, and particularly, the range of spinel compositions, has been an object of intensive research for solid state chemists, crystal chemists, mineralogists, and metallurgists during the last decades. Of the wealth of resulting publications only a relatively small group - satisfying the demand for low oxygen fugacity and equilibrium with metallic iron at the temperature of formation - is of direct interest for lunar petrology.

The stability of the titanate spinels has been discussed in the previous chapters. The chemistry of the chromite spinels has been reviewed by Ulmer (1971). Experimental data with direct bearing on lunar petrology have been presented by Muan *et al.* (1972) Lipin *et al.* (1974), and Lipin *et al.* (1975).

Muan *et al.* (1972) confirmed a complete solid solution series between the (Mg, Fe)Cr_2O_4 - (Mg, Fe)$_2TiO_4$ and the (Mg, Fe)Cr_2O_4 - (Mg, Fe)Al_2O_4 endmembers, at least at temperatures above 1000°C, while a miscibility gap exists in the series (Mg, Fe)Al_2O_4 - (Mg, Fe)$_2TiO_4$. Isothermal sections of the ternary systems MgO - Al_2O_3 - TiO_2 (Fig. 21), and FeO - Al_2O_3 - TiO_2 (Fig. 22) are presented, as well as for the system MgO - Cr_2O_3 - TiO_2 (Muan *et al.*, 1971). No information in literature is as yet available for the FeO - Cr_2O_3 - TiO_2 - system.

A striking feature of all ternary isothermal sections published today is the low solubility of eskolaite and of corundum in the isomorphous ilmenite and geikielite. In each system the restricted range of ilmenite (or geikielite) solid solution is cut off by a spinel + ilmenite (or geikielite) + "pseudobrookite" three phase triangle.

At this time research is being done on the ternary system $FeO - Cr_2O_3 - TiO_2$ in order to gain information that may be helpful in lunar petrology. Attention is thus concentrated on the compositional range of the system containing the spinel solid solution series $FeCr_2O_4 - Fe_2TiO_4$ and the ilmenite and pseudobrookite phases (Woermann et al. (1976).

The stability of the $FeCr_2O_4 - Fe_2TiO_4$ solid solution series: The equilibrium state of the reaction

$$2\ FeCr_2O_4 = 2\ Cr_2O_3 + 2\ Fe + O_2 \qquad (6)$$

has been investigated by Tretjakov et al. (1965) and by Sato et al. (1973). Our own emf-date agree reasonably well with those of Tretjakov et al. In Figure 23 they are compared with the data from the reaction $2\ Fe_2TiO_4 = 2\ FeTiO_3 + 2\ Fe + O_2$ (1) as presented in Figure 11. It is evident that the equilibrium oxygen fugacity for the ulvite + ilmenite + iron assemblage is considerably higher than the fo_2 for the assemblage chromite + eskolaite + metal at the same temperature and pressure. Intermediate composition of the continuous spinel solid solution series will have intermediate oxygen fugacities. The exact location of the corresponding $\Delta G^o/T$-curves can only be fixed, however, by direct experiment or by calculations if corresponding data on activity-composition relations are available.

The three-phase equilibrium spinel + ilmenite + ferropseudobrookite: Figure 24 is presenting some preliminary information on the subsolidus phase relations in the investigated part of the system $FeO - Cr_2 - TiO_2$. It demonstrated again the low solubility of Cr_2O_3 in $FeTiO_3$. At 1150°C the ilmenite phase contains only 7 mol % Cr_2O_3 while the composition of the coexisting spinel phase is $ulv_{53}chr_{47}$. An attempt is made to deduce the temperature dependence of the location of the three phase triangle.

It has been shown above that the stability of ilmenite in the system Fe - Ti - O is narrowing with rising temperature (Fig. 11). Figure 25 is schematically presenting the ensuing phase relations: A projection of curves (1) in figure 11 (ulv + ilm + Fe) and of curve (3) (ilm + Fps + Fe) will lead to a point of intersection "M" (which is located above the liquidus temperature for the composition in question). Point "M" corresponds to the theoretical isobaric invariant point ulv + ilm + fps + Fe. Upon addition of another component - e.g. Cr_2O_3 - the invariant point "M" will change to a monovariant line. "MN". The fact that the phase assemblage ulv + ilm + fps + Fe in the system $FeO - Cr_2O_3 - TiO_2$

is observed at a temperature of 1150°C suggests that the line "MN" is sloping towards lower temperatures with rising chromium activity.

Two points on the line "MN" have been located by emf-data at 1150°C and at 1200°C. The measurements have only been taken after heating the samples in the emf-cell at constant temperature for 15 hours to ensure equilibration of all phases involved. Since the spinel coexisting with ilmenite and pseudobrookite at 1150°C corresponds to $ulv_{53}chr_{47}$ (Figure 24), the point determined at this temperature must be located on the line for the corresponding spinel composition in Figure 23. The 1200°C point is displaced towards distinctly higher ulvite concentrations, i.e. lower chromium activity in the spinel, and consequently in all three coexisting oxide phases.

From this experiment a considerable shift of the three phase triangle towards lower chromium contents with rising temperature can be deduced. This is in agreement with the behavior of corresponding ulv + ilm + fps - three phase triangles in comparable systems, e.g. the system MgO - Fe_2O_3 - TiO_2 (Woermann et al., 1969). From these findings it follows that the secondary formation of ilmenite in ferropseudobrookite + spinel bearing assemblages does not require any material transport, but it can be explained by equilibrium reactions during the cooling of the lunar basaltic magma. This process, on the other hand, requires the reversal of the normal trends towards lower Ti/Ti + Cr + Al with cooling. This reversal has been observed in spinels in contact with ilmenite (El Goresy et al., 1975).

log fo_2/1/T - Diagram of the System FeO-Cr_2O_3-TiO_2 in Equilibrium with Metallic Iron

With the temperature falling continuously along the line MN finally the limit of stability of the pseudobrookite phase is reached. The chromium bearing ferropseudobrookite solid solution decomposes to chromous spinel + ilmenite + rutile. With the addition of rutile to the phase assemblage ulvite + ilmenite + ferropseudobrookite a new invariant point "O" (in Figure 26) is defined. Since today no information on the stabilizing effect of chromium on the ferropseudobrookite phase is available, it is impossible to locate point "O" in the log fo_2/1/T-diagram. It is known that Cr has a stability effect on the complex armalcolite solid solutions. However, no quantitative data are at hand about the degree of this stabilizing effect in the system FeO-Cr_2O_3-TiO_2. Figure 26 is merely a schematic diagram demonstrating the phase relationships in the system FeO - Cr_2O_3 - TiO_2 in equilibrium with metallic iron. From the invariant point "O" four monovariant curves emanate. It appears that a former ilmenite +

ferropseudobrookite + spinel assemblage follows the line ilmenite + spinel + rutile, towards still lower temperatures a phase assemblage that has well been observed in lunar rock.

Spinel-Pyroxene Assemblages in the Multicomponent System $MgO-CaO-FeO-Al_2O_3-Cr_2O_3-TiO_2-SiO_2$

It has been demonstrated above that in the system $MgO-FeO-TiO_2-SiO_2$ the phase assemblage spinel + pyroxene is incompatible at temperatures above about $1000^{\circ}C$. Since the liquidus surface of the system $CaMgSi_2O_6-FeO-TiO_2$ (Lipin et al. (1975)) does not show a common field boundary between the primary phase fields for ulvite and diopside,a compatibility of $(Fe, Mg)_2TiO_4$-spinels and $(Ca, Mg, Fe)SiO_3$-pyroxenes in the whole compositional range seems improbable.

From the $MgO-Al_2O_3-SiO_2$ phase diagram (Levin et al. (1964) Fig. 712) it appears that the spinel + pyroxene assemblage in this system is unstable with respect to cordierite + spinel. Only after decomposition of cordierite above 8 kb (Schreyer et al. (1960)) a stable phase assemblage spinel + pyroxene is observed (MacGregor (1974), Newton (1976)). No information is available yet for intermediate and iron rich bulk compositions.

According to Keith (1954), however, picrochromite and enstatite coexist in the system $MgO-Cr_2O_3-SiO_2$. This relation will probably persist in the $MgO-FeO-Cr_2O_3-SiO_2$ system as long as the pyroxene phase remains stable.

It appears that in the combined system $MgO-FeO-Cr_2O_3-TiO_2-SiO_2$ chromium rich spinels coexist with pyroxenes, while titanium rich spinels do not. Thus a limiting composition along the continuous chromite-ulvite solid solution series must exist for spinel-pyroxene tie lines, which will probably react sensitively to changes in temperature and pressure.

It appears that still much information is contained in the spinel-pyroxene assemblage, deserving further attention.

REFERENCES

Anderson A.T.,Bunch T.E., Cameron E.N., Haggerty S.E., Boyd F.R. Finger L.W., James O.B., Keil K., Prinz M., Ramdohr P., and El Goresy A. Proc. Apollo 11 Lunar Sci. Conf., Geochim. Cosmochim. Acta, Suppl. 1, 1, 55-63 (1970)

Bowen N.L. and Schairer J.F., Amer. J. Sci. 229, 151-217 (1935)

El Goresy A., Ramdohr P., and Taylor L.A. Proc. Lunar Sci. Conf. 3rd, 333-349 (1972)

El Goresy, A., Ramdohr P., Medenbach O., and Bernhardt H.-J., Proc. Lunar Sci. Conf. 5th, 627-652 (1974)

El Goresy, A. and Ramdohr, P. Proc. Lunar Sci. Conf. 6th, 729 (1975).
El Goresy, A., Prinz, M. and Ramdohr, P. Proc. Lunar Sci. Conf. 7th, (1976) in press.
Eugster, H.P. and Wones, D.R. Journ. Petrol., 3, 82 (1962).
Grey, I.E., Reid, A.F. and Jones, D.G. Trans. Hist. Min. Met., 83, 105 (1974).
Haggerty, S.E. Proc. Lunar Sci. Conf. 3rd, 305 (1972a).
Haggerty, S.E. Earth Planet. Sci. Lett., 13, 328 (1972b).
Haggerty, S.E. Meteorites, 7, 3, 353 (1972c).
Haggerty, S.E. EOS Trans. Amer. Geophys. Union, 54, 6, 593 (1973).
Harzman, M.J. and Lindsley, D.H. Annual Meeting Geol. Soc. Amer., 5, 7, 653 (1973).
Jacob, K.T. and Alcock, C.B. High Temp. High Press., 7, 433 (1975).
Johnson, R.E., Woermann, E. and Muan, A. Amer. Journ. Sci., 271, 278 (1971).
Keith, M.L. J. Am. Ceram. Soc., 37, 491 (1954).
Kitayama, K. and Katsura, T. Bull. Chem. Soc. Japan, 41, 525 (1968).
Kitayama, K. and Katsura, T. Bull. Chem. Soc. Japan, 41, 1146 (1968).
Larimer, S.W. Geochim. Cosmochim. Acta, 32, 1187 (1968).
Levin, E.M., Robbins, C.R. and McMurdie, H.F. American Ceram. Soc. Columbus, Ohio (1964).
Lindsley, D.H., Kesson, S.E., Hartzman, M.J. and Cushman, M.K. Proc. Lunar Sci. Conf. 5th, 1, 521 (1974).
Lipin, B.R. and Muan, A. Proc. Lunar Sci. Conf., 1, 535 (1974).
Lipin, B.R. and Muan, A. Abstracts of Papers VI Lunar Sci. Conf. 510 (1975).
MacChesney, J.B. and Muan, A. Amer. Mineral., 46, 572 (1961).
MacGregor, I.D. Am. J. Sci., 267A, 342 (1969).
MacGregor, I.D. Amer. Min., 59, 110 (1974).
Mao, H.K., El Goresy, A. and Bell, P.M. Proc. Lunar Sci. Conf. 5th, 673 (1974).
Massazza, F. and Sirchia, E. La chimica e l'industria, 40, 460 (1958).
Medaris, L.G. Am. J. Sci., 267, 945 (1968).
Merritt, R.R. and Turnbull, A.G. Journ. Solid State Chem., 10, 252 (1974).
Muan, A., Hauck, J., Osborn, E.F. and Schairer, J.F. Proc. Lunar Sci. Conf. 2nd, 1, 497 (1971).
Muan, A., Hauck, J. and Löfall, T. Proc. Lunar Sci. Conf. 3rd, 1, 185 (1972).
Nafziger, R.H. and Muan, A. Amer. Min., 52, 1364 (1967).
Navrotsky, A. Calculation of effect of cation disorder on silicate spinel phase boundaries. This volume (1976).
Nehru, C.E., Prinz, M., Dowty, E. and Keil, K. Amer. Min., 59, 1120 (1974).

Newton, R.C. Thermochemistry of garnets and aluminous pyroxenes in the CMAS system. This volume (1976).

Papike, J.J., Bence, A.E. and Lindsley, D.H. Proc. Lunar Sci. Conf. 5th, 1, 471 (1974).

Saha, P. and Biggar, G.M. Indian J. of Earth Sciences, 1, 43 (1974).

Sato, M. and Hickling, N. Abstracts of Papers IV, Lunar Sci. Conf., 650 (1973).

Sato, M., Hickling, N.L. and McLane, J.E. Proc. Lunar Sci. Conf. 4th, 1, 1061 (1973).

Schreyer, W. and Yoder, H.S. Carnegie Inst. Wash. Yearbook, 59, 90 (1960).

Simons, B. and Woermann, E. The stability of iron titanium oxides and iron silicates in equilibrium with metallic iron. In preparation (1976).

Simons, B. Diploma Thesis, Rheinisch-Westfälische Technische Hochschule Aachen (1974).

Taylor, L.A., Williams, R.W. and McCallister, R.H. Earth Planet. Sci. Lett., 16, 282 (1972).

Tretjakov, J.D. and Schmalzried, H. Ber. Bunsenges. Phys. Chem., 69, 396 (1965).

Ulmer, G.C. 'High Temperature Oxides', Ed. A.M. Alper, Academic Press Inc., New York, 251 (1970).

Usselman, T.M. and Lofgren, G.E. Abstracts of Papers VII, Lunar Sci. Conf., 88 (1976).

Wechsler, B.A., Prewitt, C.T. and Papike, J.J. Abstracts of Papers VI, Lunar Sci. Conf., 860 (1975).

Williams, R.J. Amer. J. Sci., 270, 334 (1917).

Woermann, E., Brežny, B. and Muan, A. Am. J. Sci., 267A, 463 (1969).

Woermann, E. and Ender, A. The stability of the pseudobrookite phase in the system Fe-Ti-O in equilibrium with metallic iron. In preparation (1976).

Woermann, E., Knecht, B. and Simons, B. The ulvite-ilmenite-ferropseudobrookite phase assemblage in the system FeO-Cr_2O_3-TiO_2: a possible lunar thermometer. In preparation (1976).

THERMODYNAMIC PROPERTIES OF MOLTEN SALT SOLUTIONS

O.J. Kleppa

The James Franck Institute and
The Departments of Chemistry and Geophysical Sciences,
The University of Chicago, Chicago, Illinois 60637,
U.S.A.

INTRODUCTION

Among the solutions formed by two simple molten salts we clearly must consider at least three different basic types:

(i) Mixtures of two salts with similar charge structure containing a common ion. The common ion may be an <u>anion</u> (NaCl-KCl) or a <u>cation</u> (NaCl-NaBr).

(ii) Mixtures of salts with different charge structure, but with a common ion: (NaCl-$MgCl_2$ and Na_2CO_3-NaCl).

(iii) Mixtures of two salts which have no common ion: These may, again, be charge symmetrical (NaCl-KBr) or charge unsymmetrical (NaCl-$MgBr_2$).

This list outlines, in order of increasing complexity, a series of fundamental problems in the solution chemistry of molten salts: How do the thermodynamic properties of each type of system depend on the size, charge and structure of the participating ions? Much work has been devoted to providing answers to these problems. Some of this work will be reviewed below.

The Liquid Ionic State

On theoretical grounds we are led to recognize in simple ionic liquids several contributions to the pair potential acting between two ions, i and j, of charge z_i and z_j, and of polarizability α_i and α_j, as a function of the interionic distance r[1]

D. G. Fraser (ed.), Thermodynamics in Geology, 279-299.

$$\phi_{ij}(r) = A_{ij}e^{-B_{ij}r} + z_i z_j e^2/r - [(z_i e)^2 \alpha_j + (z_j e)^2 \alpha_i]/2r^4$$

$$-C_{ij}/r^6 + \text{higher order terms.} \qquad (1)$$

The terms on the right hand side represent:

(a) Short range repulsive potential.
(b) Charge-charge interaction.
(c) Charge-dipole interaction.
(d) Dipole-dipole interaction. The coefficient C_{ij} may be approximated by the London formula for the dispersion energy[2]

$$C_{ij} = (3/2)\alpha_i \alpha_j I_i I_j/(I_i + I_j) \qquad (2)$$

where I_i is the ionization potential of the ion.
(e) Higher order terms, such as dipole-quadrupole terms, etc.

In order to obtain the total cohesive energy of the ionic melt a summation must be carried out over all the pair interactions in the system. The summation of terms (a), (b) and (d) is expected to be essentially additive. However, the charge-dipole interaction (c) involves the field created by the charges and induced dipoles surrounding a given ion. Thus this quantity has the character of a many-body term. This introduces an element of great complexity in any rigorous theoretical treatment of even a very simple molten salt. In approximate treatments this difficulty is usually sidestepped through the introduction of a dielectric constant, κ.

Additional complexity arises if the salt in question is not a simple ionic medium. This presumably is the case for many salts which have proved particularly attractive from an experimental point of view, such as, e.g. the salts of the highly polarizable ions silver, thallium, lead, cadmium, etc. Evidence for this is provided, for example, by the lattice energies of these salts, which apparently cannot be fully accounted for in terms of ionic and dispersion interactions alone. For further information, see, e.g., the review by Ladd and Lee.[3]

The rigorous theories of fused salts have not yet been developed to a point where they provide a basis for the prediction and correlation of thermodynamic information. For this purpose the most successful theoretical approach has been based on the methods of corresponding states theory.[4] For ionic liquids this approach was developed by Reiss and co-workers.[5,6]

Mixtures of Molten Salts

As a consequence of the electrostatic forces between the ions in a molten salt there will, at ordinary temperatures, be no mutual randomization of the cations and anions. This led Temkin[7] to postulate that in an ideal molten salt mixture the different types of anions will be randomly distributed among the anions (i.e., on the anion "sub-lattice"), while similarly the different types of cations will be randomly distributed among the cations.

In a mixture containing n_A moles of cation A, n_B moles of cation B, n_X moles of anion X and n_Y moles of anion Y, the integral entropy of mixing will, according to Temkin, be

$$\Delta S = -R(n_A \ell n\ N_A + n_B \ell n\ N_B + n_X \ell n\ N_X + n_Y \ell n\ N_Y). \qquad (3)$$

In this expression N represents the ionic fractions, defined through

$$N_A = n_A/(n_A + n_B), \text{ etc.}$$

For the ideal molten salt mixture we also have $\Delta H = \Delta U = \Delta V = 0$, while, of course, $\Delta G = -T\Delta S$. Note that for these ideal molten salt mixtures we have also

$$\Delta\bar{G}_{AX} = -T\Delta\bar{S}_{AX} = RT\ell n(N_A N_X), \qquad (4)$$

i.e., $a_{AX} = N_A N_X$.

In real mixtures deviations from these relations generally will be found. It is the general objective of experimental work on molten salt mixtures to obtain information on these deviations.

Reliable experimental information on the entropies of mixing in simple molten salt mixtures is less extensive and less reliable than the corresponding enthalpy data. However, work by Østvold[8] on charge unsymmetrical and by Kleppa and Hong[9] on charge symmetrical mixtures, clearly indicates that for noncomplex forming binary mixtures the Temkin expression represents a reasonable 0th order of approximation. While both positive and negative deviations from the Temkin formula are common, such deviations usually are small as long as the molar enthalpy of mixing is of the order of RT or less.

SYSTEMS OF TYPE AX-BX

The first comprehensive study of a complete family of mixed cation-common anion systems was the calorimetric investigation by Kleppa and Hersh[10] of all the binary mixtures formed by the alkali nitrates. As an example of the data obtained in this work we give, in Fig. 1, a plot of the molar enthalpy of mixing, ΔH^M, for the system KNO_3-$LiNO_3$. In the upper part of this figure we show also a plot of the enthalpy interaction parameter, $\lambda = \Delta H^M/N_1N_2$, against mole fraction, N_2. Note that this parameter is a slowly varying function of composition; it changes by less than 10% across the complete range of the binary system. We have found it convenient to describe the behaviour of simple binary molten salt mixtures by semiempirical expressions of the form

$$\Delta H^M = N_1N_2(\underline{a} + \underline{b}N_2 + \underline{c}N_1N_2). \quad (5)$$

In the binary alkali nitrates, if N_2 represents the mole fraction of the salt with the smaller cation, the parameters $\underline{a}$, $\underline{b}$, and $\underline{c}$ are all negative quantities. For these systems the energetic asymmetry ($\underline{b}N_2$) and the term $\underline{c}N_1N_2$, which Kleppa and Hersh attributed to deviations from random mixing, are numerically small compared to $\underline{a}$. Therefore, the enthalpies of mixing can be represented to a reasonable approximation by specifying a single value of the interaction parameter, conveniently chosen at $N_1 = N_2 = 0.5$, i.e., $\lambda_{0.5} = 4\Delta H^M(0.5)$. In this approximation

$$\Delta H^M \approx N_1N_2\,\lambda_{0.5}. \quad (6)$$

For the binary alkali nitrates the values of $\lambda_{0.5}$ are all negative (exothermic), and a semiempirical linear relation exists between the values of $\lambda_{0.5}$ and the square of the parameters $\delta = (d_1 - d_2)/(d_1 + d_2)$; d_1 and d_2 are the sums of the cation and anion radii for each of the two salts, i.e., we have

$$\Delta H^M \approx N_1N_2A\delta^2. \quad (7)$$

This relation is illustrated in Fig. 2. The numerical value of the constant A was found to be about -150 kcal mole^{-1}, i.e., of the order of the lattice energy of the salts.

Originally, these results were rationalized on the basis of crude hard sphere linear models, previously proposed by Førland[11] and by Blander.[12] In these models the negative enthalpies of mixing mainly result from the reduction in the Coulomb repulsive energy between next-nearest neighbour cations. A different point of view has been advanced by Lumsden,[13] who argued that the major

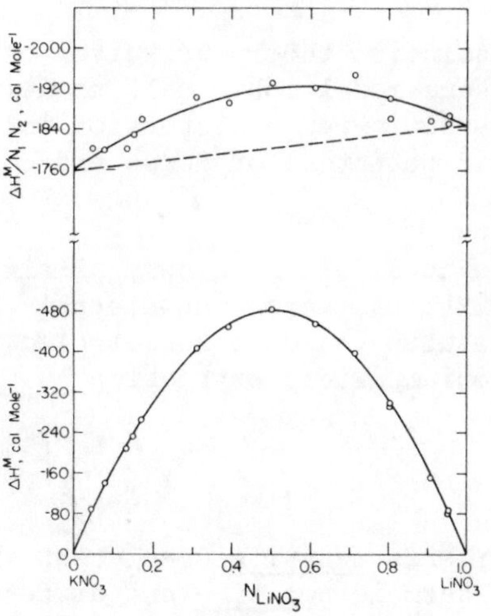

FIG. 1. Enthalpies of mixing, ΔH^M, and enthalpy interaction parameters, $\Delta H^M/N_1N_2$, in liquid potassium nitrate-lithium nitrate mixtures at 345°C. (From Kleppa and Hersh[10]).

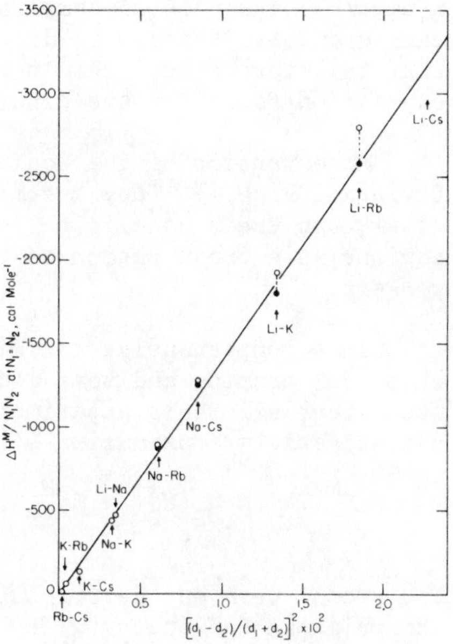

FIG. 2. Plot of enthalpy interaction parameters (evaluated at $N_1 = N_2 = 0.5$) against $\delta^2 = [(d_1 - d_2)/(d_1 + d_2)]^2$ for binary liquid alkali nitrate mixtures. (From Kleppa and Hersh[10]).

contribution to the mixing enthalpy is due to polarization forces.

Later a conformal solution theory for the enthalpies of mixing of simple, charge symmetrical molten salts was developed by Reiss, Katz and Kleppa[14] (RKK), based on the hard sphere model ionic melt of Reiss, Mayer and Katz.[6] Application of the methods of conformal solution theory (Longuet-Higgins,[15] Brown[16]) gives the following results: The first order term cancels. The second order term yields a functionally simple expression for the enthalpy of mixing

$$\Delta H^M \approx N_1 N_2 \Omega(T,P)[(d_1 - d_2)/(d_1 d_2]^2$$

$$= N_1 N_2 \Omega \delta_{12}^2. \qquad (8)$$

Here $\Omega(T,P)$ is a so far unevaluated complicated integral which is a negative function of temperature and pressure; d_1 and d_2 are characteristic interionic distances in the two pure salts. Note that this theoretical result is closely related to the empirical expression found for the binary alkali nitrates (Eq. 7).

An extension of the conformal solution theory was given by Davis and Rice.[17] They assume the same model ionic melt as Reiss et al, but their model also includes short-range dispersion interactions as a perturbation to the pair potential of Reiss and coworkers.

In a comprehensive calorimetric study of the binary alkali chloride, bromide and some iodide mixtures, Hersh and Kleppa[18] found that all their experimental results could be represented to a reasonable approximation by the semiempirical expression

$$\Delta H^M \approx N_1N_2(U_0^{++} + \Omega\delta_{12}^2). \qquad (9)$$

The second term on the right hand side is <u>negative</u> consistent with the original RKK theory. The first term is <u>positive</u> and differs from binary system to binary system. It represents a theoretical estimate of the contribution to the mixing enthalpy which arises from the London-van der Waals interaction between second nearest neighbour cations.

Hersh and Kleppa found that to a first approximation the results for the binary alkali nitrate, chloride and bromide mixtures could be represented by the same coefficient Ω in Eq. (9). This was interpreted to indicate that the principal contribution to this term is Coulombic in origin, and that in these salts nearest-neighbour cation-anion polarization makes only a modest contribution to the enthalpy of mixing.

More recent work by Holm and Kleppa[19] on the alkali fluorides, and by Melnichak and Kleppa[20] on the alkali iodides indicates that the alkali halide mixtures actually are somewhat more complex. Thus one finds that the coefficient Ω in Eq. (9) tends to increase numerically in the sequence $F^- < Cl^- < Br^- < I^-$. This is consistent with increasing importance of cation-anion polarization interaction as the polarizability of the common anion increases.

For the binary alkali fluorides the situation is further complicated by the fact that the experimental data cannot readily be expressed by the simple form of Eq. (9), but requires a more general form previously suggested by Davis and Rice (loc. cit.)

$$\Delta H^M \approx N_1N_2(U_0 + U_1\delta_{12} + U_2\delta_{12}^2). \qquad (10)$$

The new linear term in this expression is attributed to the difference in the nearest neighbour cation-anion polarization interaction in the two salts in the mixture.

A similar linear term was found by Østvold and Kleppa[21] for the alkali sulfates, and very recently by Andersen and Kleppa[22] for the alkali carbonates; however, it was not observed by Ko and Kleppa[23] in the alkali metaphosphates.

We give in Fig. 3 a graph of the experimentally determined interaction parameters for a number of AX-BX (and A_2X-B_2X) mixtures studied during the past 10-15 years. Note that this diagram indicates that for a given pair of cations, A and B, the variation in λ is relatively modest as we go from one common anion to another. This statement applies both to the order of magnitude of λ and to the extent of energetic asymmetry.

In the present discussion we have so far completely disregarded any possible dependence of the enthalpy of mixing on temperature. At the present time there is very little solid information on this point; in fact, some data published during the past year obviously are in error.[24] However, in a very recent study by Hong and Kleppa[25] it has been suggested that a correlation may well exist between $-\Delta H^M$ and ΔC_p^M. For LiF-KF at 1360 and 1181K, these authors found a maximum value of ΔC_p^M of about +0.8 cal K^{-1} mol^{-1} in the middle of the system; the corresponding value of ΔH^M is about -1.0 kcal mol^{-1}. These numbers may be useful as a guide to the order of magnitude of the temperature dependence. While we clearly are dealing with a second order phenomenon, it is one which deserves closer study.

SYSTEMS OF TYPE AX-AY

More than 20 years ago, Flood, Førland and Motzfeldt,[26] in a pioneering application of the oxygen electrode for the study of fused salt mixtures, found that Na_2CO_3-Na_2SO_4 form very nearly ideal liquid solutions. For many other mixed anion-common cation systems near-ideal behaviour similarly can be inferred from the published phase diagrams. Other related studies, such as, e.g. the heterogeneous exchange equilibrium work of Toguri, Flood and Førland[27,28] on HCl(g), HBr(g), Me(Cl,Br)(ℓ) have supported this conclusion. Hence, it is generally to be expected that the enthalpies of mixing in mixed anion-common cation systems should be numerically small. This was confirmed in the early calorimetric work of Kleppa and Meschel[29,30] on binary alkali nitrate-halides and related systems, and also in several calorimetric investigations of alkali halide-halide mixtures.[31-34] The most recent and comprehensive of these studies is the work of Melnichak and Kleppa[34] on the alkali chloride-bromide, bromide-iodide, and chloride-

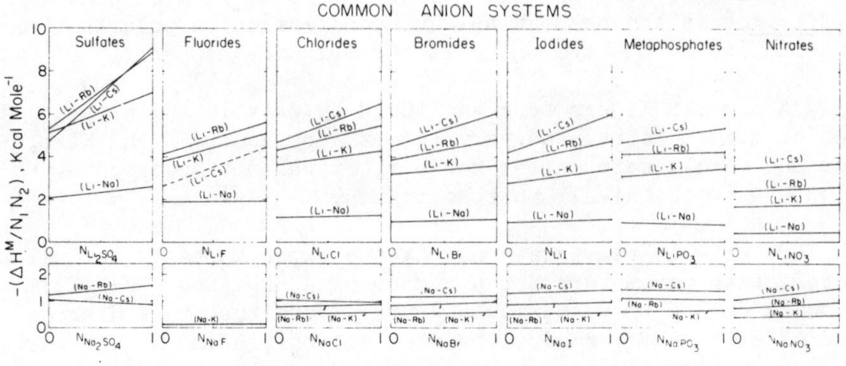

FIG. 3. Plots of enthalpy interaction parameters for lithium and sodium containing common anion mixtures.

iodide systems. A summarizing graph of their results is shown in Fig. 4.

Note that for these systems the enthalpy interaction parameters are all positive (endothermic). Note also that for a given common alkali cation, the larger the difference in the sizes of the two anions being mixed, the more endothermic is the enthalpy of mixing. All asymmetries (b) are quite small and may be positive, zero, or negative. Since the asymmetries are small, the value of the interaction parameter at the equimolar composition, $4\Delta H^M(0.5)$, well represents the whole range of compositions for each binary system.

In many respects these systems show similarities with mixed cation-common anion mixtures of the type (K-Rb)X, (K-Cs)X, and (Rb-Cs)X, which also often have quite small positive enthalpies of mixing. There is reason to believe that these similarities may be due to the fact that in both cases we are considering the mixing of two ions which differ relatively little in size, and are also fairly large and polarizable. As a consequence the small negative Coulomb contribution to the mixing enthalpy (Eq. 8) may be overshadowed by the positive London-van der Waals terms.

However, it was noted already by Kleppa and Meschel[29] in their work on the Alk(NO_3-X) solutions that the observed positive enthalpies of mixing actually are quite a bit too large to be accounted for by the calculated positive terms arising from second nearest neighbour dispersion interactions. Similar conclusions were reached by Melnichak and Kleppa.[34] It is possible that these discrepancies may be due to the use of too low values for the characteristic energy in the London expression for the dispersion

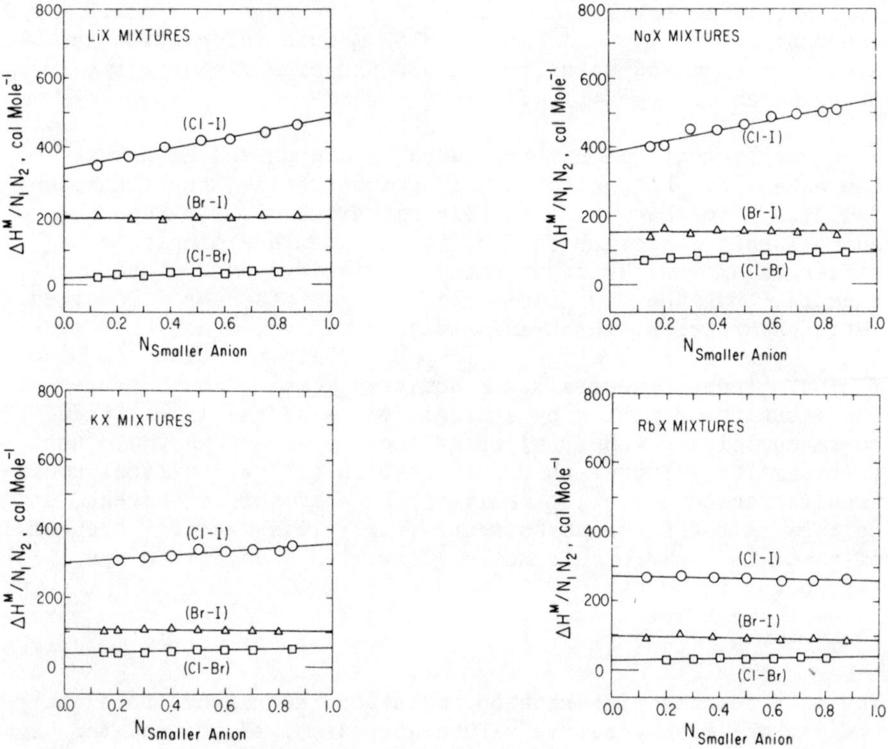

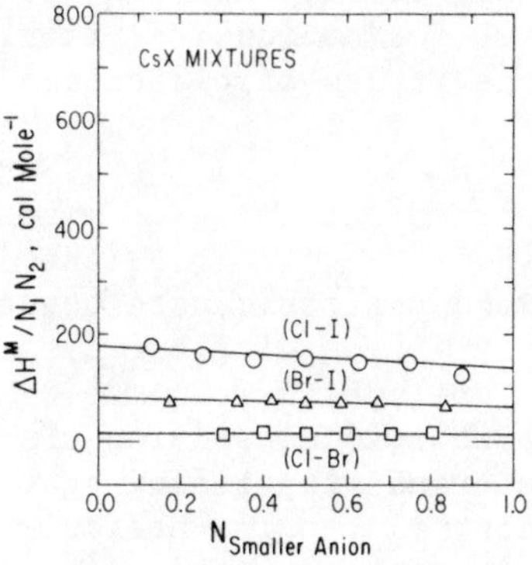

FIG. 4. Plots of enthalpy interaction parameters for chloride-bromide, chloride-iodide, and bromide-iodide liquid mixtures. (From Melnichak and Kleppa[34])

energy. However, it also seems likely that the discrepancies may be related to neglect in the calculations of the volume expansions which presumably are associated with these mixing processes. While excess volume information is now available for some mixed cation and a few mixed anion systems (Cleaver and Neil[35], Holm[36]), there is still a great lack of precise excess volume data for mixed anion-common cation systems.

Most of the thermochemical information which is presently

available on mixed anion systems is limited to salts whose anions fall within a very limited range of ionic size ($r_{A^-} \approx 2.0 \pm 0.2$Å), and until recently no detailed calorimetric information has been available on mixtures which contain the significantly smaller fluoride anion ($r_{F^-} \approx 1.4$Å).

Very recent calorimetric work by Kleppa and Melnichak[37] on the alkali fluoride-chloride, fluoride-bromide, and fluoride-iodide systems has remedied this situation. Their results are summarized graphically in Fig. 5, and indicate significant differences from the other halide-halide systems. Note, for example, that the fluoride-halide systems often have large energetic asymmetries, which are always positive, i.e., the interaction parameter is always more positive in the fluoride-rich region. Thus it generally is not very meaningful to represent the enthalpy of mixing by a single value of the interaction parameter (i.e., $4\Delta H^M(0.5)$). At the very least we would want to consider the value of this parameter in the two terminal concentration regions, i.e., $\underline{a}$ valid in the nonfluoride solvent, and ($\underline{a}$ + $\underline{b}$) in the fluoride solvent. Figure 5 shows that the values of $\underline{a}$ and ($\underline{a}$ + $\underline{b}$) for all Me(F-X) families change in a systematic way in the sequence

$$Cs < Rb < K < Na > Li.$$

The most negative interaction parameters are always found at cesium, the most positive values at sodium. Note that this systematic picture also applies for the energetic asymmetries, $\underline{b}$, which are smallest for cesium and largest for sodium. Finally, we see that for a given common cation the values of $\underline{b}$ increase systematically in the sequence

$$(F\text{-}Cl) < (F\text{-}Br) < (F\text{-}I),$$

i.e., in the order of increasing difference in ionic size between the two anions.

It is an interesting fact that the anomalous position of the lithium solutions in deviating from the trend established by the rest of the alkali systems, is common to the fluoride-halide, halide-halide,[34] and halide-nitrate systems.[29] We believe this may well be an important clue to the understanding of these mixed anion-common cation systems.

CHARGE UNSYMMETRICAL MIXTURES WHICH CONTAIN A COMMON ANION

This is a very large area of molten salt chemistry. Formally, it extends from relatively simple solutions such as

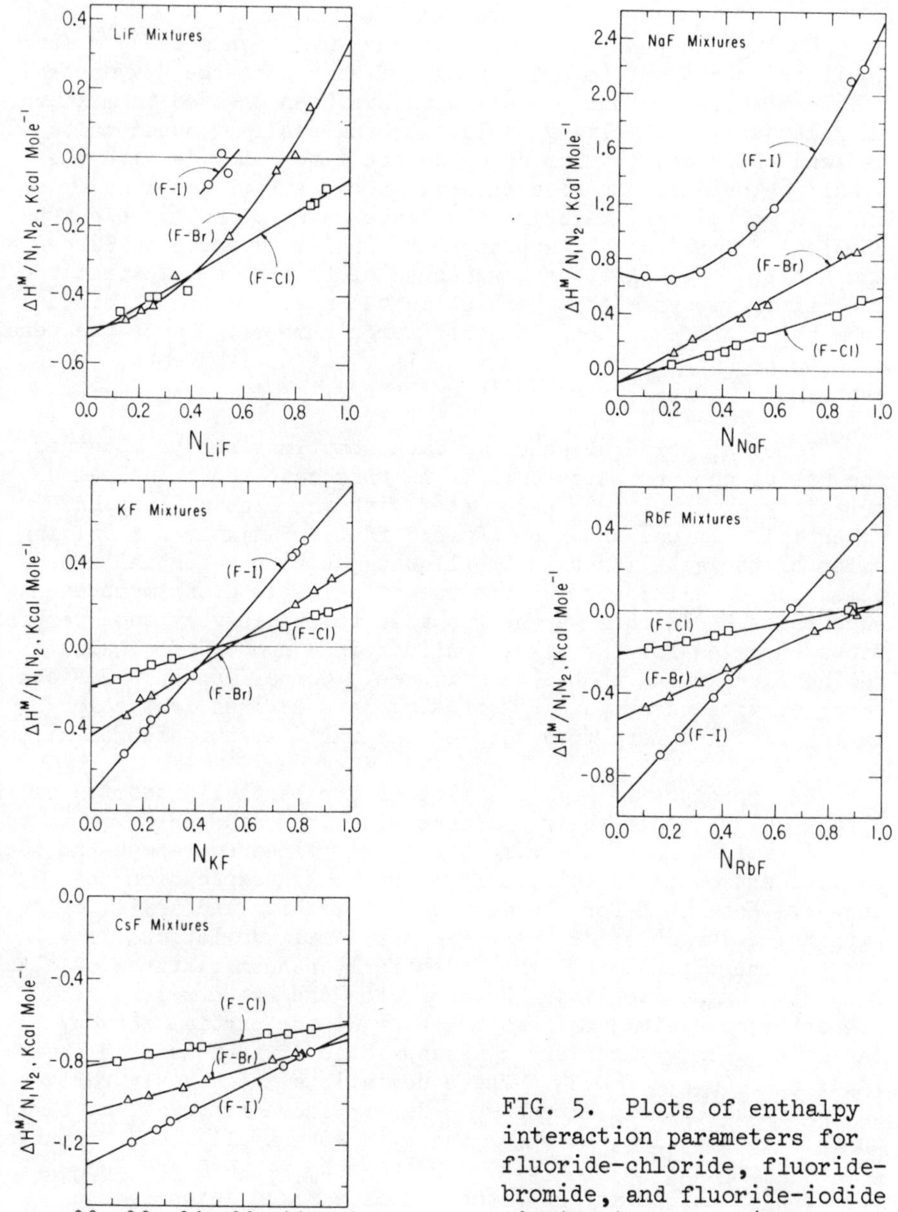

FIG. 5. Plots of enthalpy interaction parameters for fluoride-chloride, fluoride-bromide, and fluoride-iodide liquid mixtures. (From Kleppa and Melnichak[37])

those formed by alkali plus alkaline earth halides all the way to quite complex oxide mixtures involving borates, silicates and phosphates. The present discussion will be limited to the simpler types of systems.

During the past two decades there has been a large number of equilibrium and emf investigations of simple charge unsymmetrical halide mixtures. Much of this work has been devoted to mixtures of relatively low melting, volatile, and easily reduced salts such as lead chloride, cadmium chloride and zinc chloride with the alkali chlorides. Russian investigators have played a major role in this work,[38] and have in many cases made use of the electrochemical formation cell approach originally advanced by Hildebrand and Ruhle.[39] A significant weakness of this method is that the partial thermodynamic properties in the mixtures are obtained from the difference between quite large numbers. Hence the results tend to be lacking in accuracy. Also, the derived enthalpy and entropy data usually are quite uncertain.

Somewhat more reliable emf data sometimes can be obtained by the use of concentration cells. In this case the principal uncertainty often will be associated with the liquid junction potential. It was shown by Østvold that for mixtures involving salts of the alkali metals the liquid junction potential sometimes can be eliminated by the use of suitable glass membrane electrodes.[8] In this way he was able to obtain very good partial Gibbs energies for the alkali halides in a number of alkali halide-alkaline earth halide mixtures. Comparison of the Gibbs energies with corresponding enthalpy data derived from calorimetry provided very good data on the partial entropies of mixing.

We reproduce in Fig. 6 a plot of the partial entropies of the alkali chlorides in their mixtures with magnesium chloride taken from Østvold. In this figure the broken lines represent the ideal partial entropies calculated from the Temkin expression (Eq. 3, above). Note that for the mixtures of lithium and sodium chloride with magnesium chloride there are only small deviations from this curve. These presumably are essentially random mixtures of alkali ions with magnesium ions. However, for the systems involving potassium, rubidium, and cesium chloride the partial entropy curves show large and very characteristic deviations from the ideal near $N_{MgCl_2} \approx 0.33$. These deviations undoubtedly reflect the presence of very extensive order in these mixtures, centered on this composition, at which there is essentially 100% formation of the "complex" anion $MgCl_4^{2-}$. Between $N_{MgCl_2} = 0$ and $N_{MgCl_2} \approx 0.33$ the solutions may be described as a mixed anion-common cation mixture in which the two relevant anions are Cl^- and $MgCl_4^{2-}$. On the other hand, for $0.33 < N_{MgCl_2} < 1.0$ the situation is rather more complex.

Østvold also carried out similar emf and calorimetric measurements for the mixtures of the alkali chlorides with the

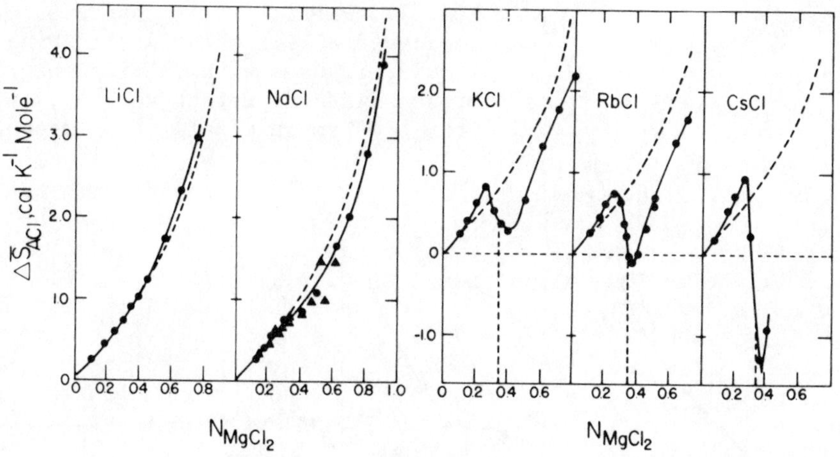

FIG. 6. Partial molar entropies of the alkali chloride in binary liquid mixtures with magnesium chloride. Broken lines represent values calculated from the Temkin model. (From Østvold[8]).

higher alkaline earth chlorides. We show in Fig. 7 his partial entropy data for the calcium chloride containing systems. Note that for these binaries there is no entropy evidence for a complex anion of compositions $CaCl_4^{2-}$. On the other hand, the entropy data for the KCl and RbCl containing systems may reflect partial order centered on $N_{CaCl_2} \approx 0.5$, i.e. the existence of a complex anion of stoichiometry, $CaCl_3^-$. Corresponding entropy data for the strontium chloride and barium chloride systems show little or no deviation from the Temkin expression.

It is of considerable interest to consider in greater detail plots of the enthalpy interaction parameters for the alkali chloride-magnesium chloride systems. These are shown in Fig. 8, taken from the work of Kleppa and McCarty.[40] Note in particular the sharp dip in λ near $N_{MgCl_2} \approx 0.33$ for the systems containing KCl, RbCl, and CsCl. This correlates with the "anomalous" partial entropy curves for these mixtures, and undoubtedly reflects the tendency of these systems to form the complex anion $MgCl_4^{2-}$. Note also that the dip is completely absent in the silver chloride and lithium chloride system, and occurs only as a very broad minimum in sodium chloride-magnesium chloride. This is consistent with the nearly ideal entropy found in these systems.

Other thermochemical investigations of systems of type ACl-BCl_2, actually predating the work of Østvold, were carried out by Papatheodorou and Kleppa.[41-43] In these investigations it was shown that chlorides of divalent transition metal ions, with ionic

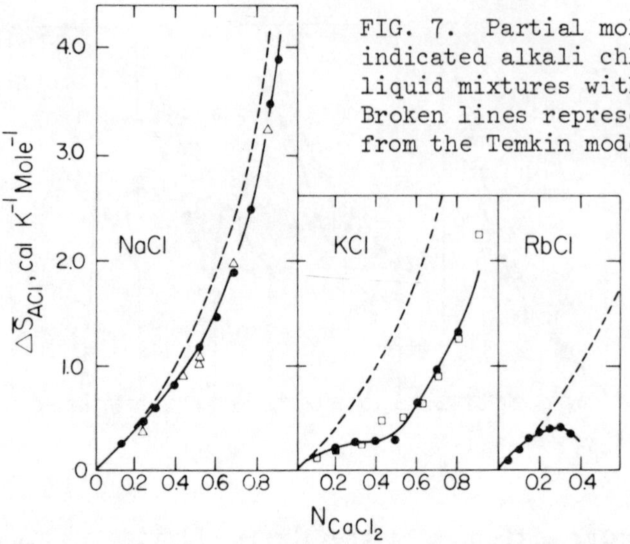

FIG. 7. Partial molar entropies of the indicated alkali chlorides in binary liquid mixtures with calcium chloride. Broken lines represent values calculated from the Temkin model. (From Østvold[8])

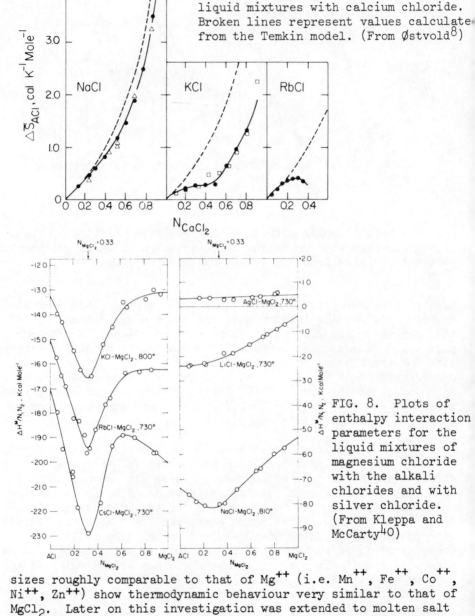

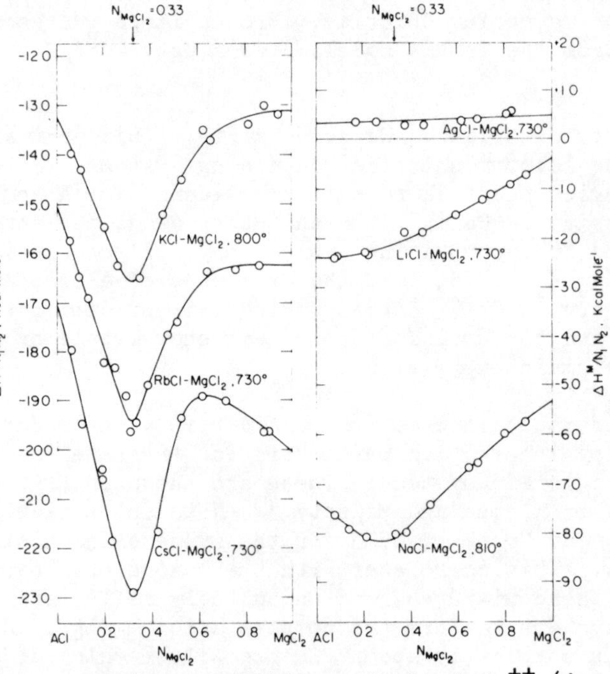

FIG. 8. Plots of enthalpy interaction parameters for the liquid mixtures of magnesium chloride with the alkali chlorides and with silver chloride. (From Kleppa and McCarty[40])

sizes roughly comparable to that of Mg^{++} (i.e. Mn^{++}, Fe^{++}, Co^{++}, Ni^{++}, Zn^{++}) show thermodynamic behaviour very similar to that of $MgCl_2$. Later on this investigation was extended to molten salt mixtures of charge structure ACl-BCl_3. From this investigation we show in Fig. 9 the experimentally determined interaction parameters for the alkali chloride-cerium chloride systems.[44] Again we find that the systems seem to fall into two distinct groups. On the one hand we have solutions involving lithium and sodium chloride.

FIG. 9. Plots of enthalpy interaction parameters for the liquid mixtures of cerium chloride with the alkali chlorides. (From Papatheodorou and Kleppa[44])

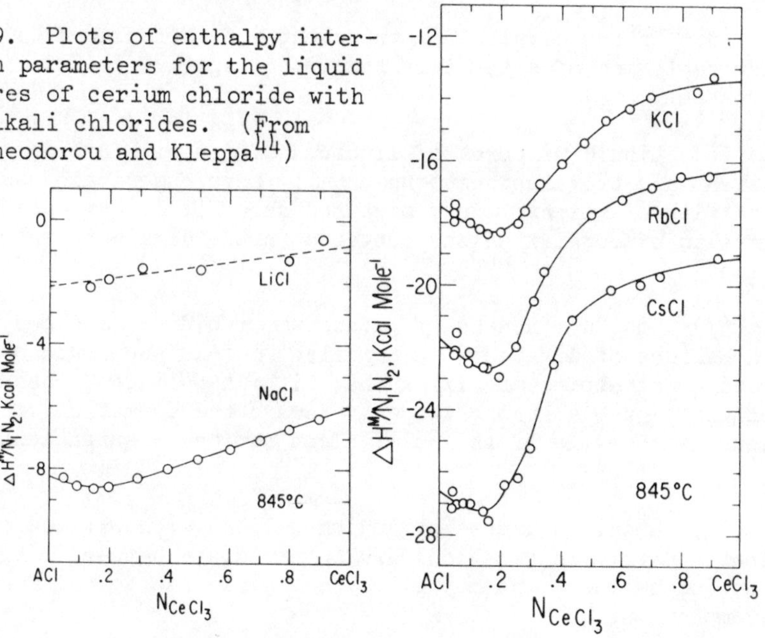

These show small (LiCl) to modest (NaCl) negative enthalpies of mixing, and no very pronounced change in the interaction parameter with composition. On the other hand, we have mixtures involving potassium, rubidium, and cesium chloride which have much larger negative enthalpies of mixing, and for which the interaction parameter shows a pronounced "dip" near $N_{CeCl_3} \approx 0.2$. There is good reason to believe that this dip may be related to the tendency of these systems to form complex anions of stoichiometry $CeCl_6^{3-}$ which would be fully "formed" at $N_{CeCl_3} = 0.25$.

It is now possible to make certain general statements regarding the thermodynamic behaviour of simple charge-unsymmetrical halide systems which contain a common anion:

(a) When a salt of a simple divalent or trivalent ion is mixed with the corresponding alkali salts, the systems usually will show increasing negative departures from ideality with increasing size of the alkali metal ion.

(b) For a given alkali metal cation the negative deviation from ideality tends to increase as the size of the divalent or trivalent ion is reduced.

(c) If the alkali ion is replaced by another singly charged ion of comparable size (i.e. Na^+ by Ag^+, or Rb^+ by Tl^+) the negative deviation from ideality is reduced (or the positive deviation increased).

(d) If the divalent or trivalent ion is replaced by a more polarizable ion of comparable size, the negative deviation is increased.

(e) Liquid mixtures of lithium halides with halides of divalent and trivalent cations usually have numerically small enthalpies of mixing. Such mixtures show little evidence of the formation of complex anions centered on the divalent or trivalent cation.

(f) Liquid mixtures of potassium, rubidium, and cesium halides with halides of divalent and trivalent metals show much more negative enthalpies of mixing than lithium halides. These mixtures frequently have a strong tendency toward the formation of complex anions if the size of the multivalent cation is sufficiently small.

(g) Liquid mixtures of sodium halides with halides of divalent and trivalent cations are intermediate between lithium and potassium halides, and may or may not give rise to the formation of complex anions.

(h) The preferred stoichiometry in complex halide anions of divalent cations is MeX_4^{2-}, although MeX_3^- is also known. The preferred stoichiometry in complex anions of trivalent cations usually is MeX_4^{3-}. However, MeX_4^- may also be formed.

In a phenomenological sense these general observations are now quite well established. Qualitatively, they clearly reflect the "competition" between the two cationic species for the common anion. This competition is aided by a small size, a high charge, and a high polarizability of the cation.

A theory for the heats of mixing of charge unsymmetrical molten salt mixtures was given by Davis.[45,46] This theory represents an extension to the charge unsymmetrical case of the conformal solution theory of Reiss et al.[14] According to the Davis theory one might expect a linear correlation between the excess thermodynamic properties of the mixture and the first power of the size parameter $\delta_{12} = (d_1 - d_2)/d_1 d_2$. Such a correlation is frequently found for a given divalent halide, and within a single halide family. However, it is common that significant deviations are found for the lithium salt. Also, it is not possible to use this theory to make quantitative predictions relating to the replacement of a certain anion by another anion.

Empirically, it is often found that replacing a larger anion by a smaller anion, i.e., Br^- by Cl^-, or Cl^- by F^-, will give rise to a more exothermic enthalpy of mixing. This situation applies, for example, in the mixtures of MgX_2 with LiX, NaX with KX, as illustrated in Fig. 10. Note, however, that in spite of the increase in exothermic character on going from KCl-$MgCl_2$ to KF-MgF_2, there is no evidence that the latter mixture has a complex anion of composition MgX_4^{2-} in analogy with the former system.

While the general rule seems to be that a smaller common anion gives rise to a more negative enthalpy of mixing, exceptions to this rule are known. Thus Østvold (loc.cit.) found that for strontium chloride, bromide, and iodide the exothermic character increases somewhat in the order Cl < Br < I. On the other hand, recent unpublished work on the strontium fluorides by Hong and Kleppa[47] shows that these latter systems are more negative than the corresponding chlorides, bromides, and iodides.

MIXTURES CONTAINING TWO CATIONS AND TWO ANIONS

As a simple illustration of the special problems posed by these reciprocal molten salt mixtures we shall consider the mixed system which contains four singly charged ions, A^+, B^+, X^-, Y^-. This is a three-component system in the sense of thermodynamics, i.e. a mixture of any composition may be prepared by bringing together suitable amounts of three of the four pure molten salts, AX, AY, BX, and BY.

Let us consider the formation (at constant P and T) of one mole of the ternary mixture $A_{N_A} B_{N_B} X_{N_X} Y_{N_Y}$ from AX, AY and BY (see Fig. 11)

$$N_X AX + (N_A - N_X) AY + N_B BY = A_{N_A} B_{N_B} X_{N_X} Y_{N_Y} \tag{11}$$

Clearly, where $N_A = N_X$ we have quasibinary mixtures of AX and BY.

If we designate the binary salt mixture which contains a common anion X by the subscript X, the binary mixture with a common cation A by the subscript A, etc., we may express the molar change in any extensive thermodynamic property R according to Eq. (11) in the following way:

$$R^M_{tern} = N_B N_X \Delta R^o_{AX+BY} + N_A \Delta R^M_A + N_B \Delta R^M_B + $$

$$+ N_X \Delta R^M_X + N_Y \Delta R^M_Y + \Delta R^E_{tern}. \tag{12}$$

In this expression ΔR^o_{AX+BY} is the standard change in the quantity R in the metathetical reaction

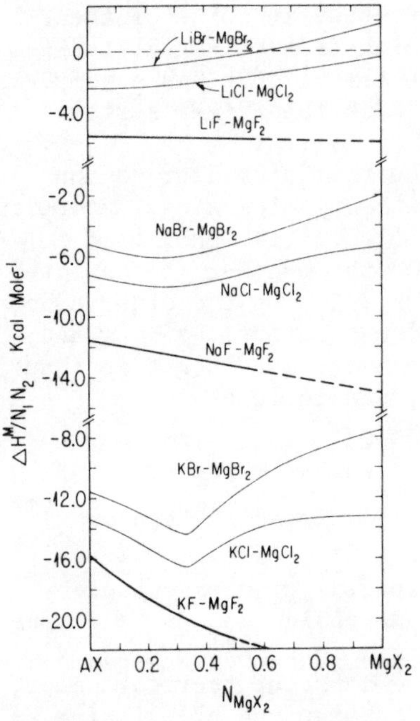

FIG. 10. Plots of enthalpy interaction parameters for liquid mixtures of magnesium fluoride, chloride, and bromide with corresponding lithium, sodium and potassium salts.

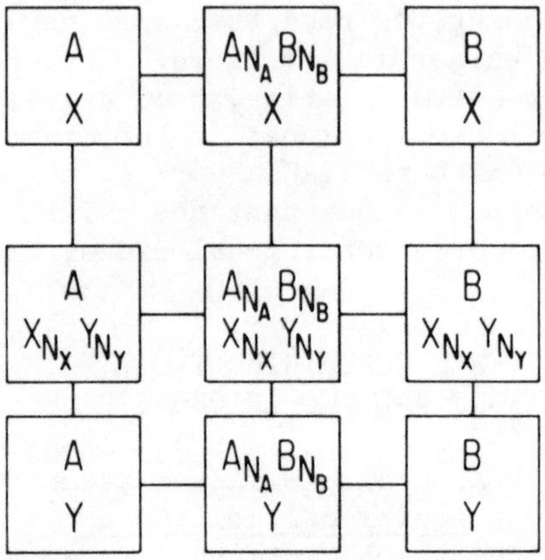

FIG. 11. Schematic diagram showing the formation of the ternary reciprocal molten salt mixture $A_{N_A}B_{N_B}X_{N_X}Y_{N_Y}$ from the pure salts AX, AY, BX, BY.

$$AX + BY = AY + BX, \tag{13}$$

i.e., this term is derived entirely from the properties of the four pure salts. Similarly the quantities ΔR_A^M, etc., represent the molar changes in R associated with the four binary mixtures, (X-Y)A, (X-Y)B, (A-B)X, and (A-B)Y. Finally, the term ΔR^E_{tern} represents a ternary excess quantity which is characteristic of the ternary system proper.

Equation (12) above can be derived directly from a simple thermodynamic cycle, first proposed by Flood, Førland and Grjotheim,[48] and has also been tested by precise calorimetric measurements aimed at the determination of ΔH^E_{tern}.[49] However, in most work on reciprocal molten salt solutions interest has been focused on the calculation of the thermodynamic activities of the constituents of interest and simplifying assumptions have been made.

For many reciprocal systems the metathetical term ΔR^{o}_{AX+BY} will be large compared to the remaining five terms. This forms the basis for the so-called first approximation of Flood and co-workers,[50] and yields the following very simple expression for the activity of AX in the mixture

$$a_{AX} = N_A N_X \gamma_{AX} \tag{14}$$

$$\Delta\bar{G}^{E}_{AX} = RT\ell n\gamma_{AX} \approx N_B N_Y \Delta G^{o}_{AX+BY} \tag{15a}$$

and correspondingly

$$\Delta\bar{G}^{E}_{BX} = RT\ell n\gamma_{BX} \approx N_A N_Y \Delta G^{o}_{AY+BX}. \tag{15b}$$

Note that $\Delta G^{o}_{AX+BY} = -\Delta G^{o}_{AY+BX}$, and that similar expressions apply for $\Delta\bar{G}^{E}_{AY}$ and $\Delta\bar{G}^{E}_{BY}$. Blander and Yosim[51] developed the conformal solution theory of Reiss et al. for reciprocal molten salt mixtures. In recent years Blander and co-workers have made a number of applications of this theory to the calculation of phase equilibria in such systems.[52-54] In these investigations they have approximated the binary mixing terms (Eq. 12) by symmetrical expressions of the form $\Delta G^{E}_{A} = N_X N_Y \lambda_A$; $\Delta G^{E}_{B} = N_X N_Y \lambda_B$; $\Delta G^{E}_{X} = N_A N_B \lambda_X$; and $\Delta G^{E}_{Y} = N_A N_B \lambda_Y$. For the activity coefficient of the component BY this yields:[53,54]

$$\begin{aligned} RT\ell n\gamma_{BY} = {} & N_A N_X \Delta G^{o}_{AX+BY} + N_A N_X (N_X - N_Y)\lambda_A \\ & + N_X (N_A N_Y + N_B N_X)\lambda_B + N_A N_X (N_A - N_B)\lambda_X \\ & + N_A (N_A N_Y + N_B N_X)\lambda_Y + N_A N_X (N_A N_Y + N_B N_X - N_B N_Y)\Lambda. \end{aligned} \tag{16}$$

The last term in this expression represents the ternary excess quantity. Λ is approximated by $-(\Delta G^{o})^2/2ZRT$ where Z is a coordination number which usually is set equal to 5 or 6. This term reflects the tendency of the ternary system to exhibit deviations from random mixing through preference for the energetically favoured ("stable") pair AX+BY or AY+BX. Equations similar to Eq. (16) for the activity coefficients of the other constituents AX, BX and AY can be obtained by suitable permutation of the relevant subscripts.[53,54]

ACKNOWLEDGEMENTS

This work has been supported by the National Science Foundation. It also has benefited from the general support of Materials Science at the University of Chicago provided by NSF-MRL.

REFERENCES

1. F.H. Stillinger, in Molten Salt Chemistry, M. Blander, Ed. (Interscience, New York - London, 1964), pp. 1-108.
2. F. London, Z. Physik. Chem. B11, 222 (1930).
3. M.F.C. Ladd and W.H. Lee, in Progress in Solid State Chemistry, H. Reiss, Ed., Vol.1 (Macmillan, New York, 1964), pp.37-82.
4. K.S. Pitzer, J. Chem. Phys. 7, 583 (1939).
5. H. Reiss and S.W. Mayer, J. Chem. Phys., 34, 2001 (1961).
6. H. Reiss, S.W. Mayer and J.L. Katz, J. Chem. Phys., 35, 820 (1961).
7. M. Temkin, Acta Physicochemica U.S.S.R., 20, 411 (1945).
8. T. Østvold, "A Thermodynamic Study of Some Fused Salt Mixtures containing Alkali and Alkaline Earth Chlorides, Bromides, and Iodides", Thesis, N.T.H. Trondheim, Tapir Trykk, 1971.
9. O.J. Kleppa and K.C. Hong, unpublished work.
10. O.J. Kleppa and L.S. Hersh, J. Chem. Phys., 34, 351 (1961).
11. T. Førland, "On the Properties of Some Mixtures of Fused Salts" Norges Tekn. Vitenskapsakademi, Series 2, No.4 (1957).
12. M. Blander, J. Chem. Phys., 34, 697 (1961).
13. J. Lumsden, Discussions Faraday Soc. 32, 138 (1961).
14. H. Reiss, J.L. Katz, and O.J. Kleppa, J. Chem. Phys., 36, 144 (1962).
15. H.C. Longuet-Higgins, Proc. Roy. Soc. (London) A205, 247 (1951).
16. W.B. Brown, Proc. Roy. Soc. A240, 561 (1957).
17. H.T. Davis and S.A. Rice, J. Chem. Phys., 41, 14 (1964),
18. L.S. Hersh and O.J. Kleppa, J. Chem. Phys., 42, 1309 (1965).
19. J.L. Holm and O.J. Kleppa, J. Chem. Phys., 49, 2425 (1968).
20. M.E. Melnichak and O.J. Kleppa, J. Chem. Phys., 57, 5231 (1972).
21. T. Østvold and O.J. Kleppa, Acta Chem. Scand., 25, 919 (1971).
22. B.K. Andersen and O.J. Kleppa, Acta Chem. Scand. (in press).
23. H.C. Ko and O.J. Kleppa, Inorg. Chem., 10, 771 (1971).
24. A.C. Macleod and J. Cleland, J. Chem. Thermodynamics, 7, 103 (1975).
25. K.C. Hong and O.J. Kleppa, J. Chem. Thermodynamics, 8, 31 (1976).
26. H. Flood, T. Førland and K. Motzfeldt, Acta Chem. Scand., 6, 257 (1952).
27. J.M. Toguri, H. Flood and T. Førland, Acta Chem. Scand., 17, 1502 (1963).
28. T. Førland, B. Saether, O. Fykse, A. Block-Bolten, J.M. Toguri, and H. Flood, Acta Chem. Scand., 16, 2429 (1962).
29. O.J. Kleppa and S.V. Meschel, J. Phys. Chem., 67, 668 (1963).
30. O.J. Kleppa and S.V. Meschel, J. Phys. Chem., 67, 2750 (1963).
31. O.J. Kleppa, L.S. Hersh and J.M. Toguri, Acta Chem. Scand., 17, 2681 (1963).

32. I.G. Murgulescu and D.I. Marchidan, Rev. Roumaine Chim., 9, 793 (1964).
33. I.G. Murgulescu, D.I. Marchidan, and C. Telea, Rev. Roumaine Chim., 11, 1031 (1966).
34. M.E. Melnichak and O.J. Kleppa, J. Chem. Phys., 57, 5231 (1972).
35. B. Cleaver and B.C.J. Neil, Trans. Faraday Soc., 65, 2860 (1969).
36. J.L. Holm, Acta Chem. Scand., 25, 3609 (1971).
37. O.J. Kleppa and M.E. Melnichak, 4th International Conference on Chemical Thermodynamics, Montpellier, France, August 1975.
38. A review of much of this work is given by G.J. Janz and C.G.M. Dijkhuis, in Molten Salts, Vol.2, Section 1: "Electrochemistry of Molten Salts: Gibbs Free Energies and Excess Free Energies From Equilibrium Type Cells", N.S.R.D.S. - NBS 28, 1969.
39. J.H. Hildebrand and G.C. Ruhle, J. Am. Chem. Soc., 49, 722 (1927).
40. O.J. Kleppa and F.G. McCarty, J. Phys. Chem., 70, 1249 (1966).
41. G.N. Papatheodorou and O.J. Kleppa, J. Inorg. Nucl. Chem., 32, 889 (1970).
42. G.N. Papatheodorou and O.J. Kleppa, J. Inorg. Nucl. Chem., 33, 1249 (1971).
43. G.N. Papatheodorou and O.J. Kleppa, Z. anorg. allg. Chem., 401, 132 (1973).
44. G.N. Papatheodorou and O.J. Kleppa, J. Phys. Chem., 78, 178 (1974).
45. H.T. Davis, J. Chem. Phys., 41, 2761 (1964).
46. H.T. Davis, J. Phys. Chem., 76, 1629 (1972).
47. K.C. Hong and O.J. Kleppa, to be published.
48. H. Flood, T. Førland and K. Grjotheim, Z. anorg. allg. Chem., 276, 289 (1954).
49. See, e.g., S.V. Meschel, J.M. Toguri and O.J. Kleppa, J. Chem. Phys., 45, 3075 (1966), and references therein.
50. B. Scrosati, H. Flood and T. Førland, Det Kgl. Norske Videnskabers Selskabs Forhandlinger, Vol. 37, Nr. 21-22, 1964.
51. M. Blander and S. Yosim, J. Chem. Phys., 39, 2610 (1963).
52. M. Blander and L.E. Topol, Inorg. Chem., 5, 1641 (1966).
53. M.L. Saboungi and M. Blander, High Temp. Sci., 6, 37 (1974).
54. M.L. Saboungi, H. Schnyders, M.S. Foster and M. Blander, J. Phys. Chem., 78, 1091 (1974).

THERMODYNAMIC PROPERTIES OF SILICATE MELTS

Donald G. Fraser

Department of Geology & Mineralogy,
University of Oxford,
Parks Road, Oxford OX1 3PR.

INTRODUCTION

One of the major current problems in experimental igneous petrology concerns the influence of additional components on liquidus boundaries observed in simple synthetic systems or in given natural rock compositions. The petrological significance of such effects may be substantial. For example, the addition of H_2O to the anhydrous system diopside-forsterite-silica causes both the forsterite and enstatite primary phase volumes to move to more silica-rich compositions (Kushiro, 1972). Adding CO_2 has the opposite effect (Eggler, 1974) as has increasing total pressure (Kushiro, 1969). These results are summarized in Fig. 1.

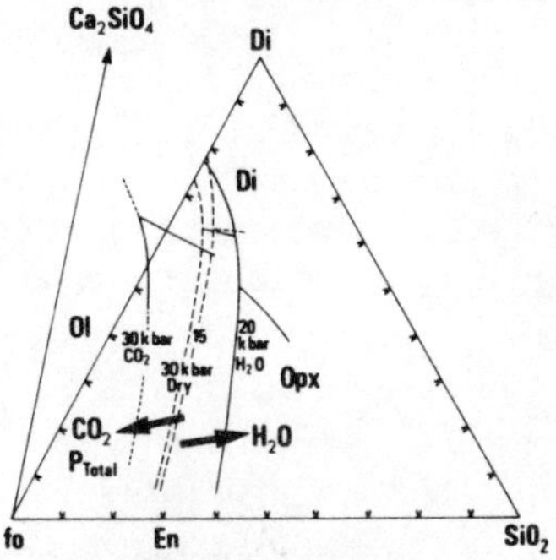

Fig. 1. Effects of added H_2O and CO_2, and of increasing total pressure, on liquidus boundaries in the system Di-Fo-SiO_2.

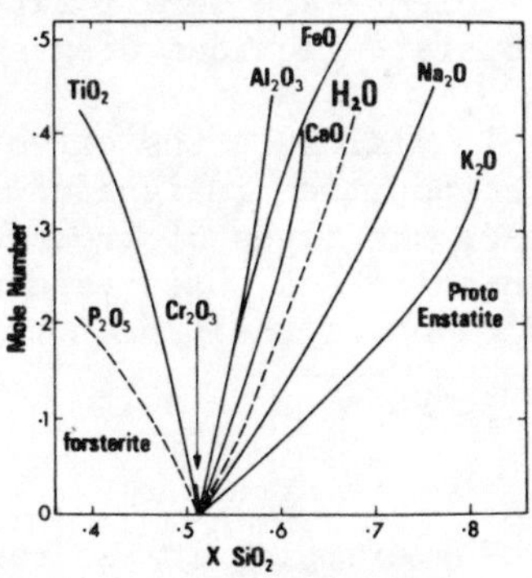

Fig. 2. Shifts in position of the forsterite-enstatite cotectic caused by adding oxides of various acid-base properties (after Kushiro, 1973).

D. G. Fraser (ed.), Thermodynamics in Geology, 301-325.

Less volatile oxide components can also cause significant changes in observed phase relations and Kushiro (1973, 1975) has summarized some of these effects as projections into the binary systems CaO-SiO_2 and MgO-SiO_2 as shown in Fig. 2. These results show that depending on the nature of the added oxide, cotectic crystallization curves may move to more, or less silica-rich compositions.

In natural systems, similar effects may be observed. For example many of the olivine-rich lavas from Nuanetsi have almost identical CaO:Al_2O_3:Na_2O ratios, and therefore lie close to a plane in the ternary join with MgO and SiO_2. Melting experiments on these rocks (Cox and Jamieson, 1974) have shown that increases in K_2O content from approximately 1% to 3% cause an extension of the crystallization range of olivine by up to 130°C before the entry of orthopyroxene. This corresponds to an increased primary phase volume of olivine relative to orthopyroxene with increasing K_2O content (cf. Fig. 2).

These observations indicate that it may prove difficult to develop detailed models of the genesis of an igneous rock suite from a limited number of melting experiments. The importance of polybaric, polythermal crystallization paths has been noted by O'Hara (1968). However the more recent studies referred to above make it clear that the effects of changing composition in open systems must be considered as well. Changing amounts of H_2O, CO_2 and other oxides present during partial melting or crystallization processes may alter the topology of phase diagrams and hence affect the evolution of a magmatic system. Similar constraints may well be important in detailed trace element studies since distribution coefficients are likely to be temperature, pressure and composition dependent. It is often forgotten that trace element distribution coefficients are merely alternative representations of the appropriate liquidus diagrams.

To deal with the effects of changing temperature, pressure and composition in igneous systems, it may be useful to attempt to develop the types of treatment described in preceding chapters for solid-solid and solid-vapour equilibria by considering equilibrium conditions of the type:

$$\mu_i^{crystal} = \mu_i^{melt} \qquad (1)$$

For example, in the case of olivine and orthopyroxene in equilibrium with an n-component silicate melt, the olivine-enstatite cotectic may be defined by (n-1) equations of the form:

$$\mu^o_{Mg_2SiO_4,ol} + RT\ln a^{ol}_{Mg_2SiO_4} = \mu^o_{Mg_2SiO_4,melt} + RT\ln a^{melt}_{Mg_2SiO_4} \qquad (2)$$

and

$$\mu^{o}_{MgSiO_3,opx} + RT\ln a^{opx}_{MgSiO_3} = \mu^{o}_{MgSiO_3,\,melt} + RT\ln a^{melt}_{MgSiO_3} \qquad (3)$$

Values of μ^{o} can, in principle, be determined calorimetrically and the expression of activity-composition relationships for components in mineral solid solutions has been discussed in previous chapters. Moreover many of the components which have been shown to influence the crystallization behaviour of rock-forming minerals (e.g. H_2O, CO_2, K_2O, P_2O_5) have such small solubilities in the solid phases that the activities of the mineral components in the solid state will be almost unaffected by varying amounts of H_2O, CO_2, K_2O etc. in a system.

The changes in liquidus boundaries caused by their addition must reflect the influence of these oxides on the activities of mineral components in the melt phase.

The purpose of this chapter is to review what is known of the mixing properties of molten silicates, to examine the relative properties of different oxide components and to suggest a simple scheme for predicting the crystallization behaviour of multi-component melts from phase relations observed in simpler systems.

NATURE OF SILICATE MELTS

Measurements of the electrical conductivities and transport properties of silicate melts (Bockris et al., 1952a, 1952b; Waffe and Weill, 1975) have shown that, with the exception of melts containing transition-metal ions in which significant charge transfer processes may operate, conduction in the melts is entirely ionic with Faraday's law being obeyed. Moreover the conductance is unipolar (Bockris et al., 1952b; Bockris and Mellors, 1956) occurring by the transport of relatively mobile cations while the anions remain stationary. Silicate melts are thus ionic liquids like other molten salts (cf. Chapter 15).

The thermodynamic properties of silicate melts however, are complicated by the presence of a large number of different silicate anions. While silicate minerals are usually monodisperse, containing only a single type of anion (e.g. SiO_4^{4-} in olivine), molten silicates contain a distribution of different polymeric silicate anions of different molecular weights and are thus polydisperse systems. The presence of a distribution of silicate anions in the melt can be inferred from the mixing properties of silicate melts (Richardson, 1956; Masson, 1965). However, it has recently become possible to separate some of the different species in silicate melts using chromatographic techniques (Lentz, 1964; Götz and Masson, 1970).

The chromatographic separation of polyanionic species in quenched oxy-acid melts was first successfully applied to polyphosphate systems (Ebel, 1953; Crowther, 1954) and the results of these and other studies (see Van Wazer, 1958) have been of importance in developing an understanding of the properties of molten silicates. Inorganic phosphates contain tetrahedral PO_4 groups analogous to the SiO_4 units in silicate minerals. By sharing oxygen atoms, the PO_4 groups may polymerize to form chain, ring or sheet structures, and by choosing cations with similar radii, phosphate compounds may be readily prepared which are excellent chemical and structural analogues of silicate systems of geological importance. For example $LiPO_3$ and $MgSiO_3$ both have infinite pyroxene chains, and $NaPO_3$ has a polymorph with 'dreierketten' similar to those in the β-wollastonite structure (Hilmer and Dornberger-Schiff, 1956; Jost, 1963). By using Lewis acids like Al^{3+} and Be^{2+}, even three-dimensionally cross-linked phosphate analogues can be produced. BeP_2O_6 (≡ $3SiO_2$) has the structure of SiO_2 - keatite (Schultz and Liebau, 1973) while $AlPO_4$ (≡ $2SiO_2$) shows all eight of the low-pressure SiO_2 polymorphs (Kleber and Winkhaus, 1949).

One of the advantages of working with phosphate analogues is that because the ionic charges are lower they act as softened systems with lower melting temperatures and greater water solubility than the M^{II} silicates of geological importance. Moreover the presence of the P = O double bond makes phosphate tetrahedra sensitive to their structural positions in polyanions (cf. Fig. 4 below). Alkali phosphate melts can be quenched to give water soluble glasses and the constituents of these can be separated by paper chromatography. The distributions of anions in solutions of quenched melts in the system Na_2O - P_2O_5 are shown plotted as functions of mean chain length in Fig. 3, (after Westman and Gartaganis, 1957). Each melt compositions can be seen to contain a distribution of phosphate polymers.

The infra-red spectra of quenched glasses in the same system also show the changing proportions of end-groups and middle groups with changing composition as shown in Fig. 4 (after Fraser, 1975a). The existence of polymeric units in the melts themselves is indicated by the observation of P-O-P bands in the Raman spectrum of molten $NaPO_3$ (Bues and Gehrke, 1956).

More recently it has become possible to separate out polymeric anions in silicate glasses. Since most silicates are sparingly soluble in water, the inorganic silicate anions are first converted to volatile organic methyl-silyl derivatives using

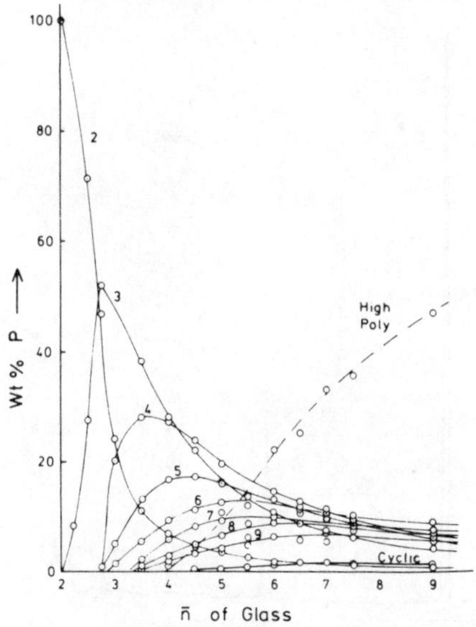

Fig. 3. Molecular weight distribution in some Na_2O-P_2O_5 glasses. Each curve represents the observed distribution of an individual polymer determined by paper chromatography. Data from Westman and Gartaganis (1957).

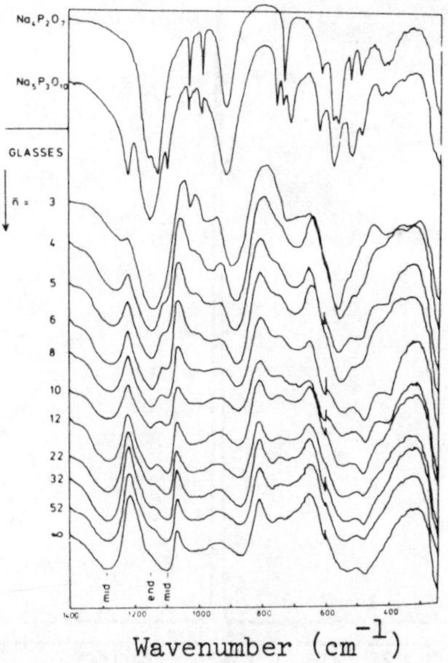

Fig. 4. Infra-red spectra of some Na_2O-P_2O_5 glasses. Note increase in relative intensities of P-O stretch bands from middle groups as mean chain length ($\bar{n}$) increases. Data from Fraser (1975a).

hexamethyl disiloxane, $(CH_3)_3Si - O - Si\ (CH_3)_3$ (Lentz, 1964; Götz and Masson, 1970). These volatile methyl siloxanes may then be separated by gas-liquid chromatography. It has been shown (Götz and Masson, 1970) that this technique does not change the molecular weight distribution greatly since, for example, crystalline hemimorphite treated in the same way gives a chromatogram containing almost exclusively peaks corresponding to the dimer (Si_2O_7) of the crystalline phase (Fig. 5).

In contrast, even a glass with as simple a composition as Pb_2SiO_4 contains a number of different anions (Fig. 6).

There is thus good experimental evidence that silicate melts are ionic liquids containing relatively free cations and mixtures of polymeric silicate anions. In a previous chapter Kleppa has reviewed what is known of the mixing properties of simple molten salts. The applications of these principles to melts containing a large number of different polyanions requires the introduction of methods developed by organic polymer chemists (Flory, 1936, 1952). Before describing the polymer models which have been applied to silicate melts it will be useful to review briefly the use of the terms acidic and basic as applied to oxides or melts.

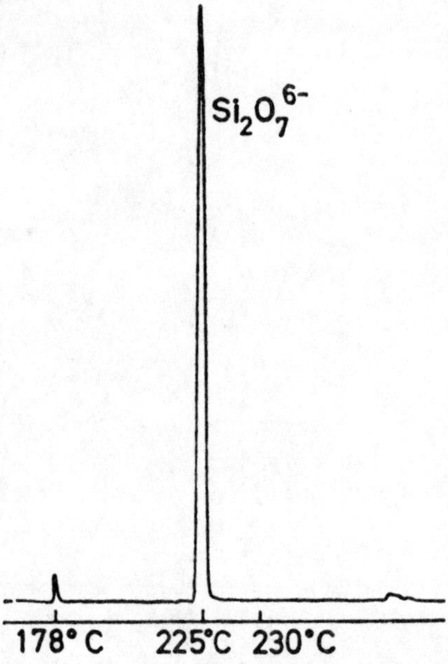

Fig. 5. Gas chromatogram of methyl-silyl derivatives of crystalline hemimorphite. Only a single major peak is present (after Götz and Masson, 1970).

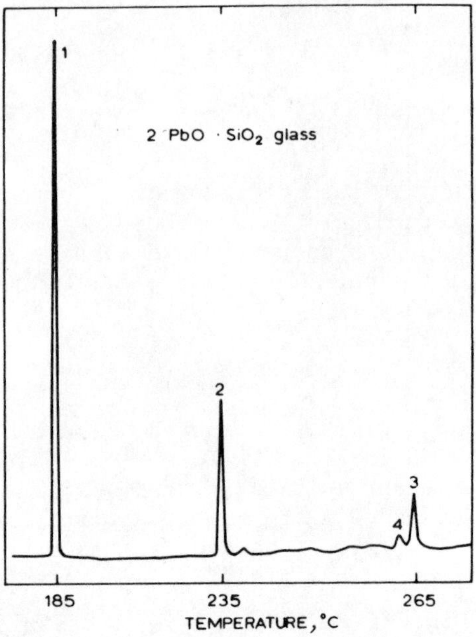

Fig. 6. Gas chromatogram of methyl-silyl derivatives from quenched Pb_2SiO_4 glasses. Glass contains several anions. Peaks corresponding to SiO_4 (1), Si_2O_7 (2), Si_3O_9 (3) and Si_4O_{12} (4) can be identified (after Masson, 1972).

ACIDS AND BASES

The concept of acidic and basic oxides is familiar to geologists and dates back to Berzelius. A more exact definition, however, was proposed by Flood and Förland (1947) following the work of Lux (1935). In aqueous systems, acidic or basic behaviour is conveniently treated using the conjugate acid-base formalism of Brönsted and Lowry:

$$\text{Acid} = \text{Base} + H^+ \quad (4)$$

$$\text{e.g. } H_2SO_4 = HSO_4^- + H^+ \quad (5)$$

In non-protonic solvents like molten oxides and silicates, such a definition is clearly unsuitable, and in the case of molten oxides or oxy-acids the Lux-Flood acid-base definition is frequently used. In this, oxide-ion, O^{2-}, takes the place of protons in the Brönsted-Lowry scheme. Thus, a basic oxide is a substance capable of furnishing oxide ions, and an acidic oxide is one which reacts with O^{2-}:

$$\text{Base} = \text{Acid} + O^{2-} \tag{6}$$

$$\text{e.g. } SiO_3^{2-} = SiO_2 + O^{2-} \tag{7}$$

It should, therefore, be possible to investigate the mixing properties of silicate melts by titrating the melts with oxide ion, just as protonic acid-base systems are investigated by pH titration.

BINARY SILICATE MELTS

"Titration" curves for a number of binary silicate melts are shown in Fig. 7. These were obtained by adding successive amounts of each basic oxide to SiO_2. The first amounts of basic oxide, MO, added react almost entirely so that activity, a_{MO}, remains at a low level. However as the orthosilicate composition is approached the possibilities for the oxide to react with the melt by depolymerization virtually disappear, and further additions of basic oxide cause a sharp rise in oxide activity.

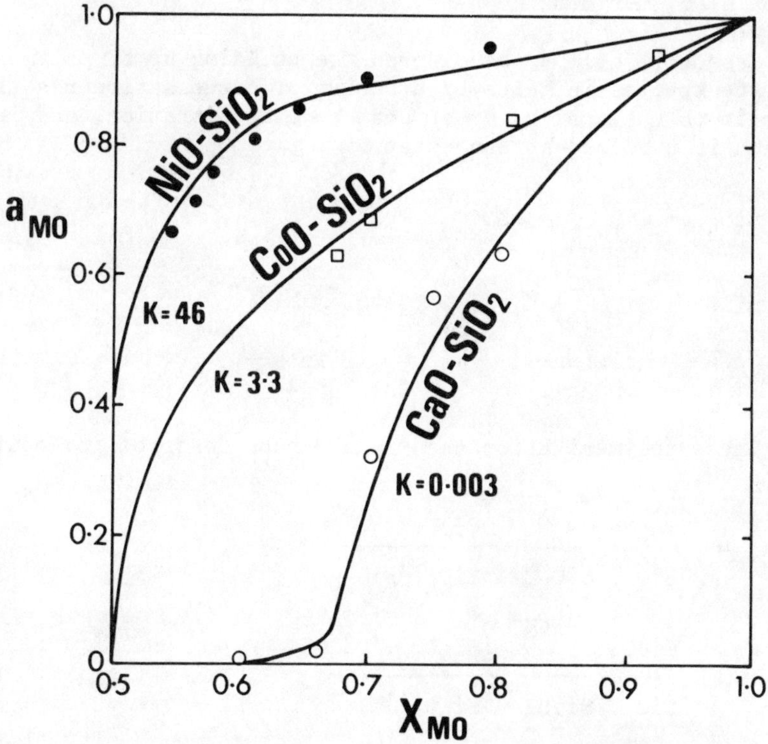

Fig. 7. Oxide activities in binary silicate melts compared with theoretical curves calculated using equation (10) after Masson, (1968).

The thermodynamic activity of the oxide component may be determined for each composition by a number of methods. For example, the melts may be equilibrated with a pure metal so as to allow the volatile oxide component to be absorbed by the molten metal until equilibrium is achieved (Richardson and Webb, 1956). Alternatively, activities may be extracted from phase diagrams, or measured directly by making e.m.f. measurements using CaO/ZrO_2 solid electrolytes (e.g. Sato in Ulmer, 1971).

The activity-composition curves shown in Fig. 7 may be interpreted if a suitable reaction model involving the basic oxide is available. Since silicate melts appear to contain large numbers of different molecular species, reaction models have been developed which are based on the concepts of organic polymer theory. Two such models will be described below:

1) Masson's polymer models

2) The Toop and Samis (quasi-chemical) model.

Masson's Polymer Models

Masson (1965) has considered the building up of polymeric silicate species in melts by a series of condensation reactions opposite to the process described above as titration, and in which O^{2-} ion is produced at each step.

$$SiO_4^{4-} + SiO_4^{4-} \rightleftarrows Si_2O_7^{6-} + O^{2-} \qquad (8.1)$$

$$SiO_4^{4-} + Si_2O_7^{6-} \rightleftarrows Si_3O_{10}^{8-} + O^{2-} \qquad (8.2)$$

$$SiO_4^{4-} + Si_nO_{3n+1}^{(2n+2)-} \rightleftarrows Si_{n+1}O_{3n+4}^{(2n+4)-} + O^{2-} \qquad (8.n)$$

These polymerization reactions can be described by equilibria of the form:

$$K_1 = \frac{X_{Si_2O_7} . X_{O^{2-}}}{X_{SiO_4} . X_{SiO_4}} \cdot \frac{\gamma_{Si_2O_7} . \gamma_{O^{2-}}}{\gamma_{SiO_4} . \gamma_{SiO_4}} \qquad (9.1)$$

$$\frac{X_{Si_3O_{10}} . X_{O^{2-}}}{X_{SiO_4} . X_{Si_2O_7}} \quad \frac{\gamma_{Si_3O_{10}} . \gamma_{O^{2-}}}{\gamma_{Si_2O_7} . \gamma_{O^{2-}}} \qquad (9.2)$$

It has been established by polymer chemists that as polymers increase in size, the reactivity of the condensing functional groups rapidly becomes independent of chain length. Meadowcroft and Richardson (1965) and Cripps-Clark et al. (1972) have shown that this becomes true in polyphosphate systems for chains above two or three PO_4-units long. This is the result of the poor transmission of electronic bonding information along polymer chains.

Masson applied these principles to silicate melts (Masson, 1965) by considering the condensing SiO_4 tetrahedra to be bifunctional so that the silicate polymers formed by reactions (8) must be unbranched chains with functional groups at the ends. Assuming equal reactivity of functional groups, the equilibrium constants K_1, K_2 etc., are identical. Masson assumed further that the silicate polyanions mix ideally so that ion fractions may be used instead of activities in equations (9). This assumption is most unlikely to be true in reality because of the different sizes of the mixing species. However its success in describing the mixing properties of silicate melts and reasons for this success will be discussed below.

Making these assumptions the infinite set of equilibria (9) can be simplified to yield an expression for the variation of oxide activity with bulk composition (see Masson, 1965, 1972):

$$\frac{1}{X_{SiO_2}} = 2 + \frac{1}{1 - a_{MO}} - \frac{1}{1 + a_{MO}(\frac{1}{K} - 1)} \tag{10}$$

It should be noted that this expression assumes:

1) Only bifunctional condensation
2) Ideal mixing of silicate polymers
3) Ideal Temkin mixing of cations and anions on the independent cation and anion matrices
4) Complete dissociation of MO in the standard state to give $M^{2+} + O^{2-}$

Activity-composition curves calculated from this model are in reasonable agreement with experimental data when appropriate values are given to K (Fig. 7). More recently this model has been extended to allow for chain branching, but not self-condensation to form ring or network structures (Whiteway et al., 1970). The expression analogous to (10) is:

$$\frac{1}{X_{SiO_2}} = 2 + \frac{1}{1 - a_{MO}} - \frac{3}{1 + a_{MO}(\frac{3}{K} - 1)} \tag{11}$$

The predictions of the branched chain model agree better with experiments for the systems $SnO-SiO_2$, $FeO-SiO_2$, $MnO-SiO_2$ and $PbO-SiO_2$, while for $CoO-SiO_2$ the simple unbranched chain model gives a better fit (Masson, 1972).

Toop and Samis (quasi-chemical) model

An alternative treatment of the mixing properties of silicate melts has been proposed by Toop and Samis (1962a, 1962b) following the work of Fincham and Richardson (1954). These authors suggested that the chemical properties of silicate melts can be interpreted by considering the anions to be composed of just three different quasi-chemical oxygen species:

O^{o} = bridging oxygen atom
O^{-} = singly-bound oxygen
O^{2-} = free oxide ion

By assuming the reactivities of bridging oxygen groups towards oxide ion to be independent of molecular size (cf. polymer theory above) the depolymerization reactions of a basic oxide with a silicate melt can be represented by the single reaction:

$$O^{2-} + O^{o} \rightleftarrows 2O^{-} \qquad (12)$$

The use of quasi-chemical species to describe the mixing properties of silicate melts is equivalent to a first order refinement of the regular solution model. In a truly regular solution there is a heat of mixing, but ΔS_{mix} is considered to be configurational. This is unlikely to be true in any case, but in particular when large heats of mixing are observed as in the case of silicate melts with different silica contents, because the heats of mixing correspond to attractions or repulsions among the mixing components. The quasi-chemical treatment allows for these effects by considering the production of 'reacted species' in the mixture as in the non-ideal mixing of O^{2-} and O^{o} in (12). Thus, the non-ideal mixing of O^{2-} and O^{o} is expressed in terms of <u>ideal</u> mixing of the quasi-chemical species O^{2-}, O^{o} and O^{-}, the proportions of each being determined by an equilibrium constant (cf. Guggenheim, 1952, p.38).

To apply this quasi-chemical model to describe the mixing properties of silicate melts, Toop and Samis assumed that:

1. The Temkin model holds in binary silicate melts, $MO\text{-}SiO_2$. All the free energy of mixing is then anionic since the metal ion fraction in the cation matrix ($X^{c.m.}_{M^{2+}}$) is 1.0.

2. Metal oxide components, MO, are 100 per cent dissociated.

3. The reactivities of oxo-bridges are independent of molecular configuration.

If the quasi-chemical species mix ideally, then the

equilibrium condition for the reverse of (12), i.e. polymerization, is:

$$K = \frac{X_{O^{2-}} . X_{O^o}}{X^2_{O^-}} \tag{13}$$

Since one mole of O^{2-} reacts with O^o to produce two moles of O^-, the integral free energy of mixing per mole of melt is given by:

$$\Delta G_{mix} = \frac{n_{O^-}}{2} RT\ln K \tag{14}$$

Values of ΔG_{mix} can be calculated using this model for any values of K if values of n_{O^-} can be obtained for a given composition. This is done by assuming that:

1) 1 mole MO gives 1 mole O^{2-}, and
1 mole SiO_2 gives 2 moles O^o

2) Charge balance for Si^{4+} so that in a binary

$$4X_{SiO_2} = 2n_{O^o} + n_{O^-}$$

3) Mass balance so that

$$n_{O^o} = 2X_{SiO_2} - \tfrac{1}{2}n_{O^-}$$

Applying these constraints and the equilibrium condition (13), n_{O^-} is given by the quadratic:

$$(n_{O^-})^2 (4K-1) + n_{O^-} (2+2X_{SiO_2}) + 8X_{SiO_2} (X_{SiO_2} - 1) = 0 \tag{15}$$

Solving for n_{O^-}, it is then possible to calculate ΔG_{mix} for any bulk composition for given values of K using equation (14). Curves calculated in this way are compared with experimental data in Fig. 8 and it can be seen that the calculated curves agree closely with the experimental data if appropriate values of K are used.

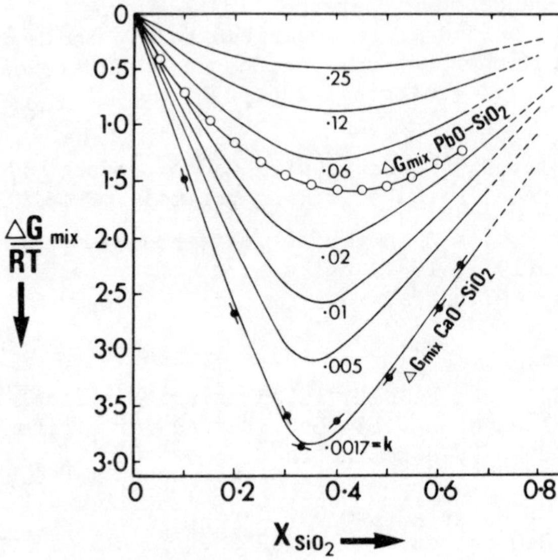

Fig. 8. Integral free energies of mixing in binary silicate melts. Data for systems $PbO-SiO_2$ and $CaO-SiO_2$ compared with curves calculated from equations (14) and (15) using different values of K (after Toop and Samis, 1962).

APPLICATION OF MIXING MODELS

Polymer models of the type described above have been used wit considerable success to interpret the mixing properties of certain binary silicate melts. Although the polymerization constants for melts of geological complexity are as yet unknown and the application of the models outside binary systems is not straightforward partly because of differences in the degree of dissociation of different basic oxides as is described below, the models may already be used to interpret some aspects of silicate melt behaviour which are of geological interest.

1) H_2O, CO_2 and SO_2 Solubilities

The solubility of H_2O in albite melts has been studied by Burnham and Davis (1971, 1974) and some of their results are shown in Fig. 9.

The results shown in Fig. 9 indicate that water solubility in albite melts increases linearly with the square root of water fugacity. This has been observed for a large number of different silicate melt compositions. The linear dependence of water solubility on the square root of f_{H_2O} implies that H_2O dissociates in the melt to produce two independent species. Two equivalent models were proposed to explain this behaviour (Burnham and Davies, 1974; Burnham, 1975).

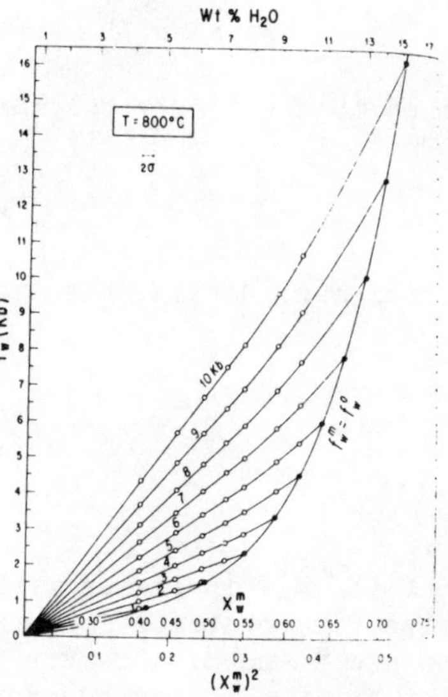

Fig. 9. Water solubility of albite melts at 800°C (after Burnham and Davies, 1974).

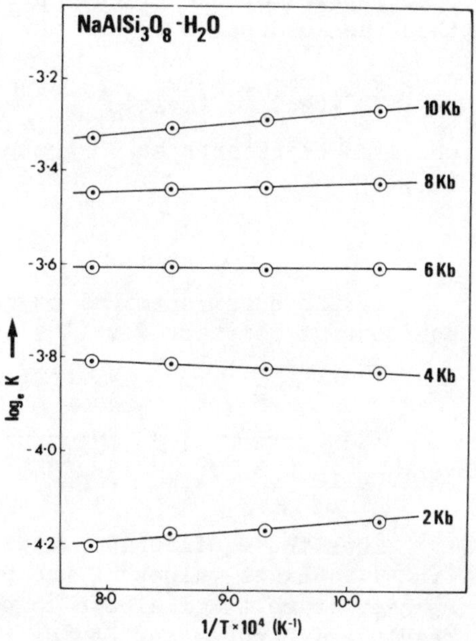

Fig. 10. Water solubility of albite melts from 700°C-1000°C and 2-10 kbar expressed as linear plots of lnk vs. 1/T using quasi-chemical reaction model and equation (19) (after Fraser, 1975a).

1) $NaAlSi_3O_8 + H_2O = AlSi_3O_7OH + NaOH$ (16)

2) $NaAlSi_3O_8 + H_2O = AlSi_3O_7OH + NaO^- + H^+$ (17)

These reactions seem unlikely to be correct since melts containing 50 mol per cent H_2O (cf. Fig. 9) would contain 50 mol per cent of molten NaOH in equilibrium with the acidic species $AlSi_3O_7OH$, or alternatively the same proportion of free protons in an oxide melt.

It is important to emphasize that no thermodynamic data can prove the validity of a structural theory. Thus in the present case, any mechanism involving the dissociation of H_2O can account for the observed dependence of solubility on f_{H_2O}. One way of doing this is by adapting the Toop and Samis model.

Since the viscosities of silicate melts are decreased markedly by the absorption of H_2O, it is likely that, at least in acidic melts, water is absorbed in the melt by reaction with oxo-bridges causing depolymerization - i.e. as a basic oxide. If the reactivities of oxo-bridges are independent of molecular size, then the absorption of H_2O may be expressed by:

$$H_2O_{(vap)} + O^o_{(melt)} = 2\ OH_{(melt)} \tag{18}$$

where OH represents an OH group attached to a silicate polymer, i.e.

-Si-OH

If the quasi-chemical oxygen species mix ideally, then the equilibrium constant for (18) is

$$K = \frac{X^2_{OH}}{f_{H_2O}\ X_{O^o}} = \frac{(2X^{melt}_{H_2O})^2}{f_{H_2O} \cdot X_{O^o}} \tag{19}$$

Thus the equilibrium water content ($X^{melt}_{H_2O}$) increases with $(f_{H_2O})^{\frac{1}{2}}$. Exact values of the polymerization constants for oxygen species in melts of albite composition are unknown. However the results of Burnham and Davies (1974) on albite-H_2O may be expressed by the simple equilibrium relation (19) if it is assumed that one mole of $NaAlSi_3O_8$ melt contains eight moles of O^o. Values of lnK calculated from (19) for different pressures are plotted against 1/T in Fig. 10.

It should be noted that deviations from the linear relationship between water solubility and f_{H_2O} occur at higher water contents. These deviations may be the result of

(1) differences in the reactivities of Al-O-Si and Si-O-Si oxo-bridges towards basic oxides. The net reaction of H_2O with the melt may therefore be due to the two reactions

$$H_2O + O^o_{Al,Si} \rightleftarrows 2\ OH \qquad \Delta G^o_1 \tag{20}$$

$$H_2O + O^o_{Si,Si} \rightleftarrows 2\ OH \qquad \Delta G^o_2 \tag{21}$$

In describing the properties of aluminosilicate melts, it may be necessary to modify the simple Toop and Samis model so as to distinguish different quasi-chemical O^o species in this way. This was found to be necessary in order to interpret the H_2O solubilities of some alkali phosphate melts (Fraser, 1975a) in which the reactivities of oxo-bridges are known to vary according to their co-ordination with end groups, middle groups etc. (Van Wazer and Holst, 1950). To test this possibility for aluminosilicate melts, water solubilities should be determined for more compositions in the binary system $NaAlO_2$-($NaAlSi_3O_8$)-SiO_2. It is interesting to note that in the system SiO_2-$NaAlSi_3O_8$, the primary phase volume of quartz increases with increasing H_2O contents of the melt (Tuttle and Bowen, 1958) whereas in the system diopside-forsterite-silica the primary phase volume of the SiO_2 polymorphs decreases with increasing H_2O content. This may indicate that in qz-ab-an, Al-O-Si bonds are broken in preference to Si-O-Si oxo-bridges by added H_2O, thus increasing the activity of SiO_2 relative to aluminosilicate in the melt.

(2) increasing amounts of molecular H_2O dissolved in the melt, in which case $X^{melt}_{H_2O}$ increases with the <u>square</u> of $f^{\frac{1}{2}}_{H_2O}$ (e.g. Anfilogov and Kadik, 1973). The observed deviations from linearity may, therefore, be due simply to increasing amounts of dissolved molecular H_2O in the melt. It should also be noted that in very basic melts H_2O may dissolve as an acidic oxide according to the reaction:

$$H_2O + O^{2-} \rightleftarrows 2\ OH^- \tag{22}$$

Like other oxides, H_2O can be expected to show amphoteric properties depending on the nature of the other components present.

The solubility of CO_2 in silicate melts may be expressed in an analogous way by considering CO_2 to be an acidic oxide.

$$CO_2 + O^{2-} \rightleftarrows CO_3^{2-} \tag{23}$$

$$CO_{2(vap)} \rightleftarrows CO_{2(melt)} \tag{24}$$

Mysen et al. (1975) found infra-red evidence to support the existence of both molecular CO_2 and carbonate ion in quenched silicate melts.

The solubility of sulphur in silicate melts has been investigated by Fincham and Richardson (1954) who inferred the operation of two mechanisms depending on the magnitude of f_{O_2}.

$$\tfrac{1}{2}S_2 + O^{2-}{}_{(melt)} \rightleftarrows \tfrac{1}{2}O_2 + S^{2-}{}_{(melt)} \tag{25}$$

$$\tfrac{1}{2}S_2 + \frac{3}{2}O_2 + O^{2-}{}_{(melt)} \rightleftarrows SO_4^{2-}{}_{(melt)} \tag{26}$$

With increasing f_{O_2} the sulphur solubility in silicate melts decreases according to (25) before increasing again as reaction (26) becomes effective with sulphur dissolving as SO_4^{2-}.

Oxidation states of transition metal ions in silicate melts

The oxidation states of transition metal ions in silicate melts are determined not only by f_{O_2}, T and P, but also by the composition of the melt. For example the Fe(III)/Fe(II) ratios of silicate melts increase with increasing melt basicity at constant f_{O_2}, T and P both in natural rock compositions (Fudali, 1965) and in synthetic melts (Paul and Douglas, 1965a). This observed tendency of Fe(III)/Fe(II) ratios to increase in alkaline silicate liquids has been called the "alkali-ferric-iron effect" (Carmichael and Nicholls, 1967). Similar results have been observed for Ce^{3+}/Ce^{4+} (Paul and Douglas, 1965b), Mn^{2+}/Mn^{3+} (Paul and Lahiri, 1966), Cr^{3+}/Cr^{6+} (Nath and Douglas, 1965) and Eu^{2+}/Eu^{3+} (Morris and Haskin 1974a). These results are clearly important in affecting the distribution behaviour of these elements.

It has been shown experimentally that at constant oxygen fugacity, the melts become increasingly oxidized as the basicity increases. This cannot be interpreted on the basis of simple redox reactions of the type

$$4Eu^{3+} + 2O^{2-} \rightleftarrows 4Eu^{2+} + O_2 \tag{27}$$

To interpret the redox behaviour of altervalent components in silicate melts it is necessary to write a reaction in terms of oxide components:

$$Eu_2O_3 \rightleftarrows 2EuO + \tfrac{1}{2}O_2 \tag{28}$$

In silicate melts the activities of the oxide components in

(28) will be determined by their reactions with the melts. Thus EuO is likely to dissociate as a basic oxide like CaO

$$EuO \rightleftarrows Eu^{2+} + O^{2-} \qquad (29)$$

However Eu_2O_3 may well show amphoteric behaviour like Al_2O_3. Its net behaviour may be considered to result from two reactions.

$$Eu_2O_3 \rightleftarrows 2Eu^{3+} + 3O^{2-} \qquad \text{'basic'} \qquad (30)$$

$$Eu_2O_3 + O^{2-} \rightleftarrows 2EuO_2^- \qquad \text{'acidic'} \qquad (31)$$

The species of EuO_2^- represents Eu-based structural units of polymers in the melt analogous to AlO_2^- groups. Similar reactions may be written for other transition metal oxides. Proceeding in this way it can be shown (Fraser, 1975b) that if equilibrium constants for the basic and acidic reactions are represented by K_b and K_a, then the ratio Eu(III)/Eu(II) is given by:

$$\frac{Eu(III)}{Eu(II)} = \text{const.}\ f_{O_2}^{\frac{1}{4}} \ \{K_a (a_{O^{2-}})^{\frac{3}{2}} \left[\frac{1-X_{MO}}{X_{MO}+\frac{1}{4}(1-X_{MO})}\right] + \frac{K_b}{(a_{O^{2-}})^{\frac{1}{2}}}\} \qquad (32)$$

This equation predicts that the ratio Eu(III)/Eu(II) should increase with $(fO_2)^{\frac{1}{4}}$ as is observed experimentally (Morris and Haskin, 1974b). If Eu_2O_3 behaves predominantly as a basic oxide, then $K_a \rightarrow 0$ and the Eu(III)/Eu(II) ratio would decrease with increasing oxide activity as is predicted erroneously by the simple equation (27). However if Eu_2O_3 is amphoteric, the acidic reaction (31) must also be considered. As a limiting case it is instructive to consider the behaviour of Eu_2O_3 as an entirely acidic oxide. In this case $K_b \rightarrow 0$ and the Eu(III)/Eu(II) ratio is seen to increase with increasing melt basicity as was observed by Morris and Haskin (1974a).

Oxide activities calculated using equation (32) with $K_b = 0$ from the observed Eu(III)/Eu(II) ratios in the system $CaAl_2Si_2O_8$ - Mg_2SiO_4 - Ca_2SiO_4 are shown in Fig. 11. The calculated activities define curves similar to the titration curves determined experimentally in silicate melts (Fig. 7) despite the assumption of the completely acidic behaviour of Eu_2O_3 and differences between SiO_2 and anorthite as end member components. It should also be noted that (31) is only one of a number of reactions which can be written to represent the acidic behaviour of Eu_2O_3.

By expressing the acid-base properties of different oxides in this way it is possible to construct a geochemical acid-base scale:

Cs_2O, Rb_2O	$\rightarrow$	Fe_2O_3, Al_2O_3	$\rightarrow$	TiO_2, SiO_2, P_2O_5
Basic		Amphoteric		Acidic
$K_a \rightarrow 0$		K_a, K_b		$K_b \rightarrow 0$

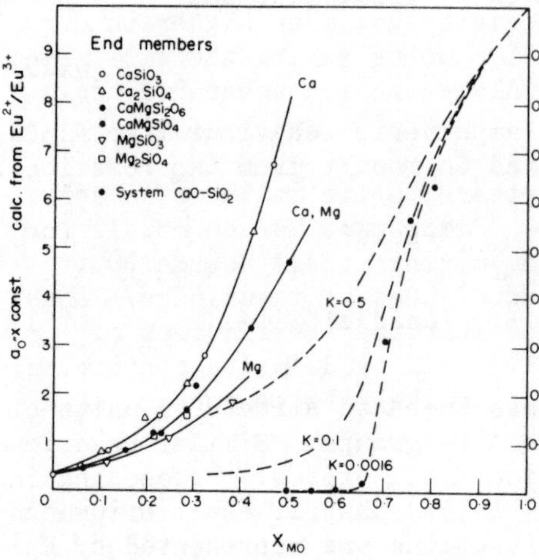

Fig. 11. Curves of relative oxide activity calculated from observed Eu(III)/Eu(II) ratios in the binary systems shown with $CaAl_2Si_2O_8$ as the other end member (after Fraser, 1975b).

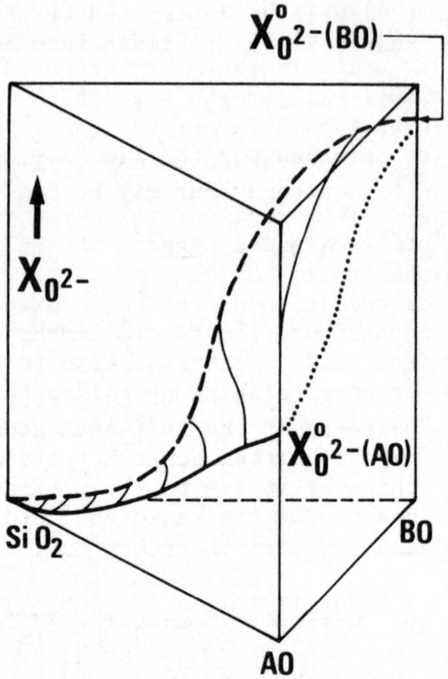

Fig. 12. Schematic diagram illustrating different standard state values of X^o_{O2} for two oxides AO and BO (eg. MgO and CaO).

STANDARD STATES OF OXIDE COMPONENTS

The curves shown in Fig. 11 are of some interest because the cation effect is the opposite of that which might be predicted from studies of the effects of cation polarizing power on the polymerization constants (Hess, 1971). Hess has pointed out that values of K tend to increase as the polarizing power of the cation (Z/r^2) increases. Referring to the polymerization reactions:

$$2O^- \rightleftarrows O^o + O^{2-} \qquad (33)$$

$$\text{or} \quad SiO_4^{4-} + SiO_4^{4-} \rightleftarrows Si_2O_7^{6-} + O^{2-} \qquad (34)$$

it would seem that the more polarizing Mg^{2+} cations should stabilize higher oxide activities than should the larger Ca^{2+} whereas the opposite was found from the experimental data using equation (32). (cf. Fig. 11).

This is, in fact, quite reasonable since in the presence of cations of still higher polarizing power, e.g. Si^{4+} or Ti^{4+}, it is unlikely that the free oxide activity would be higher in e.g. $TiO_2 - SiO_2$ melts than in $Rb_2O - SiO_2$ melts having the same X_{SiO_2} even though larger values of the polymerization constants would be inferred for the $TiO_2 - SiO_2$ system.

The origin of this dilemma appears to lie in the standard states assumed by both the Toop - Samis and Masson models for the "basic" oxide component. These mixing models assume 100% dissociation of the basic oxide constituent of the binary. Thus in the pure liquid end-member oxide melt the ion fraction of free oxide $X^o_{O^{2-}}$ is assumed to be 1.0. In oxide melts containing strongly polarizing cations this is unlikely to be correct since the dissociation reactions e.g.

$$TiO_2 \rightleftarrows Ti^{4+} + 2O^{2-} \quad (35)$$

$$SiO_2 \rightleftarrows Si^{4+} + 2O^{2-} \quad (36)$$

(cf. (33))

are unlikely to proceed to completion.

The polymer models assume ideal mixing and so the ion fractions are related to fugacities (via Raoult's Law)

$$X^o_{O^{2-}} \equiv f^o_{O^{2-}} \quad (37)$$

However oxide activities are by definition ratios of the fugacities of O^{2-} in the melt to those in the pure liquid oxide standard state:

$$a_{O^{2-}} = f_{O^{2-}}/f^o_{O^{2-}} = X_{O^{2-}}/X^o_{O^{2-}} \quad (38)$$

The oxide activities shown in Fig. 11 for the Mg^{2+} and Ca^{2+} systems refer to a common (hypothetical) standard state value of $X_{O^{2-}} = 1.0$ for both systems so that the activities can be compared. Note that if either oxide fails to dissociate completely then $X^o_{O^{2-}}$ in the pure liquid oxide will be <u>less than</u> 1.0. The departure of $X^o_{O^{2-}}$ from unity is likely to increase with cation polarizing power (cf. SiO_2).

Experimentally, activities are measured relative to the empirical properties of the end-member oxide so that by definition $a_{MO} = 1.0$ in pure liquid MO. The binary titration curves used to fit values of K are therefore each normalized to a value of 1.0 for a_{MO} when $X_{MO} = 1.0$. However in this standard state, values of $X^o_{O^{2-}}$ for different oxides may be quite different.

It is therefore clear that unless the different standard state values of $X^o_{O^{2-}}$ are calibrated, oxide ion activities in different binary systems obtained using the Temkin equation ($a_{MO} = a_{M^{2+}} . a_{O^{2-}}$) cannot be compared since the experimental activities relate to different ionic standard states ($X^o_{O^{2-}}$).

The Masson and Toop and Samis mixing models assume complete dissociation of the basic oxide components. Thus, while these models may allow the calculation of oxide activities within any binary, activities in different binary systems cannot be compared since they relate to different standard states. The magnitude of the polymerization constant for a system is a measure of the shape of the titration curve not its absolute position. Thus with decreasing K the curves become steeper, or more sharply inflected, reflecting strong interactions between O^{2-} and O^o or Si-O-Si. As pointed out by Hess (1971) values of K decrease as Z/r^2 decreases. So the activity curves for Ca^{2+} systems should be more steeply inflected than in the case of the equivalent Mg^{2+} compositions and these effects are shown in Fig. 11.

In attempting to use the polymer or quasi-chemical models to calculate the mixing properties of ternary and more complex melts, the differences in the standard state values of $X^o_{O^{2-}}$ assumed for the end-member oxide components may be very important. In the absence of any additional cross-interactions the polymerization constants in a multicomponent system containing a mols of AO, b mols BO etc. will be given by

$$\bar{K} = (K^a_{AO} . K^b_{BO} \; .....)^{1/(a + b + ...)} \qquad (39)$$

However it will also be necessary to calibrate the reference values of $X^o_{O^{2-}}$ for each oxide relative to a common standard state. This is illustrated schematically in Fig. 12.

EFFECTS OF MELT STRUCTURE ON LIQUIDUS BOUNDARIES

It has been shown above that silicate melts contain distributions of silicate polyanions of different sizes. These polymeric anions can all be constructed from five structural units:

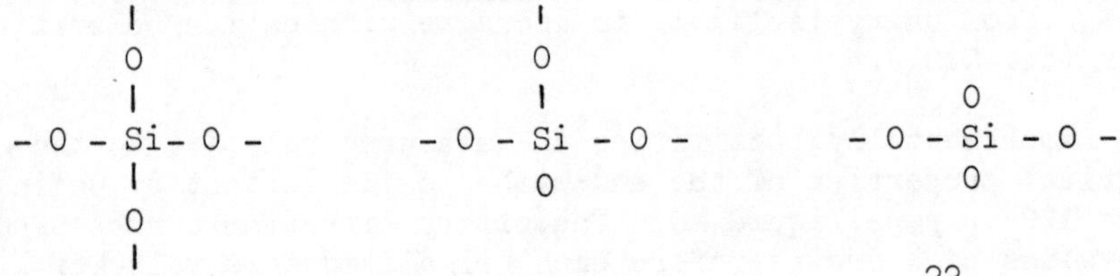

^{04}Si ^{13}Si ^{22}Si

```
         O                   O
 - O - Si O              O  Si O
         O                   O

      31Si                 40Si
```

These structural groups may be given the notation ^{ij}Si where i is the number of singly bound oxygens and j is the number of bridging oxygens in the group (Fraser, 1974).

When silicate minerals crystallize from these melts, only particular structural units can be accommodated by the growing mineral. For example pyroxene chains contain only middle groups (^{22}Si) and olivines contain only orthosilicate (^{40}Si) anions.

Masson (1965) assumed that the discrete silicate polyanions of different sizes mix ideally. However the activity coefficients (γ) in equation (9) will also disappear if they lie in a geometrical series and cancel out (Masson, 1972). This is equivalent to allowing the ideal mixing of the structural units of the polymers. It should be of interest, therefore, to treat silicate melts as ideal solutions of structural units or structons (Huggins, 1954) and cations.

This concept may be used to derive simple solubility expressions for silicate minerals in melts. Consider, for example, the crystallization of enstatite from a melt. Since the structure of enstatite can only accommodate middle groups the "solubility product" for enstatite in a silicate melt can be written in the form

$$MgSiO_3 = Mg^{2+} + {}^{22}Si \qquad (40)$$

For forsterite the equivalent expression is

$$Mg_2SiO_4 = 2Mg^{2+} + {}^{40}Si \qquad (41)$$

Referring to equations (2) and (3) it can be seen that (40) and (41) suggest in the melt that the activities of the $MgSiO_3$ and Mg_2SiO_4 components are given by the expressions

$$a^{melt}_{MgSiO_3} = X_{Mg^{2+}} \cdot X_{^{22}Si} \qquad (42)$$

$$a^{melt}_{Mg_2SiO_4} = X^2_{Mg^{2+}} \cdot X_{^{40}Si} \qquad (43)$$

The ion fractions of different structons may be easily calculated from the equilibrium proportions of the Toop and Samis oxygen species O^- and O^o in a melt. Thus the probabilities of forming the different structons are given by the combinatorial expressions in Table 1.

$$p_{04} = X_{O^o}^4 \times 1 = X_{O^o}^4$$
$$p_{13} = X_{O^o}^3 . X_{O^-} \times \frac{4!}{3!\,1!} = 4X_{O^o}^3 . X_{O^-}$$
$$p_{22} = X_{O^o}^2 . X_{O^-}^2 \times \frac{4!}{2!\,2!} = 6X_{O^o}^2 . X_{O^-}^2$$
$$p_{31} = X_{O^o} . X_{O^-}^3 \times \frac{4!}{1!\,3!} = 4X_{O^o} . X_{O^-}^3$$
$$p_{40} = X_{O^-}^4 \times 1 = X_{O^-}^4$$

Table 1. Calculation of relative proportions of different structural groups in silicate melts. (after Fraser, 1975a).

The variation of polymerization constants with temperature is known for the system 'FeO'- SiO_2 (Distin et al., 1971; Masson, 1972). The proportions of different structons can thus be calculated at different temperatures using the combinatorial formulae in Table 1. Activities of fayalite in the melt can therefore be calculated from the expression:

$$a_{Fe_2SiO_4}^{melt} = \frac{X_{Fe}^2 . X_{40(melt)}}{X_{Fe}^2 . X_{40}(\text{pure } Fe_2SiO_4 \text{ melt})}$$

$$= \frac{X_{40(melt)}}{X_{40(pure)}} \qquad (44)$$

This calculated activity may be compared with activities obtained from the depression of freezing-point equations (see Wood and Fraser, 1976):

$$\int_1^{a_i} \partial \ln a_i = \int_{T_o}^{T} \frac{\Delta H^o}{RT^2} . dT \qquad (47)$$

where ΔH^o is the heat of fusion at temperature, T.

Using the data in Kubaschewski et al. (1967) for Fe_2SiO_4 we have

	Tm(K)	Teutectic(K)	ΔH^o_{fus}(kcal)	ΔCp(cal.mol.K^{-1}
Fe_2SiO_4	1479.5	1455	22.0	21.0

From these data the activity of Fe_2SiO_4 in the melt at, for example, the eutectic, is

$$a^{melt}_{Fe_2SiO_4} = 0.882.$$

This can be compared with the value calculated from equation (44). Taking the data of Distin et al. (1971) and Masson (1972) a value of the polymerization constant for FeO-SiO_2 melts can be obtained at 1455°K from a regression of lnK versus 1/T. From this, K = 0.574. Using this value the activity of Fe_2SiO_4 in the melt calculated from equation (44) is

$$a^{melt}_{Fe_2SiO_4(calc.)} = 0.893.$$

This agrees very well with the empirical value.

There are insufficient data available at present to allow more extensive tests of the model on other systems. However by combining this model with the acid-base scale described above, the shifts in liquidus boundaries noted by Kushiro (1973, 1975) can be interpreted.

The effects of different oxides on the forsterite-enstatite cotectic are shown in Fig. 2. For the pure mineral phases, the cotectic is defined by

$$K_{s.p.(Fo)} = a^{melt}_{Mg_2SiO_4} = X^2_{Mg^{2+}} \cdot X_{40_{Si}} \quad (46)$$

and

$$K_{s.p.(En)} = a^{melt}_{MgSiO_3} = X_{Mg^{2+}} \cdot X_{22_{Si}} \quad (47)$$

Consider the effects of adding a basic oxide (e.g. K_2O). The added oxide will disturb the polymerization equilibrium according to:

$$O^{2-} + O^{o} \rightleftarrows 2O^{-} \quad (48)$$

This will increase X_{40} $(=X^4_{O^-})$ relative to $X_{22}(= 6X^2_{O^o}.X^2_{O^-})$ and so the activity of Mg_2SiO_4 in the melt will increase and the activity of $MgSiO_3$ will decrease. This will cause enstatite to dissolve and forsterite to crystallize. To maintain both minerals in equilibrium with the melt, the cotectic must move to more SiO_2-rich compositions or otherwise add O^o to the system.

Thus the addition of basic oxides like K_2O, Na_2O, H_2O will cause the forsterite-enstatic cotectic to move to more silica-rich compositions. The addition of acidic oxides will have the opposite effect as is observed for P_2O_5 and TiO_2 (Kushiro) and also for CO_2 (Eggler, 1974).

Thus, by combining what is already known of the nearly ideal mixing properties of silicate structural units in the melt (Masson 1972) with the requirement that crystallizing mineral phases can only accommodate certain of those present in the polymers, it may be possible to construct tables of approximate solubility products for silicate minerals in melts similar to those for salts in aqueous solution.

REFERENCES

Anfilogov, V.N. and Kadik, A.A., Geokhimiya, 9, 1396 (1973).

Bockris, J.O'M., Kitchener, J.A., Ignatowicz, S. and Tomlinson, J.W., Trans. Faraday Soc., 48, 75 (1952a).

Bockris, J.O'M., Kitchener, J.A. and Davies, A.E., Trans. Faraday Soc., 48, 536 (1952b).

Bockris, J.O'M. and Mellors, G.W. J. Phys. Chem., 60, 1321 (1956).

Bues, W. and Gehrke, H.-W., Z. anorg. Chem., 288, 292 (1956).

Burnham, C.W., Geochim. Cosmochim. Acta, 39, 1077 (1975).

Burnham, C.W. and Davies, N.F., Amer. J. Sci., 270, 54 (1971).

Burnham, C.W. and Davies, N.F., Amer. J. Sci., 274, 902 (1974).

Cox, K.G. and Jamieson, B.G., J. Petrology, 15, 269 (1974).

Cripps-Clark, C.J., Sridhar, R., Jeffes, J.H.E. and Richardson, F.D. in J.H.E. Jeffes and R.J. Tait (Eds.), "Physical chemistry of process metallurgy", Inst. Mining. Met., 1974.

Crowther, J. Anal. Chem., 26, 1383 (1954).

Distin, P.A., Whiteway, S.G. and Masson, C.R. Canad. Metall. Omart., 10, 73 (1971).

Ebel, J.P. Bull. Soc. Chim. Fr., Mem. 20, 1089 (1953).

Eggler, D.H. Carnegie Inst. Wash. Yearbook, 73, 215 (1974).

Fincham, C.J.B. and Richardson, F.D. Proc. Roy. Soc. 223A, 40 (1954).

Flood, H. and Förland, T. Acta Chem. Scand., 1, 592 (1947).

Flory, P.J. J. Amer. Chem. Soc., 58, 1877 (1936).

Flory, P.J. "Principles of polymer chemistry", Cornell University Press (1953).

Fraser, D.G. Faraday Symposium Chem. Soc. No.8, 64 (1974).

Fraser, D.G. D.Phil. Thesis, University of Oxford (1975a).

Fraser, D.G. Geochim. Cosmochim. Acta, 39, 1525 (1975b).

Fudali, R.F. Geochim. Cosmochim. Acta, 29, 1063 (1965).

Götz, J. and Masson, C.R. J. Chem. Soc. A, 2683 (1970).

Guggenheim, E.A. Mixtures, Clarendon Press, Oxford (1952).

Hess, P.C. Geochim. Cosmochim. Acta, 35, 289 (1971).

Hilmer, W. and Dornberger-Schiff, K. Acta Cryst., 9, 87 (1956).

Huggins, M.L. J. Phys. Chem., 58, 1141 (1954).
Jost, K.H. Acta Cryst., 16, 428 (1963).
Kleber, W. and Winkhans, B. Fortschr. Mineral., 28, 175 (1949).
Kubaschewski, O., Evans, E.L. and Alcock, C.B. Metallurgical Thermochemistry, Pergamon, Oxford (1967).
Kushiro, I. Amer. J. Sci., 267A, 269 (1969).
Kushiro, I. J. Petrology, 13, 311 (1972).
Kushiro, I. Carnegie Inst. Wash. Yearbook, 72, 497 (1973).
Kushiro, I. Amer. J. Sci., 275, 411 (1975).
Lentz, C.W. Inorg. Chem., 3, 574 (1964).
Lux, H. Z. Elektrochemie, 45, 303 (1939).
Masson, C.R. Proc. Roy. Soc., 287A, 201 (1965).
Masson, C.R. J. Iron Steel Inst., 192, 89 (1972).
Meadowcroft, T.R. and Richardson, F.D. Trans. Faraday Soc., 61, 54 (1965).
Morris, R.V. and Haskin, L.A. Geochim. Cosmochim. Acta, 38, 1435 (1974a).
Morris, R.V., Haskin, L.A., Biggar, G.M. and O'Hara, M.J. Geochim. Cosmochim. Acta, 38, 1447 (1974b).
Mysen, B.O., Eggler, D.H., Seitz, M.G. and Holloway, J.R. Amer. J. Sci., 276, 455 (1976).
Nath, P. and Douglas, R.W. Phys. Chem. Glasses, 6, 197 (1965).
O'Hara, M.J. Earth Sci. Rev., 4, 69 (1968).
Paul, A. and Douglas, R.W. Phys. Chem. Glasses, 6, 207 (1965a).
Paul, A. and Douglas, R.W. Phys. Chem. Glasses, 6, 212 (1965b).
Paul, A. and Lahiri, D. J. Amer. Ceram. Soc., 49, 565 (1966).
Richardson, F.D. Trans. Faraday Soc., 52, 1312 (1956).
Richardson, F.D. and Webb, L.E. Trans. I.M.M., 64, 529 (1955).
Sato, M.
Schulz, E. and Liebau, F. Naturwiss., 60, 429 (1973).
Toop, G.W. and Samis, C.S. Trans. Met. Soc. A.I.M.E., 224, 878 (1962a).
Toop, G.W. and Samis, C.S. Canad. Met. Quartz., 1, 129 (1962b).
Tuttle, O.F. and Bowen, N.L. Geol. Soc. Amer., Mem. No. 74 (1958).
Van Wazer, J.R. "Phosphorus and its compounds", Interscience, New York (1958).
Van Wazer, J.R. and Holst, K.A. J. Amer. Chem. Soc., 72, 639 (1950).
Waffe, H.S. and Weill, D.F. Earth Planet. Sci. Letters, 28, 254 (1975).
Westman, A.E.R. and Gartaganis, P.A. J. Amer. Ceram. Soc., 40, 293 (1957).
Wood, B.J. and Fraser, D.G. "Elementary thermodynamics for geologists", Oxford University Press, Oxford (1976).

THE ACTIVITIES OF COMPONENTS IN NATURAL SILICATE MELTS

J. Nicholls

Department of Geology, University of Calgary, Calgary
Canada T2N 1N4

INTRODUCTION

In a series of papers, Carmichael and his co-workers have applied the thermodynamics of heterogeneous equilibria to igneous processes for two ends. First, thermodynamics can be used to describe the activity or effective concentration of components in silicate melts. This in turn, provides a quantitative basis for characterizing or classifying igneous rocks. This topic has been discussed in detail by Carmichael et al. (1970, 1974). Second, the thermodynamics of heterogeneous equilibria can be used to estimate pressures and temperatures of equilibration between melt and solid phases (Nicholls et al., 1971; Nicholls and Carmichael, 1972; Bacon and Carmichael, 1973; and Carmichael et al., in press).

This paper will review these two topics. The first topic is easily understood by petrologists as it can be interpreted in terms of existing classifications such as the CIPW normative classification or the classification based on silica saturation developed by Shand (1943). The second topic requires a discussion of the sources of error in the estimates of temperature and pressure. In general there are three sources. First, errors due to an approximation to the complete thermodynamic theory. Second there are possible errors due to inadequacy in the basic thermodynamic data such as heats of formation, volumes, entropies, and heat capacities. The third sources of error are the analytical uncertainties of the constituent phases in the rock. As most igneous minerals are zoned, the problem is to determine which zone of mineral A crystallized contemporaneously with mineral B. The uncertainty in the estimates of T and P due to the first

D. G. Fraser (ed.), Thermodynamics in Geology, 327-348.

two sources of error has been discussed by Carmichael et al. (in press) and only the results of the discussion will be quoted here. Maximum uncertainties due to the third source of error can be determined if the compositional range of the constituent minerals is ascertained. This problem will be discussed in some detail.

THERMODYNAMIC CHARACTERIZATION OF SILICATE MELTS

The activity or effective concentration of SiO_2 in a melt which is in equilibrium with olivine and lime-poor pyroxene can be defined in terms of the reaction:

$$\underset{\text{olivine}}{Mg_2SiO_4} + \underset{\text{melt}}{SiO_2} = \underset{\text{pyroxene}}{Mg_2Si_2O_6} \tag{1}$$

The thermodynamic equation for the Gibbs energy change of this reaction is:

$$\Delta G/RT = \Delta G°/RT - \ln a^{melt}_{SiO_2} - \ln a^{olivine}_{Mg_2SiO_4} + \ln a^{pyroxene}_{Mg_2Si_2O_6} \tag{E1}$$

where a^j_i is the activity of component i in phase j, $\Delta G°$ is the Gibbs energy of reaction with both reactants and products in their standard states. T is the temperature on the Kelvin scale and R is the gas constant. At equilibrium $\Delta G/RT$ is zero and equation (E1) becomes:

$$\ln a^{melt}_{SiO_2} = \Delta G°/RT + \ln a^{pyroxene}_{Mg_2Si_2O_6} - \ln a^{olivine}_{Mg_2SiO_4} \tag{E2}$$

The standard states which will be used in this paper are the pure stable solids and pure liquids where possible. The one exception will be SiO_2. The standard will be silica glass for the component in the melt as data for silica liquid are not available. The solid SiO_2 standard state will be β-quartz as this is the most common polymorph of silica found as a primary phase in igneous rocks.

Equation (E2) can be used to determine the activity of silica in a melt if both olivine and pyroxene coexist in equilibrium with the same melt. The data required to make the calculations are temperature, pressure, compositions of the solid solutions and an expression relating the activities of the solid phases to their compositions. For the moment, the progressive substitution of Fe for Mg in both olivine and pyroxene will be ignored and silicate melts will be characterized in terms of silica activity defined by the pure end members (i.e., at unit activity). During the crystallization of the groundmass of volcanic rocks, the pressure should approach one bar. With these approximations, equation (E2) becomes an expression relating the two variables temperature and silica activity. The variation of the activity of silica with temperature according to equation (E2) is shown in Figure 1.

Obviously there are several mineral associations which can be used to define silica activity. Several reactions which correspond to some associations of importance in igneous rocks are listed in Table 1. Among these associations is the single mineral quartz which is characteristic of the oversaturated rocks in Shand's classification. It is noteworthy that because quartz is nearly pure SiO_2 melts in equilibrium with this mineral will have silica activities that, for all practical purposes, lie exactly on the curve labeled Q in Figure 1. Reaction (1) characterizes the tholeiite basalts (Carmichael et al., 1974; Tilley, 1950) As rocks belonging to the tholeiite suite generally contain normative quartz and, in the more siliceous varieties, modal quartz as well, melts corresponding to lavas in this suite should have silica activities between curves 1 and 2 (Figure 1).

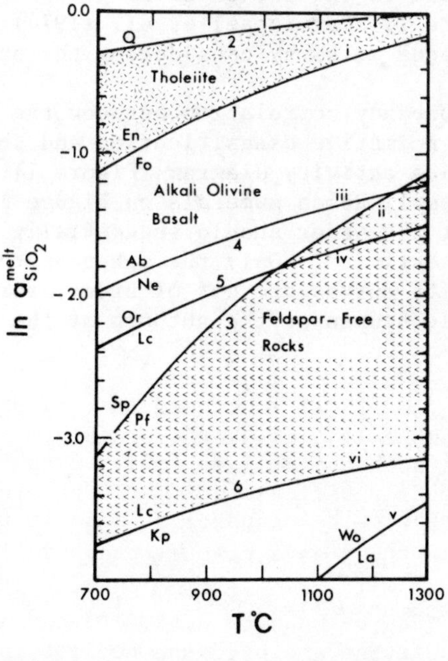

Figure 1: Variation of silica activity with temperature for several mineral assemblages. Arabic numerals correspond to the reactions listed in Table 1. Roman numerals correspond to the order of desilication steps in the CIPW normative classification. Curves calculated from data tabulated in Nicholls et al. (1971) and Carmichael et al. (in press).

The perovskite-sphene curve (labeled 3, Figure 1) is of relevance because perovskite is rarely, if ever, found with feldspar in terrestrial rocks. In other words, curve 3 approximates the lower limit to silica activity in feldspar-bearing igneous melts. Melts with silica activities between curves 1 and 3 will correspond to the alkali olivine basalts, basánites, trachytes, phonolites and other feldspar-bearing moderately silica undersaturated rocks. Below curve 3 will fall melts with silica activities corresponding to the feldspar-free igneous rocks which are the bane of petrography students. The lower limit to silica activity in igneous rocks will probably approximate closely the kaliophillite-leucite curve (number 6, Figure 1). Only for silicate melts with mineral assemblages containing kaliophillite (or kalsilite) but lacking leucite could one expect that the silica activity was less than that defined by curve 6.

To summarize, silica activity can be used to characterize four large groups of igneous silicate rocks: Quartz-bearing varieties, the tholeiite suite, alkali olivine basalts and their associates, and the feldspar-free rocks. The fields of silica activity appropriate to the latter three groups of rocks are indicated on Figure 1. Carmichael et al. (1974) have further characterized igneous melts by considering the activity of alumina.

There is an uncanny correlation between the desilication steps of the CIPW normative classification and the order of the curves on the silica activity diagram (Figure 1). The CIPW steps are labeled with small Roman numerals on Figure 1. If the correlation were perfect the order should sequentially increase with decreasing silica activity. Only the sphene-perovskite and the wollastonite-larnite curves are out of order, surely a tribute to the petrographic acumen of the authors of the CIPW classification.

Table 1: Some reactions which can define the activities of components in silicate melts.

No.	Reaction	Component
1	$Mg_2SiO_4 + SiO_2 = Mg_2Si_2O_6$	SiO_2
2	$SiO_2 = SiO_2$ melt β-quartz	SiO_2
3	$CaTiO_3 + SiO_2 = CaTiSiO_5$	SiO_2
4	$1/2\ NaAlSiO_4 + SiO_2 = 1/2\ NaAlSi_3O_8$	SiO_2
5	$KAlSi_2O_6 + SiO_2 = KAlSi_3O_8$	SiO_2
6	$KAlSiO_4 + SiO_2 = KAlSi_2O_6$	SiO_2
7	$Ca_2SiO_4 + SiO_2 = 2CaSiO_3$	SiO_2
8	$Fe_2SiO_4 = Fe_2SiO_4$ melt olivine	Fe_2SiO_4
9	$CaMgSi_2O_6 = CaMgSi_2O_6$ melt pyroxene	$CaMgSi_2O_6$
10	$NaAlSi_3O_8 = NaAlSi_3O_8$ melt feldspar	$NaAlSi_3O_8$
11	$KAlSi_3O_8 = KAlSi_3O_8$ melt feldspar	$KAlSi_3O_8$
12	$CaMgSi_2O_6 + Al_2O_3 + 1/2\ SiO_2 = CaAl_2Si_2O_8$ melt melt $+ 1/2\ Mg_2SiO_4$	Al_2O_3, SiO_2
13	$Mg_2SiO_4 + Al_2O_3 = 1/2\ Mg_2Si_2O_6 + MgAl_2O_4$	Al_2O_3
14	$3CaMgSi_2O_6 + Al_2O_3 = Ca_3Al_2Si_3O_{12}$ $+ 3/2\ Mg_2Si_2O_6$	Al_2O_3
15	$NaAlSi_3O_8 + Mg_2SiO_4 = Mg_2Si_2O_6 + NaAlSi_2O_6$ melt	$NaAlSi_3O_8$
16	$2/3\ Fe_3O_4 + SiO_2 = Fe_2SiO_4 + 1/3\ O_2$	SiO_2
17	$NaAlSi_3O_8 = SiO_2 + NaAlSi_2O_6$ melt melt	$NaAlSi_3O_8$, SiO_2

ESTIMATION OF PRESSURES AND TEMPERATURES OF EQUILIBRATION

The estimation of equilibration conditions is a straightforward application of the thermodynamics of heterogeneous equilibria. An equation such as (E2) has as variables temperature, pressure, composition of the solid phases, and composition of the melt. In theory then, if the composition of all the phases were known for two such independent equations, they could be solved simultaneously for unique values of pressure and temperature. Conditions of equilibration between melt and solids could be calculated for several cases. Examples include equilibration between melt and phenocrysts, melt and megacrysts, and melt and residual minerals in the source region. The important requirement is the prior knowledge of the compositions of the melt and solids. Many of these compositions can be estimated from chemical analyses of the rock and the constituent minerals. An exception could be the composition of the residual minerals in the source region. Quite often however, one may wish to calculate possible equilibration conditions to help test hypotheses concerning the source region. In these cases, mineral compositions may be estimated from xenoliths brought to the surface by lavas or possibly experimental data on natural minerals which constitute plausible source materials.

The practical difficulties with this approach are two. First, there is a general lack of quality data needed to express the several functions of $\Delta G°/RT$ for the reactions as explicit functions of pressure and temperature. This is especially true for compounds which can be used as components in the melt. A set of consistent data for several convenient reactions has been compiled by Carmichael et al. (in press). For conditions at one bar these data have been reduced to the form:

$$\Delta G°/RT = A(\ln T - 1) + BT + CT^{-1} + DT^{-2} + E \qquad (E3)$$

where A, B, C,....., are constants. These reduced data are given in Table 2. For original sources of data the reader is referred to the paper cited above (Carmichael et al., in press). In order to extend these data to higher pressures the following expression can be used:

$$\Delta G°(P,T) = \Delta G°(1\text{ bar}, T) + \int_1^P \Delta v_s^o \, dP - \int_1^P v_m^o \, dP \qquad (E4)$$

where the left hand side of the equation is the value of $\Delta G°$ at a pressure P and temperature T. The first term on the left hand side is the value of $\Delta G°$ at a pressure of 1 bar and the same temperature T (Table 2). The term under the first integral sign is the volume change of the solid components of the reaction in their standard state (sum of products minus sum of reactants).

The term under the second integral sign is the volume of the melt component in its standard state. As written, equation (E4) assumes that the components in the melt are treated as reactants (e.g. reaction 1). The data needed to evaluate the first integral for the free energy functions listed in Table 2 are given in Table 3. As will be seen later, for our purposes here, the second integral need not be evaluated. The second practical difficulty lies in finding suitable expressions for the activities of the components in the solid solutions and the melt. These topics are considered next.

Table 2: Coefficients for the expressions $G°/RT = A(\ln T - 1) + BT + CT^{-1} + DT^{-2} + E$ corresponding to the reactions given in Table 1.

No.	A	$Bx10^3$	C	$Dx10^{-5}$	E
1	0.045	0.186	-2198.2	0.148	0.6433
2	0.00	0.00	- 715.4	0.00	0.439
8	10.462	-2.335	- 952.4	1.690	- 61.9063
9	14.920	0.662	10562.0	3.960	-103.3363
10	12.072	-3.497	1301.1	3.777	- 71.5698
11	9.928	-3.246	- 35.2	4.290	- 57.8935
12	8.834	-1.834	339.2	0.927	- 59.5084
13	7.438	-1.157	-3025.2	1.897	- 47.2212
14	16.737	-5.724	3127.5	-2.790	- 97.8648
15	12.197	-3.613	-1570.1	4.227	- 65.3606
16	0.00	0.00	19042.0	0.00	- 6.47
17	12.152	-3.799	628.1	4.129	- 66.0039

The activity-composition relations of solid solutions

The state of the art of expressing the activities of components of mineral solid solutions is in flux (cf. Kerrick and Darken, 1975; Wood and Strens, 1971, Wood and Banno, 1973; Carmichael et al., in press). Consequently, the relationships suggested here are used without apology but with the realization that alterations, perhaps extensive, may be required in light of future work. The suggested expressions for the activities of solid components are given in Table 4.

Activity-composition relations in silicate melts

The data which are required, but which are lacking, for a rigorous formulation of the activities of components in natural silicate melts are enthalpies and partial molar enthalpies. Because of this deficiency, it has been assumed that to a first approximation silicate liquids can be adequately treated as symmetric, regular solutions (Thompson, 1967). This model has

Table 3: Volumes of the solid components in their standard state expressed in the form $v° = (aT + b)(1 - \beta P)$. Units are cal/bar.

Solid	$a \times 10^6$	b	$\beta \times 10^6$
Mg_2SiO_4	35.78	1.0349	0.79
Fe_2SiO_4	31.43	1.0996	0.91
$Mg_2Si_2O_6$	45.08	1.4888	1.01
$Fe_2Si_2O_6$	40.73	1.5655	1.13
$CaMgSi_2O_6$	46.27	1.5644	1.07
$NaAlSi_2O_6$	38.75	1.4314	0.75
$CaAl_2Si_2O_8$	34.34	2.3991	1.0
$NaAlSi_3O_8$	67.66	2.3777	1.48
$KAlSi_3O_8$	70.94	2.5795	1.76
$Ca_3Al_2Si_3O_{12}$	70.46	2.9725	0.60
$MgAl_2O_4$	26.04	0.9408	0.41
$SiO_2(\beta\text{-q})$	-3.82	0.5703	1.776

adequately accounted for the departures from ideality in several systems with a limited number of components (Lumsden, 1961; Somerville et al., 1973; Nicholls and Carmichael, 1972). Following Kudo and Weill (1970) who treated activity coefficients as functions of composition alone, Nicholls and Carmichael (1972) used the following expression for the activity of component i in a silicate melt at a pressure of one bar:

$$\ln a_i = \ln x_i + \phi_i/T \qquad \text{(E5)}$$

where ϕ_i is function of composition whose value is to be determined, but is independent of pressure and temperature. It can be shown that this formulation is consistentwith the regular solution model (Carmichael et al., in press) but it is important to note that equation (E5) is more general than the analogous equation for strictly regular, symmetric solutions. All that is required is that ϕ be a function of composition alone. At constant pressure, the only other variable that may affect the activity is temperature. Obviously, over a sufficiently small temperature range equation (E5) must be a reasonable approximation for a melt of constant composition. It is worth noting here that an expression for the activity which satisfies equation (E5) has as a consequence that there be no excess heat capacity (Carmichael et al., in press) or, in other words, that the heat capacity of a complex melt be a linear function of the mole fractions of the

Table 4: Expressions for the activities of components in solid solutions.

Phase	Component	Activity
Olivine	Mg_2SiO_4	$x^2_{Mg_2SiO_4}$
	Fe_2SiO_4	$x^2_{Fe_2SiO_4}$
Pyroxene	$CaMgSi_2O_6$	MgCa(Fe+Mg+Ca+Na-1)/(Fe+Mg)
	$Mg_2Si_2O_6$	Mg^2 (Fe+Mg+Ca+Na-1)/(Fe+Mg)^2
		where Ca, Mg, etc. are the number of Ca, Mg, etc. cations in the standard pyroxene formula based on 6 oxygens (modified from Nicholls, in press).
Plagioclase	$CaAl_2Si_2O_8$	$x_{CaAl_2Si_2O_8}$
	$NaAlSi_3O_8$	$x_{NaAlSi_3O_8}$
Magnetite	Fe_3O_4	$1 - x_{usp}$ (Carmichael, 1967)

For additional and alternative expressions, the reader is referred to Waldbaum and Thompson (1969) and Carmichael _et al_., (in press).

components. Over a large range in composition, with the oxides as components, this has been established to within experimental error ($\pm$0.3%)(Carmichael _et al_., in press).

The pressure dependence of the activity is related to the partial molar volume and standard state volume by:

$$(\partial \ln a_i/\partial P)_{T,x_i} = (\bar{v}_i - v_i^o)/RT \qquad \text{(E6)}$$

where $\bar{v}_i$ is the partial molar volume of component i and v_i^o is the volume of component i in its standard state. Integration of (E6) between the limits of one bar and pressure P gives upon substitution of (E5):

$$\ln a_i = \ln x_i + \phi_i/T + \left(\int_1^P \bar{v}_i \, dP - \int_1^P v_i^o \, dP\right)/RT \qquad \text{(E7)}$$

A knowledge of a_i at one known pressure and temperature will suffice for the evaluation of ϕ_i if the liquid remains constant in composition. This evaluation is most conveniently done with extrusive rocks wherein the groundmass crystallized at pressures of approximately one bar. Quite often the temperature (Tq) of the crystallization of the groundmass can be estimated from the compositions of the iron-titanium oxides (Buddington and Lindsley, 1963). Substitution of Tq and the requisite mineral compositions into equations similar to (E2) but for the appropriate reaction in Table 1 provides an estimate of a_i. Whence:

$$\phi_i = Tq(\ln a_i - \ln x_i) \qquad \text{(E8)}$$

There remains only the problem of evaluating x_i for the components in the melt. From the point of view of the analytical chemist, the most convenient set of components are the oxides. In this case the x_i are the mole fractions of the constituent oxides, SiO_2, etc. For multioxide compounds such as Fe_2SiO_4, the calculation of x_i is, perhaps, not so straight forward. An example will serve to illustrate the method. An olivine in equilibrium with a melt can be represented, in part, by the reaction:

$$\underset{\text{melt}}{2FeO} + \underset{\text{melt}}{SiO_2} = \underset{\text{olivine}}{Fe_2SiO_4}$$

This equilibria can be written in terms of the chemical potentials as:

$$\mu^{\text{olivine}}_{Fe_2SiO_4} = 2\mu^{\text{melt}}_{FeO} + \mu^{\text{melt}}_{SiO_2} \qquad \text{(E9)}$$

Subsitution of the appropriate expressions for the activities of the components in the melt (equation E5), assuming a pressure of one bar gives:

$$\mu^{\text{olivine}}_{Fe_2SiO_4} = (2\mu^{o}_{FeO} + \mu^{o}_{SiO_2}) + RT\ln(x^2_{FeO}x_{SiO_2})$$

$$+ 2R\phi_{FeO} + R\phi_{SiO_2} \qquad \text{(E10)}$$

where μ^o_i are the liquid and glass standard state chemical potentials of FeO and SiO_2 respectively. Consider the case of x_{FeO} equal to 2/3 and x_{SiO_2} equal to 1/3; that is the case of the melt being pure Fe_2SiO_4. Obviously, the only olivine which can be in equilibrium with such a melt is pure fayalite. Along the fusion curve the standard state chemical potential (Gibbs energy) of Fe_2SiO_4 liquid ($\mu^o_{Fe_2SiO_4}$) can be defined in terms of the standard state chemical potential of pure crystalline fayalite (μ^o_{Fa}) (Carmichael <u>et al</u>., in press). Substituting into equation (E10) gives:

$$\mu^o_{Fa} = \mu^o_{Fe_2SiO_4} = (2\mu^o_{FeO} + \mu^o_{SiO_2}) + RT\ \ln(4/9 \cdot 1/3) + 2R\phi^o_{FeO} + R\phi^o_{SiO_2} \quad (E11)$$

where ϕ^o_{FeO} and $\phi^o_{SiO_2}$ are the values of ϕ appropriate to pure Fe_2SiO_4 liquid. Rearrangement gives:

$$(2\mu^o_{FeO} + \mu^o_{SiO_2}) = \mu^o_{Fe_2SiO_4} - (2R\phi^o_{FeO} + R\phi^o_{SiO_2}) - RT\ \ln(4/27) \quad (E12)$$

Substitution of (E12) into (E10) gives:

$$\mu^{olivine}_{Fe_2SiO_4} = \mu^o_{Fe_2SiO_4} + RT\ \ln(27/4\ x^2_{FeO}\ x_{SiO_2}) + R\phi_{Fe_2SiO_4} \quad (E13)$$

where 27/4 is a stoichiometric factor needed to adjust the oxide component mole fractions to a multioxide basis. The last term results from setting:

$$\phi_{Fe_2SiO_4} = 2(\phi_{FeO} - \phi^o_{FeO}) + (\phi_{SiO_2} - \phi^o_{SiO_2}) \quad (E14)$$

which simply measures the difference in the ϕ between the multicomponent melt and pure Fe_2SiO_4; again a constant if the melt remains constant in composition. It is perhaps worth noting that from equation (E13) one obtains:

$$\ln a^{melt}_{Fe_2SiO_4} = \ln(27/4\ x^2_{FeO}\ x_{SiO_2}) + \phi_{Fe_2SiO_4}/T \quad (E15)$$

a formulation exactly parallel to (E5). A similar exercise will reveal the stoichiometric factor for the other multioxide compounds. The ideal contributions to the activities (the terms under the logarithm sign) for the multioxide compounds considered in this paper are set down in Table 5.

Table 5: Ideal constributions to the activities of multioxide compounds in silicate melts.

Compound	Ideal Contribution
Fe_2SiO_4	$27/4\ x^2_{FeO}\ x_{SiO_2}$
$CaMgSi_2O_6$	$64\ x_{CaO}\ x_{MgO}\ x^2_{SiO_2}$
$NaAlSi_3O_8$	$512/27\ x^{1/2}_{Na_2O}\ x^{1/2}_{Al_2O_3}\ x^3_{SiO_2}$
$KAlSi_3O_8$	$512/27\ x^{1/2}_{K_2O}\ x^{1/2}_{Al_2O_3}\ x^3_{SiO_2}$

Substitution of the equation for the activity of silica (of the form E7) into equation (E2), taking into account (E4), gives:

$$\ln x^{melt}_{SiO_2} + \phi_{SiO_2}/T + \left(\int_1^P \bar{v}_{SiO_2}\, dP - \int_1^P v^o_{SiO_2}\, dP\right)/RT$$

$$= G^o/RT + \left(\int_1^P v^o_s\, dP - \int_1^P v^o_{SiO_2}\, dP\right)/RT \qquad \text{(E16)}$$

$$+ \ln a^{pyroxene}_{Mg_2Si_2O_6} - \ln a^{olivine}_{Mg_2SiO_4}$$

For this equation, one should note that the terms involving $v^o_{SiO_2}$ cancel. The partial molar volume data needed for the evaulation of the remaining integral terms are given in Table 6.

Equation (E16) is an example of a general expression involving equilibria between components in the melt and components in the minerals of interest. The only two unknown variables remaining, given the compositions of the melt, solids, and a value of ϕ_i, are the temperature T and pressure, P. This equation is most easily solved for P as a function of T by computer. A FORTRAN program written for the CDC 6400 computer is given in the appendix. This program incorporates the data in Tables 2, 3, and 6.

Table 6: Partial molar volumes of components in silicate melts, given in the form $\bar{v} = aT + b - cP$. units are cal/bar. Data compiled from Bottinga and Weill, (1970), Carmichael et al. (in press).

Component	$a \times 10^6$	b	$c \times 10^6$
Fe_2SiO_4	101.75	1.0822	1.94
$CaMgSi_2O_6$	148.65	1.7035	7.26
$NaAlSi_3O_8$	112.79	2.5322	7.49
$KAlSi_3O_8$	195.39	2.5958	8.86
SiO_2	6.15	0.6304	2.66
Al_2O_3	24.37	0.8664	0.47

EXAMPLE AND DISCUSSION

Carmichael et al. (in press) have compared the estimated pressures and temperatures of olivine and augite crystallization with the experimentally determined conditions for very basic melts studied by Bultitude and Green (1971). The average difference between observed and calculated conditions was 5.3 kilobars in the pressure and 38° in the temperature for a total of 32 data points. In this comparison the possibility of any error in the experimental data was assumed to be nil. The values quoted above are thought typical of the possible error in the theoretical approximations (e.g. the activity expressions) and the basic thermodynamic data (Tables 2, 3, and 6). The numbers quoted above will be termed a measure of the accuracy of the method. Given the theoretical formulations and the basic thermodynamic data, any further error will be introduced through the possible errors in the mineral and melt compositions. The measure of this last uncertainty will be termed the precision. This quantity is best discussed by means of example, the details of which are given in Table 7.

Table 7: Example problem.

A basanite lava from a cinder cone in the Itcha Mountains of British Columbia, Canada contains small phenocrysts of olivine in a groundmass of olivine, augite, plagioclase, alkali feldspar, nepheline, leucite and Fe-Ti oxides. Analytical data on the rock and enclosed minerals are given below. Combining these data with the equations in the Tables and text, estimate the pressures and temperatures at which a melt of the composition of the rock could be in equilibrium with source materials having the following thermodynamic characteristics:

Component	$Mg_2Si_2O_6$	$CaMgSi_2O_6$	Mg_2SiO_4	Fe_2SiO_4	$MgAl_2O_4$	
Activity	0.5601	0.4129	0.7435	0.0190	0.7107	(a)
	0.7170	0.4493	0.8036	0.0107	0.6116	(b)

Component	$NaAlSi_2O_6$	$Ca_3Al_2Si_3O_{12}$	
	0.1016	0.0038	(a)
	0.0563	0.0038	(b)

Lava and enclosed mineral compositions:

Olivine Phenocrysts; Average Fo 70.1, range 71.3-69.4
Groundmass; Average Fo 69.4, range 70.0-68.4

Plagioclase Average An 50, range 65-40

Pyroxene formula		Lava Composition	Wt.%	Mole %
Si	1.788	SiO_2	45.18	48.66
Al^{iv}	0.215	TiO_2	2.99	2.42
Al^{vi}	0.031	Al_2O_3	13.88	8.81
Ti	0.088	Fe_2O_3	3.89	1.58
Fe	0.271	FeO	8.77	7.90
Mn	0.005	MnO	0.16	0.15
Mg	0.710	MgO	9.48	15.22
Ca	0.873	CaO	8.93	10.30
Na	0.050	Na_2O	3.79	3.96
		K_2O	1.48	1.02

Fe-Ti oxide temperature 965°C
$\ln f_{O_2} = -26.94$ Magnetite ss Mt 39.86

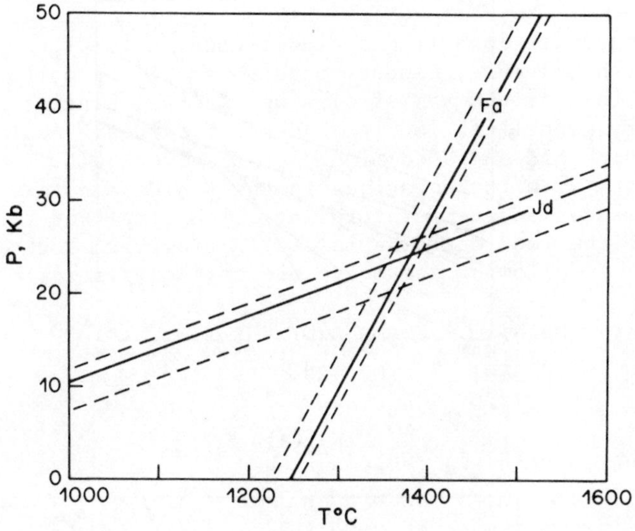

Figure 2: Equilibration curves for reactions (8) and (17), Table 1, using data in Table 7. Solid lines correspond to average mineral compositions in the lava. The dashed line correspond to maximum and minimum fayalite and albite contents in the respective minerals in the lava.

In Figure 2 are shown the equilibration curves for reactions (8) and (17), Table 1. The curves were calculated using the program listed in the appendix. The solid lines correspond to the maximum and minimum fayalite and albite contents of the minerals in the lava, including groundmass. The source material was that designated (b) in Table 7. The approximate parallellogram outlined by the dashed lines marks the maximum uncertainty in the precision of the estimates; approximately 7 kilobars and 65 degrees. In other words, the uncertainty in the precision is slightly larger than the accuracy of the method.

The results of solving seven of the equations for P in terms of T are shown in Figure 3 for two different source materials. The first (Figure 3a) is for bulk composition more iron-rich than is usual in lherzolite xenoliths found in basanites (Bacon and Carmichael, 1973; Fiesinger and Nicholls, 1976). The second is for bulk composition which is similar to lherzolite xenoliths. This difference is apparent in olivine and enstatite activities (Table 7). As is appreciated from Figure 3, this makes for a considerable difference in the intersection of the curves. Presumably one would conclude that the more iron-rich material is the more plausible residua of the source region.

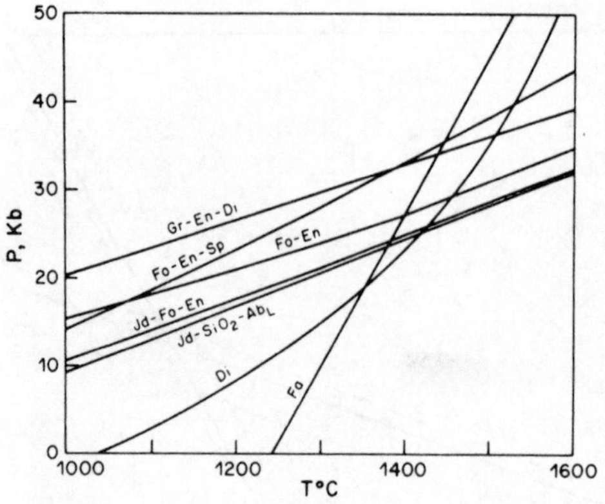

(a)

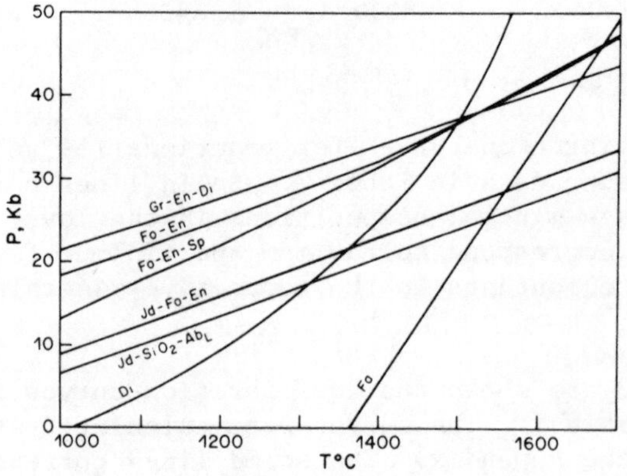

(b)

Figure 3: Calculated equilibration curves between a basanite melt and two possible source materials. See Table 7 and text for details.

ACKNOWLEDGEMENTS

Material support was provided by NRC operating grant A7372 and by the Department of Geology and Geophysics, University of California Berkeley for computer costs. This paper obviously owes a debt to Professor I. Carmichael which is beyond words and just as obviously lacks his clarity of presentation.

APPENDIX: Program for solving equations for P and T similar to equation (E16) in text.

```
      PROGRAM MAGMA(INPUT,OUTPUT,TAPE5=INPUT,TAPE6=OUTPUT)
      DIMENSION NAME(7),ABAR(10),BBAR(10),CBAR(10),A(10),B(10),C(10)
      DIMENSION D(10),E(10),AS(10),BS(10),AB(10),BB(10),T(10),
     +P(10,10)
      DIMENSION X(10),XS(10),FT(10),FP(10),REAC(10),PSI(10),IN(10),
     +F(10)
      DIMENSION RC(10)
      DATA A/10.462,14.920,12.072,9.928,8.834,0.045,7.438,16.737,
     +12.197,12.152/
      DATA B/-2.335E-03,0.6617E-03,-3.497E-03,-3.246E-03,-1.834E-03,
     +0.186E-03,-1.157E-03,-5.724E-03,-3.613E-03,-3.799E-03/
      DATA C/-952.412,10562.0,1301.12,-35.2072,339.197,-2198.24,
     +-3025.22,3127.47,-1570.14,628.10/
      DATA D/1.690E05,3.960E05,3.777E05,4.290E05,0.9272E05,0.145E05,
     +1.897E05,-2.7904E05,4.2270E05,4.129E05/
      DATA E/-61.9063,-103.3363,-71.5698,-57.8935,-59.5084,0.6433,
     +-47.2212,-97.8648,-65.3606,-66.0039/
      DATA ABAR/101.7522E-06,148.6498E-06,112.7898E-06,195.3872E-06,
     +27.4419E-06,6.15E-06,2*24.37E-06,112.80E-06,106.64E-06/
      DATA BBAR/1.0822,1.7035,2.5322,2.5985,1.1816,.6304,2*.8664,
     +2.5322,1.9018/
      DATA CBAR/1.94E-06,7.26E-06,7.49E-06,8.86E-06,13.96E-06,
     +2.66E-06,2*0.4696E-06,7.49E-06,4.83E-06/
      DATA AS/31.43E-06,46.27E-06,67.66E-06,70.94E-06,23.85E-06,
     +9.30E-06,12.80E-06,-0.73E-06,48.05E-06,38.75E-06/
      DATA BS/1.0996,1.5644,2.3777,2.5795,1.8696,0.4539,.6503,.5125
     +,1.8853,1.4314/
      DATA AB/28.6013E-12,49.5089E-12,100.1368E-12,124.8544E-12,
     +13.0973E-12,17.2646E-12,5.1756E-12,-37.9545E-12,46.3271E-12,
     +29.0625E-12/
      DATA BB/1.0006E-06,1.6739E-06,3.5190E-06,4.5399E-06,
     +,1.5428E-06,.6861E-06,.32E-06,-.9827E-06,1.7597E-06,1.0736E-06/
      DATA REAC/8HFAYALITE,8HDIOPSIDE,8HALBITE  ,8HSANIDINE
     +8HAN FO DI,8HFO EN SI,8HSP FO EN,8HGR EN DI,8HJD EN FO,
     +8HAB SI JD/
 1000 FORMAT(I2,7A8)
 1001 FORMAT(10I2,F6.0,F6.2)
 1002 FORMAT(10F6.4,/,10F6.4)
 1003 FORMAT(10F8.2)
 2000 FORMAT(1H1)
 2001 FORMAT(26X,9H*********,/,2X,7A8,/)
 2002 FORMAT(2X,44HEQUILIBRATION TEMPERATURES AT A PRESSURE OF ,
     +F6.2,3H KB,//,2X,41HREACTION   T DEG C   ACTIVITY PRODUCT    X
     +,16H LIQUID       PSI,4X,13HVALUE OF F(T))
 2003 FORMAT(2X,A8,3X,F6.0,8X,F7.4,8X,F6.4,3X,F8.2,2X,E13.4)
 2004 FORMAT(//,2X,29HEQUILIBRATION PRESSURES IN KB,/)
 2005 FORMAT(2X,7HT DEG C,4X,10(A8,2X))
```

```
2006 FORMAT(1H-)
2007 FORMAT(2X,F7.1,2X,10(F8.2,2X))
     LC=55
     KT1=5
     KT2=6
     R=1.9872
     WRITE(KT2,2000)
   1 READ(KT1,1000)N,NAME
     IF(EOF,KT1)999,10
  10 READ(KT1,1001)IN,T0,P0
     T0=T0+273.15
     P0=P0*1000.0
     DO 15 I=1,N
  15 T(I)=T0
     READ(KT1,1002)X,XS
     READ(KT1,1003)PSI
     DO 30 I=1,N
     J=IN(I)
     DO 20 KK=1,20
     F(I)=A(J)*(ALOG(T(I))-1.0)+B(J)*T(I)+C(J)/T(I)+
    +D(J)/(T(I)*T(I))+E(J)+ALOG(XS(I))
     F(I)=F(I)+((AS(J)*T(I)+BS(J))*P0-0.5*(AB(J)*T(I)+BB(J))
    +P0*P0)/(R*T(I))
     F(I)=-F(I)+ALOG(X(I))+PSI(I)/T(I)
     F(I)=F(I)+((ABAR(J)*T(I)+BBAR(J))*P0-0.5*CBAR(J)*P0*P0)/
    +(R*T(I))
     IF(ABS(F(I)).LE.1.0E -10) GO TO 30
     FT(I)=ALOG(X(I))+(ABAR(J)/R)*P0-A(J)*ALOG(T(I))-2.0*B(J)*
    +T(I)
     FT(I)=FT(I)+D(J)/(T(I)*T(I))-E(J)-(AS(J)*P0-0.5*AB(J)*
    +P0*P0)/R-ALOG(XS(I))
     FT(I)=(FT(I)-F(I))/T(I)
     T(I)=T(I)-F(I)/FT(I)
  20 T(I)=ABS(T(I))
  30 CONTINUE
     DO 35 I=1,N
  35 T(I)=T(I)-273.15
     P0=P0/1000.0
     LC=LC-24-N
     IF(LC.GE.0) GO TO 40
     LC=55-24-N
     WRITE(KT2,2000)
  40 IF(LC+24+N.LT.55) WRITE(KT2,2006)
     WRITE(KT2,2001)NAME
     WRITE(KT2,2002)P0
     DO 45 I=1,N
     J=IN(I)
     WRITE(KT2,2003)REAC(J),T(I),XS(I),X(I),PSI(I),F(I)
  45 CONTINUE
```

```
      T(1)=1273.15
      DO 50 I=1,9
      K=I+1
   50 T(K)=T(I)+100.0
      DO 70 K=1,N
      J=IN(K)
      DO 70 I=1,10
      F(I)=A(J)*(ALOG(TI))-1.0)+B(J)*T(I)+C(J)/T(I)+
     +D(J)/(T(I)*T(I))+E(J)+ALOG(XS(K))
      F(I)=(-F(I)+ALOG(X(K))+PSI(K)/T(I))*R*T(I)
      FT(I)=(ABAR(J)-AS(J))*T(I)+BBAR(J)-BS(J)
      FP(I)=0.5*(AB(J)*T(I)+BB(J)-CBAR(J))
      DISC=FT(I)*FT(I)-4.0*FP(I)*F(I)
      IF(DISC.LT.0.0) GO TO 60
      P(I,K)=(-FT(I)+SQRT(DISC))/(2000.0*FP(I))
      GO TO 70
   60 P(I,K)=0.0
   70 CONTINUE
      WRITE(KT2,2004)
      DO 80 I=1,N
      J=IN(I)
   80 RC(I)=REAC(J)
      WRITE(KT2,2005)(RC(J),J=1,N)
      DO 90 I=1,10
      T(I)=T(I)-273.15
      WRITE(KT2,2007)T(I),(P(I,J),J=1,N)
   90 CONTINUE
      GO TO 1
  999 STOP
      END
```

Table A1: Details of data input.

Card 1	Col. 1-2	N, the number of reactions for which data will be entered. N 10. Format I2.
	Col. 3-58	NAME. An identification array for the problem. Format 7A8.
Card 2	Col. 1-20	IN, an array of reaction codes to identify the reactions. Reaction, Table 1: 8 9 10 11 12 1 13 14 15 17 Code 1 2 3 4 5 6 7 8 9 10 Enter N codes, one code for each reaction every two columns. Format I2. Leave blank those columns not needed.
	Col. 21-26	T0, an estimate of the temperature of equilibration in ^{o}C, e.g. 1300. Format F6.0.

Table A1, cont.:

	Col. 27-32	PO, an approximate pressure of equilibration in kb, e.g. 20.0. Format F6.2.
Card 3	Col. 1-60	X, an array of mole fractions of components in the melt. N values required in the order corresponding to the reaction codes. Format F6.4.
Card 4	Col. 1-60	XS, an array of the contributions by the solids to the activity product of the reaction. (e.g. for reaction 1, Table 1 this value would be $a_{Mg_2Si_2O_6}/a_{Mg_2SiO_4}$) N values required in the order of the corresponding reaction codes. Format F6.4.
Card 5	Col. 1-80	PSI, an array of Ø values in the order of the corresponding reaction codes. N values required. Format F8.2.

Repeat cards 1-5 for each problem (e.g. each different lava).

REFERENCES CITED

Bacon, C.R., and Carmichael, I.S.E., Contr. Mineral. Petrol., 41, 1-22 (1973).

Buddington, A.F., and Lindsley, D.H., Jour. Petrol., 5, 310-357 (1964).

Bultitude, R.J., and Green, D.H., Jour. Petrol., 12, 121-147.

Carmichael, I.S.E., Contr. Mineral. Petrol., 14, 24-66 (1967).

Carmichael, I.S.E., Nicholls, J., Spera, F.J., Wood, B.J., and Nelson, S.A., Trans. Roy. Soc. (in press).

Carmichael, I.S.E., Nicholls, J., and Smith, A.L. Amer. Mineral., 55, 246-263 (1970).

Carmichael, I.S.E., Turner, F.J., and Verhoogen, J., Igneous Petrology, McGraw-Hill Book Co. (1974).

Fiesinger, D.W., and Nicholls, J., Geol. Assoc. Canada Spec. Paper (1976).

Kerrick, D.M., and Darken, L.S., Geochim. Cosmochim. Acta, 39, 1431-1442 (1975).

Kudo, A.M. and Weill, D.F., Contr. Mineral. Petrol., 25, 52-65 (1970).

Lumsden, J., Metal. Soc. Conf. 7, Interscience, Inc. (1961).

Nicholls, J., Contr. Mineral., (in press).

Nicholls, J. and Carmichael, I.S.E., Amer. Mineral., 57, 941-959 (1972).

Nicholls, J., Carmichael, I.S.E., and Stormer, J.C., Contr. Mineral. Petrol., 33, 1-20 (1971).

Shand, S.J., The Eruptive Rocks, 2nd Ed., Wiley and Sons Book Co. (1943).

Thompson, J.B., Researches in Geochemistry, 2, John Wiley, Inc. (1967).

Tilley, C.E., Q.J. Geol. Soc. London, 106, 37-61 (1950).

Waldbaum, D.R., and Thompson, J.B., Amer. Mineral., 54, 1274-1298 (1969).

Wood, B.J., and Banno, S., Contr. Mineral. Petrol., 42, 109-124 (1973).

Wood, B.J., and Strens, R.G.J., Earth Planet. Sci. Lett., 11, 1-6.

STUDY PROBLEM

A rhyolite from Inyo Craters, California has been analysed by Carmichael (1967) with the results as set out below. Estimate the temperatures and pressures of equilibration between minerals and melt and list the assumptions or interpretations required to make these estimates.

Mode	Vol %	Compositions
Glass	97.2	
Plagioclase	1.0	Rims = An 16, Cores = An 22
Alkali feldspar	1.0	Or 70 Ab 30
Quartz	0.2	
Ca-rich pyroxene	Tr	Ca 39.1 Mg 25.5 Fe 35.4
Ca-poor pyroxene	Tr	Ca 3.0 Mg 30.8 Fe 66.2
Fe-Ti oxides	0.3	mt 50 usp 50 il 90 hm 10

Rock Analysis

	Wt%	Mole %
SiO_2	71.25	76.93
TiO_2	0.26	0.21
Al_2O_3	14.60	9.29
Fe_2O_3	0.65	0.26
FeO	1.35	1.22
MnO	0.06	0.05
MgO	0.27	0.43
CaO	1.06	1.23
BaO	0.12	0.05
Na_2O	4.57	4.78
K_2O	5.08	3.50
P_2O_5	0.06	0.03
H_2O+	0.56	2.02

THE THERMODYNAMICS OF TRACE ELEMENT DISTRIBUTION †

R.K. O'Nions[1] and R. Powell[2]

1. Lamont-Doherty Geological Observatory of Columbia University, Palisades, New York 10964.
2. Department of Earth Sciences, University of Leeds, Leeds.

INTRODUCTION

Conventionally, elements are categorized as major, minor or trace depending upon their relative abundances in the system of interest. The divisions between these three categories have not been rigidly fixed, but the term trace element has frequently been used for any element whose abundance is 1000 ppm or less. At very low concentrations of a trace solute constituent, the thermodynamic behaviour of the solution may be expected to approach that of an ideal dilute solution. The concentration range over which Henry's Law is obeyed by a given trace constituent cannot as yet be predicted and must be determined by experiment. It is unfortunate that the term trace element as currently used encompasses concentration ranges where departures from Henry's Law may occur. As more data become available on the solution behaviour of trace elements some qualifications of the term may prove worthwhile.

The present article is primarily concerned with a discussion of the thermodynamics of typical solid-liquid and some solid-gas equilibria which may have been important in controlling the distribution of trace elements in the earth, moon and meteorites. In this respect attention is drawn to the limitations of the conventional Berthelot-Nernst and Henderson-Kracek distribution coefficients in geochemistry. Particular attention is paid to the thermodynamics of gas-solid reactions since these are becoming increasingly important as evidence accrues for fractionation of rare-earth elements by solid-gas reactions prior to meteorite formation.

† Lamont-Doherty Geological Observatory Contribution 2457.

D. G. Fraser (ed.), Thermodynamics in Geology, 349-363.

HENRY'S LAW

In the majority of discussions of trace element variations in Nature it is implicitly assumed that Henry's Law is obeyed. For solid and liquid silicates Henry's Law is illustrated in Figure 1, where a_i is the activity of the trace component i in the solid or liquid phase and X_i is the mole fraction of i present.

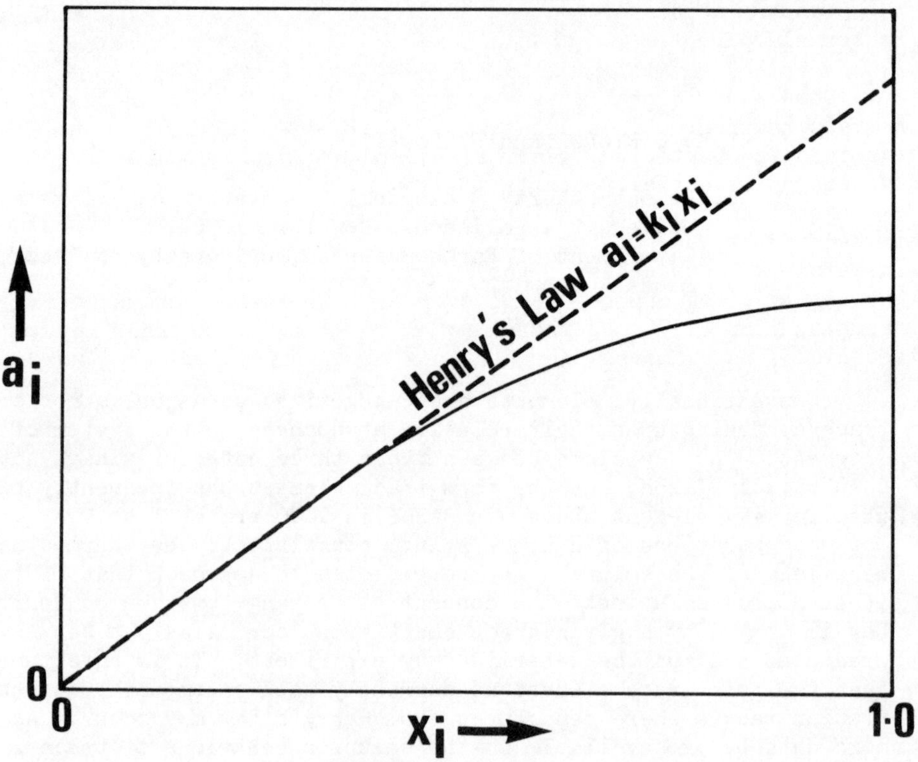

Fig. 1. Illustration of the possible solution behaviour of a component i.

As the amount of X_i becomes small and approaches infinite dilution ($X_i \to 0$) then a_i becomes proportional to X_i:

as $X_i \to 0$

$$a_i \to k_iX_i \tag{1}$$

where k_i is the Henry's Law constant for the component i and is independent of X_i, but usually dependent on P and T. For a trace constituent i in a gas phase the analogous expression as $X_i \to 0$ is

$$p_i \to k_i X_i \tag{2}$$

where p_i is the partial pressure of i in the gas.

A more detailed discussion of the nature of Henry's Law and the standard states of trace constituents is given in Wood and Fraser (1976).

Some experimental data are now available which bear on the Henry's Law region in silicates. Drake and Weill (1975) have examined the partitioning of Sr, Ba and Sm in plagioclase-liquid systems of essentially constant Ab/An ratios. The electron probe techniques employed by these authors did not permit them to examine in detail the very dilute solutions which are commonly encountered in Nature. The results of their experiments (Figure 2) suggest that Henry's Law may be obeyed over a considerable concentration range. Mysen (1976), using a much more sensitive β-track mapping technique has examined the behaviour of dilute solutions of Sm and Ni in olivine-liquid and orthopyroxene-liquid systems at 20 kb and 1025°C (Figure 2). In each case examined, the data can approximate to a straight line through the origin at low solute concentrations, but deviate considerably at higher concentrations, which in the case of Ni in olivine includes the range of geological interest. In general, the straight line portion through the origin should represent the region in which both liquid and solid phases obey Henry's Law (see Mysen, 1976). In most instances examined by Mysen (1976) it is reasonable to assume that Henry's Law is obeyed by both phases at low concentration. It is conceivable however that the behaviour of Sr and Ba in plagioclase could be more complex than indicated in Figure 2, since data for very low concentrations have not been obtained.

FORMULATION OF DISTRIBUTION COEFFICIENTS

Many attempts to model trace element abundance variations in natural systems have employed the so-called Berthelot-Nernst distribution coefficient or distribution coefficients following the Henderson and Kracek treatment (see review of McIntire, 1963). Both McIntire (1963) and Banno and Matsui (1973) have previously discussed the relative advantages and disadvantages of these distribution coefficients and reference is made to these works for additional discussion.

For the distribution of a trace element i between a solid and a liquid phase, the Berthelot-Nernst distribution coefficient (D_i)

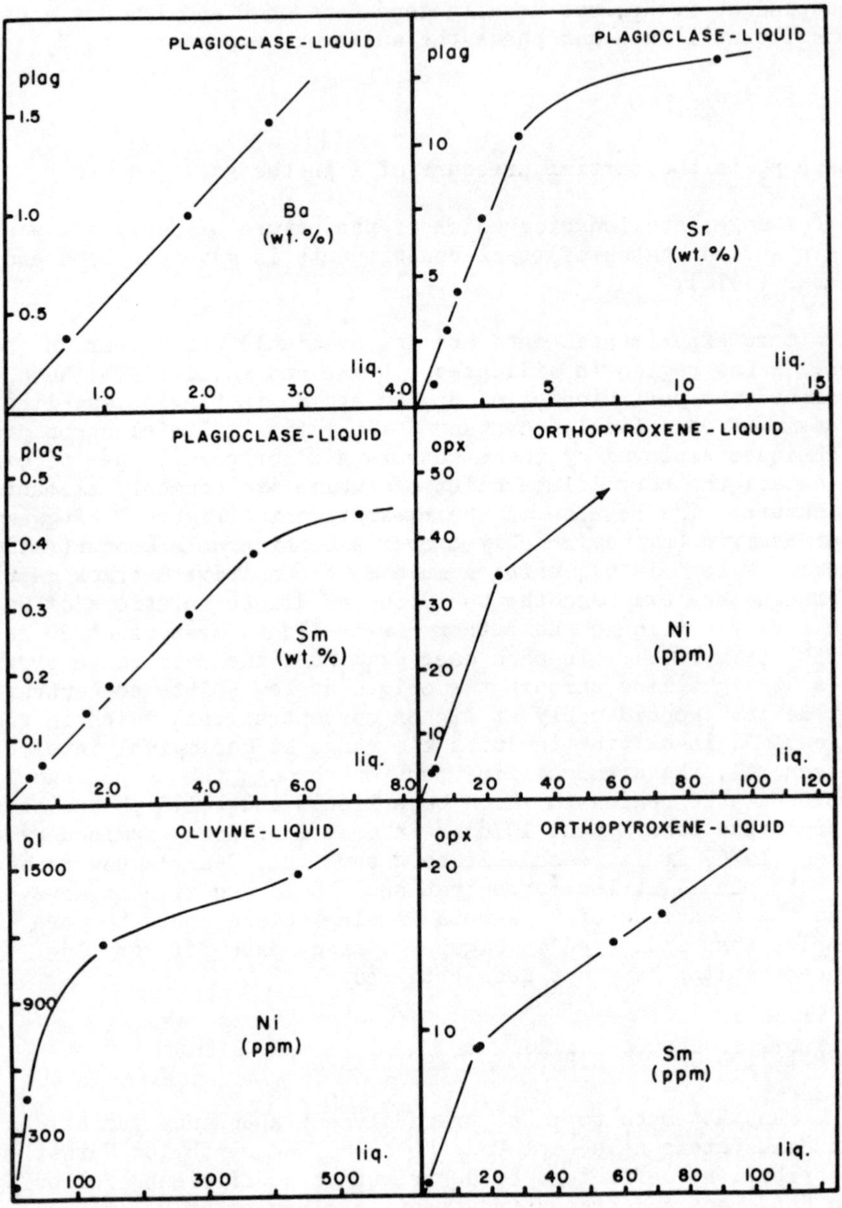

Fig. 2. Partitioning of Sr, Ba and Sm between plagioclase and liquid after Drake and Weill (1975). Partitioning of Sm and Ni between olivine, orthopyroxene and liquid after Mysen (1976).

is defined as:

$$D_i = \frac{C_i^s}{C_i^l} \tag{3}$$

where C_i^s and C_i^l are the concentrations of i in the solid and liquid phases respectively. For two solid phases α and β, for example, we have

$$D_i = \frac{C_i^\alpha}{C_i^\beta} \tag{4}$$

An alternative form of distribution coefficient (Henderson and Kracek Formula) which has been comparatively little used in geochemistry, introduces the concept of 'carrier' elements which can be thought of as the major elements for which the trace element substitutes. Thus:

$$D_i' = \frac{C_{Tr}^s}{C_{Cr}^s} \qquad \frac{C_{Tr}^l}{C_{Cr}^l} \tag{5}$$

where Tr and Cr indicate trace and carrier constituent respectively.

RELATIONSHIP BETWEEN DISTRIBUTION COEFFICIENTS AND EQUILIBRIUM CONSTANTS

Despite the apparently over-simplified form of the Berthelot-Nernst distribution coefficient, it has been extensively employed in trace element geochemistry often with considerable success. In this respect it is instructive to compare D_i with the true equilibrium constant for a given reaction. As an example the partition of Ni between olivine and clinopyroxene (Häkli and Wright, 1967; Broecker and Oversby, 1971; Banno and Matsui, 1973; Carmichael et al., 1974). Consider the following exchange reaction:

$$2CaMgSi_2O_6 + Ni_2SiO_4 = 2CaNiSi_2O_6 + Mg_2SiO_4 \tag{6}$$

for which

$$-\frac{\Delta G^o}{RT} = \ln \frac{\left(a_{CaNiSi_2O_6}^{cpx}\right)^2 \cdot a_{Mg_2SiO_4}^{ol}}{\left(a_{CaMgSi_2O_6}^{cpx}\right)^2 \cdot a_{Ni_2SiO_4}^{ol}} = \ln K \tag{7}$$

If the Ni-concentrations are low the activities of the $CaMgSi_2O_6$ and Mg_2SiO_4 components will be close to unity, and substituting from equation (1) for activities of the trace components,

$$\ln \frac{\left(X^{cpx}_{CaNiSi_2O_6}\right)^2}{X^{ol}_{Ni_2SiO_4}} = \ln K - 2 \ln k' + \ln k'' \qquad (8)$$

where k' and k" are the Henry's Law constants for the Ni component in clinopyroxene and olivine respectively and K is the equilibrium constant for reaction (6). The left hand side of (8) is related to $\ln D_i$, the Berthelot-Nernst distribution coefficient, which is not usually the same as the equilibrium constant. If the Ni components formed ideal solutions then k' = k" = 1.0 and the relationship is simplified. It is clear that D_i will vary with P and T as do the true equilibrium constant K, and the Henry's Law constants k, but must also vary when the $a^{cpx}_{CaMgSi_2O_6}$ and $a^{ol}_{Mg_2SiO_4}$ depart from unity with addition of other components such as Fe and Al to the system.

When one of the phases is a liquid phase, some insight into factors influencing D_i can be gained by considering a reaction such as the following:

$$2NiO_{(liq)} + SiO_{2(liq)} = Ni_2SiO_{4(olivine)} \qquad (9)$$

for which

$$-\frac{\Delta G^o}{RT} = \ln \frac{a^{ol}_{Ni_2SiO_4}}{a^{liq}_{NiO}\, a^{liq}_{SiO_2}}$$

Introducing Henry's Law constants for Ni-olivine (k') and NiO_{liq} (k") then,

$$\ln \frac{X^{ol}_{Ni_2SiO_4}}{\left(X^{liq}_{NiO}\right)^2} = \ln K - \ln k' + 2 \ln k'' + \ln a^{liq}_{SiO_2} \qquad (10)$$

In this situation the left hand side which is related to D_i depends not only on K and the Henry's Law constants k' and k" but also on $a^{liq}_{SiO_2}$. This point has also been made by Banno and Matsui (1973) and Drake and Weill (1975). The activity of silica in the liquid, $a^{liq}_{SiO_2}$, may be a strong function of liquid composition as is well exemplified both by the effects of alkalis, H_2O and CO_2 etc. on phase relationships in Di-Fo-Qz (Kushiro, 1972, 1974 and Fraser, this volume, p.301), and also the differences in crystallization relationships in acid and basic magmas (Carmichael et al., 1970; Nicholls et al., 1971).

The high degree of success which has been obtained with the simple Berthelot-Nernst distribution coefficient must reflect the comparatively small effect of its deficiencies compared with the wide range of values of D_i and the magnitude of trace element abundance variations in nature. Its great limitation is inevitably in the interpretation of more subtle variations, such as might arise through variations in temperature or pressure of equilibration.

There have been few attempts to date to formulate true equilibrium constants for reactions involving trace elements. One notable attempt is that of Drake and Weill (1975), who have considered plagioclase-liquid equilibria. These authors have used the reaction:

$$MAl_2Si_2O_{8(plag.)} = MAl_2O_{4(liq)} + 2SiO_{2(liq)} \tag{11}$$

where M may be Sr, Ba, Eu, etc.

$$-\frac{\Delta G^o}{RT} = \ln \frac{a^{liq}_{MAl_2O_4} \cdot \left(a^{liq}_{SiO_2}\right)^2}{a^{plag}_{MAl_2Si_2O_8}}$$

The selection of the components SiO_2 and MAl_2O_4 is justified by their role as network formers or as acidic oxides (cf. Fraser, 1975 and this volume). Their activities have been approximated by:

$$a^{liq}_{MAl_2O_4} = X_{MAl_2O_4} \Big/ \Sigma_{\text{network forming components}}$$

$$a^{liq}_{SiO_2} = X_{SiO_2} \Big/ \Sigma_{\text{network forming components}} \tag{12}$$

There are insufficient data at present to assess the advantages of this approach.

An alternative approach to the problems posed by specifying the activities of components in the liquid phase is to use exchange reactions involving only a simple MO component in the liquid. For example for the partition of Ni between olivine and liquid we can write:

$$2MgO_{(liq)} + Ni_2SiO_{4(olivine)} = 2NiO_{(liq)} + Mg_2SiO_{4(olivine)} \tag{13}$$

for which

$$-\frac{\Delta G^o}{RT} = \ln \frac{a^{ol}_{Mg_2SiO_4}\left(a^{liq}_{NiO}\right)^2}{a^{ol}_{Ni_2SiO_4}\left(a_{MgO}\right)^2} \tag{14}$$

Assuming Temkin solution:

$$\left.\begin{aligned} a_{MgO} &= a_{Mg^{2+}} . a_{O^{2-}} \\ a_{NiO} &= a_{Ni^{2+}} . a_{O^{2-}} \end{aligned}\right) \tag{15}$$

Both Mg^{2+} and Ni^{2+} will be present in the cation matrix (C.M. of the liquid. Thus,

$$-\frac{\Delta G^o}{RT} = \ln \left[\frac{a^{ol}_{Mg_2SiO_4}}{a^{ol}_{Ni_2SiO_4}}\right]_{solid} . \left[\frac{a_{Ni^{2+}}}{a_{Mg^{2+}}}\right]_{liq\ (C.M.)}$$

If it is assumed that Mg and Ni do not show any site preferen in olivine then the activities can be approximated by

$$a^{ol}_{Mg_2SiO_4} = X^2_{Mg(M)} . \gamma^2_{Mg(M)}$$

and

$$a^{ol}_{Ni_2SiO_4} = X^2_{Ni(M)} . \gamma^2_{Ni(M)}$$

If the mixing on sites in both solid and liquid is ideal (i.e the activity coefficients are unity), then,

$$\ln K = \left[\frac{X_{Mg(M)}}{X_{Ni(M)}}\right]^2_{olivine} . \left[\frac{X_{Ni^{2+}}}{X_{Mg^{2+}}}\right]^2_{liquid\ (C.M.)} \tag{16}$$

The equilibrium constant for this reaction bears a strong resemblance to the Henderson and Kracek compound distribution coefficient (equation 5). The equilibrium constant in (16) should be less dependent on liquid composition than (9), since the $a_{O^{2-}}$ in the liquid phase, which is difficult to estimate, cancels. Likewise, the compound distribution coefficients should show less compositional dependence than the Berthelot-Nernst coefficients (see also Banno and Matsui, 1973).

GEOTHERMOMETRY AND GEOBAROMETRY

The equilibrium constants for the above reactions will all depend upon temperature and pressure in the following manner.

$$\left(\frac{\partial \ln k}{\partial T}\right)_P = \frac{\Delta H^o}{RT^2} \qquad (17)$$

and

$$\left(\frac{\partial \ln k}{\partial P}\right)_T = -\frac{\Delta V^o}{RT} \qquad (18)$$

and thus both the Berthelot-Nernst and the Henderson-Kracek distribution coefficients will depend upon temperature and pressure, but will be less simply related to the ΔH^o and ΔV^o for the appropriate reactions. Also in general ΔH^o and ΔV^o are expected to be smaller for exchange reaction of the type (13) than for reactions such as (9) which should make the Henderson and Kracek coefficient less T and P dependent than the Berthelot-Nernst coefficient (c.f. Banno and Matsui, 1973; Drake and Weill, 1975).

Ideally a trace element geothermometer should be based upon a reaction with a large ΔH^o and small ΔV^o and vice versa for a geobarometer. A number of experimental studies have now demonstrated the temperature dependence of Berthelot-Nernst distribution coefficients. Noteworthy amongst these are the experiments of Shimizu (1974) on the distribution of Sr, Ba, K, Rb, and Cs between clinopyroxenes and liquid and those of Drake and Weill (1973) on the partitioning of Sr, Ba and Eu between plagioclase and liquid. In both of these instances the presentation of major element data for solid and liquid phases permit a more rigorous estimation of activities according to the treatments outlined in section 5. However until more data of a similar nature become available the value of these alternative approaches cannot be fully assessed.

Nevertheless, it is doubtful that geothermometers based upon the Berthelot-Nernst distribution coefficient will ever be useful in systems of variable composition.

TRACE ELEMENTS IN GAS-SOLID REACTIONS

Considerable interest in the partitioning of trace elements in gas-solid reactions has been generated recently. To a large extent this interest stems from the demonstration by Grossman (1973, 1976) that some inclusions in chondrites, such as the C3 chondrite Allende, may have been formed by high-temperature condensation from a gas of solar composition. In addition measurements of rare-earth element abundances in inclusions from Allende and other chondrites (e.g. Tanaka and Masuda, 1973) have revealed the presence of anomalous Ce and Yb abundances. Yb-anomalies do not occur in terrestrial, or for that matter lunar, samples and as far as is known cannot be generated by solid-liquid equilibria. Boynton (1975) in an important discussion of the origin of these rare-earth element abundance anomalies has shown how such anomalies are predictable from thermodynamic properties of the rare-earths.

In the present section we shall present a discussion of gas-solid equilibria pertinent to the interpretation of such rare-earth element anomalies, but which will also serve as a general approach to the thermodynamics of gas-solid reactions involving trace elements. The treatment here is somewhat different to that of Boynton (1975), but employs essentially the same basic thermodynamic data.

In calculating the distribution of rare-earth elements, or indeed any other trace elements, between a gas of solar composition and condensing crystalline phases there are two basic steps. Firstly, the distribution of rare-earth species in the nebula must be estimated and secondly, the partition of the REE between the gas and solid phases calculated. In the ensuing discussion it is assumed following Grossman (1974) and Boynton (1975), that equilibrium was achieved in the condensing solar nebula.

In the nebula gas finite amounts of many REE species will exist, but studies of vapours in equilibrium with REE oxides suggest that M and MO species will be the most important.

For a particular REE the proportions of these two species in the nebula are governed by

$$\tfrac{1}{2}\, O_2 + M_{(gas)} = MO_{(gas)}$$

for which the equilibrium relationship is

$$-\frac{\Delta G^{o}_{M}}{RT} = \ln\left(\frac{X_{MO}}{X_M . fO_2^{\frac{1}{2}}}\right) = RT \ln K_M \qquad (19)$$

Given thermodynamic data for gaseous M, MO and O_2 the ratio M/MO can be calculated for a given f_{O_2} and temperature. The results of these calculations are shown diagrammatically in Figure 3.

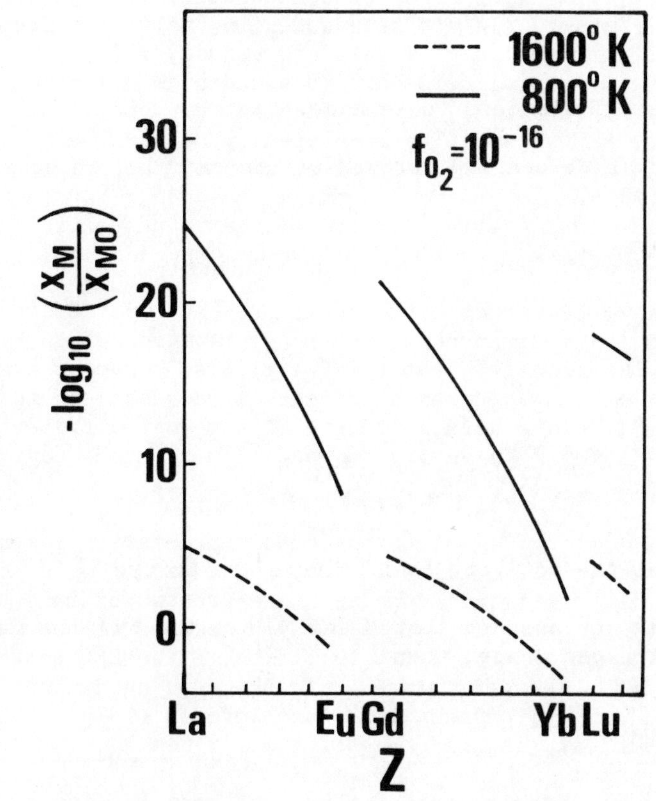

Fig. 3. X_M/X_{MO} in gas as a function of temperature and f_{O_2} for the rare earth elements.

Noteworthy is the stepped fractionation of REE species corresponding to basic changes of electronic configuration of the REE (except Ce, Gd and Lu) in going from $M_{solid} \rightarrow M_{gas}$ (involving transfer of the 5d electron to 4f) and the absence of this change for reaction $M_{solid} + \frac{1}{2} O_2 \rightarrow MO_{gas}$ (Ames, 1967). Also noteworthy is the effect of temperature on X_M/X_{MO} ratios.

The strong stepwise fractionation of $\frac{X_M}{X_{MO}}$ ratios is reflected in the partition of REE into condensing solids. The condensing solids are likely to be Y_2O_3 solid solutions or perovskite (Grossman, 1973, 1976; Boynton, 1975). In the absence of thermodynamic data for the REE-bearing end-members of the appropriate phases the assumption will be made that

$$\Delta_f G_{M' \text{ endmember}} - \Delta_f G_{M'' \text{ endmember}}$$

$$= \tfrac{1}{2} (\Delta_f G_{M'_2O_3} - \Delta_f G_{M''_2O_3})$$

which allows calculations to be made.

Consider a Nb-bearing perovskite for example, in equilibrium with a gas phase,

$$CaLaTiNbO_6 + M_{(gas)} = CaMTiNbO_6 + La_{(gas)} \tag{20}$$

$$K_c = \left(\frac{X_M}{X_{La}}\right)_{perv} \left(\frac{X_{La}}{X_M}\right)_{gas} \tag{21}$$

$$-\frac{\Delta G^o_c}{RT} = \ln \left(\frac{X_M}{X_{La}}\right)_{perv} \left(\frac{X_{La}}{X_M}\right)_{gas}$$

assuming that the activity coefficients are unity.

If it is now assumed that M and MO are the two dominant species in the gas phase, then

$$X_M + X_{MO} = \frac{2\,A_M}{A_H} \tag{22}$$

where A_M is the solar abundance of M and A_H is the solar abundance of H.

Using the identity from (19)

$$X_{MO} = X_M K_M \; fO_2^{\frac{1}{2}}$$

and substituting into (22) gives

$$X_M (1 + K_M \, fO_2^{\frac{1}{2}} = \frac{2\,A_M}{A_H} = 2\,X'_M$$

where X'_M is now the hydrogen - normalized abundance of M. Thus

$$X_M = \frac{2\,X'_M}{(1 + KfO_2^{\frac{1}{2}})} \tag{23}$$

Substituting (23) and the analogous expression for X_{La} into (21)

$$K_c = \left(\frac{X_M}{X_{La}}\right)_{Perov} \left(\frac{X'_{La}}{X'_M} \frac{(1 + K_M fO_2^{\frac{1}{2}})}{(1 + K_{La} fO_2^{\frac{1}{2}})}\right)_{gas} \quad (24)$$

where X'_{La}/X'_M is the ratio of La to M in the nebula from all species.

Exchange reactions such as (20) above are of general applicability to the prediction of trace element partitioning between a gas of solar composition and condensing solid phases.

As an example of how the patterns of REE fractionation can be predicted the above analyses can be extended to calculate enrichment factors (E) for each REE relative to La in the condensed solid,

$$E = \left(\frac{X_M}{X_{La}}\right)_{Perv} \left(\frac{X'_{La}}{X'_M}\right)_{gas}$$

$$= K_c \frac{1 + K_{La} fO_2^{\frac{1}{2}}}{1 + K_M fO_2^{\frac{1}{2}}} \approx \frac{K_c K_{La} fO_2^{\frac{1}{2}}}{1 + K_M fO_2^{\frac{1}{2}}} \quad (25)$$

The approximation on the right hand side can be made because $K_{La} >> 1.0$.

Using published thermodynamic data, values of E have been calculated for $T = 1200^oK$ and $T = 1600^oK$ and fO_2 of 10^{-16} and 10^{-8} (Table 1).

Table 1. Enrichment factors for Perovskite.

	log E*(1200 K)	log E*(1600 K)
La	0.0	0.0
Ce	2.4	2.0
Sm	0.3	0.8
Eu	-2.6(-2.6)	-0.5(1.7)
Gd	3.7	2.7
Dy	3.5	2.7
Er	4.9	3.7
Yb	2.4(3.9)	-0.5(3.0)
Lu	6.6	4.7

*E = Enrichment Factor given by (M/La) perovskite/$(M/La)_{gas}$. Fractionation of Eu and Yb are a function of fO_2. Values listed are for $fO_2 = 10^{-16}$ and values in parentheses are 10^{-8}.

It is noteworthy that the predicted values for E in perovskite show markedly reduced values for the trivalent REE at Eu and Yb and that the magnitude of the predicted anomalies are a function of f_{O_2}. Despite the fact that only trivalent REE have been considered, it is clear, as shown by Boynton (1975), that gas-solid condensation reactions of the kind considered here provide a mechanism for generating the anomalous REE patterns found in some carbonaceous chondrites. The actual patterns such as those found in Allende (Tanaka and Masuda, 1973) may be made up of several components condensing at different temperatures and possibly in different parts of the nebula. Subsequent solid-solid or solid-liquid equilibria will not be able to eradicate those anomalies. The absence of such anomalies on the earth and moon may reflect the fact that they represent well homogenized samples of the condensed material. The possibility that the earth contains a core which is an early direct condensate and a mantle which condensed at low temperatures, e.g. Anderson (1975) is rendered highly unlikely by the regularity of REE distribution patterns in terrestrial rocks.

It is hoped that these comments at least serve to illustrate the potential value of investigating gas-solid reactions such as those above.

CONCLUDING REMARKS

The interpretation and modelling of trace element abundance variations in rocks and minerals have employed Berthelot-Nernst distribution coefficients or less frequently compounded coefficients such as the Henderson-Kracek coefficients. Although from the standpoint of thermodynamics these are inadequate and may differ considerably from true equilibrium constants, their use has met with considerable success.

In situations where trace element abundance variations are large, the use of these distribution coefficients will certainly help to limit the possible ways in which the variations were generated. The fact that these distribution coefficients, particularly the Berthelot-Nernst coefficients, vary with P, T and composition, greatly limits their usefulness in situations where the variations in trace element chemistry are more subtle as could arise from P and T variations. For geothermometry and geobarometry, it is necessary to calculate true equilibrium constants and some possible approaches to this problem have been discussed here.

The distribution of trace elements between solid and gas phases during, for example, the condensation of the solar nebula, is also amenable to thermodynamic treatment, and the results of one particular approach have been presented. The current interest in trace element fractionation between the earth, moon and the various classes of meteorite provides considerable scope for further investigation of gas-solid reactions.

REFERENCES

Anderson, D.L. Bull. Geol. Soc. Amer., 86, 1593 (1975).

Ames, L.L., Walsh, P.N. and White, D. Jour. Phys. Chem., 71, 2707 (1967).

Boynton, W.V. Geochim. Cosmochim. Acta, 29, 569 (1975).

Broecker, W.S. and Oversby, V.M. "Chemical equilibria in the Earth", McGraw Hill (1971).

Carmichael, I.S.E., Turner, F.J. and Veerhoogen, J. "Igneous petrology", 739 pp. McGraw Hill (1974).

Carmichael, I.S.E. and Smith, A.L. Amer. Mineral., 55, 246 (1970).

Drake, M.J. and Weill, D.F. Geochim. Cosmochim. Acta, 39, 689 (1975).

Fraser, D.G. Geochim. Cosmochim. Acta, 39, 1525 (1975).

Grossman, L. Geochim. Cosmochim. Acta, 37, 1119 (1973).

Grossman, L. and Ganapathy, R. Geochim. Cosmochim. Acta, 40, 331 (1976).

Häkli, T. and Wright, T.L. Geochim. Cosmochim. Acta, 31, 877 (1967).

Henderson, L.M. and Kracek, F.C. J. Amer. Chem. Soc., 49, 739 (1927).

Kushiro, I. Jour. Petrology, 13, 311 (1972).

Kushiro, I. Carneg. Inst. Wash. Yearb., 73, 248 (1974).

McIntire, W.L. Geochim. Cosmochim. Acta, 27, 1209 (1963).

Mysen, B. Earth Planet. Sci. Lett., 31, 1 (1976).

Nicholls, J., Carmichael, I.S.E. and Stormer, J.C. Contr. Mineral. Petrol., 33 (1971).

Shimizu, N. Geochim. Cosmochim. Acta, 38, 1789 (1974).

Tanaka, T. and Masuda, A. Icarus, 19, 523 (1973).

Wood, B.J. and Fraser, D.G. "Elementary Thermodynamics for Geologists", 303 pp. Oxford University Press, Oxford (1976).

THE SOLUBILITY OF CALCITE IN SEA WATER*

Wallace S. Broecker and Taro Takahashi

Lamont-Doherty Geological Observatory,
Columbia University, Palisades, New York

INTRODUCTION

The mineral calcite is one of the most prominent phases in deep sea sediments. Its distribution with water depth on the flanks of the oceanic ridges everywhere in the world oceans has the same basic character. Sediments with uniformly high calcite content extend from the crest down the ridge flank to what Berger (1968) has termed the lysocline. Here a decrease in calcite content with water depth commences. This decrease continues until sediments nearly free of calcite are encountered. Below this horizon (often referred to as the calcite compensation depth) the sediments are free of calcite. Although the pattern is everywhere the same the depth of the lysocline and the width of the transition zone (i.e. distance between lysocline and compensation depth) vary from basin to basin (see Fig. 1).

It is generally agreed that this pattern is generated by calcite dissolution. The sediments above the lysocline have lost little calcite to dissolution while those below compensation depth have lost almost all their calcite to dissolution. However, there is little agreement with regard to the relationship between the lysocline (or compensation depth) and the saturation horizon (i.e. that depth at which sea water goes from calcite supersaturation to undersaturation). Some workers have concluded that it is more or less coincident with the lysocline while others conclude that it lies roughly a kilometer above the lysocline. Two uncertainties permit such a range of opinion: (1) the solubility of calcite in equilibrium with sea water cannot be

* Lamont-Doherty Geological Observatory Contribution 2439

D. G. Fraser (ed.), Thermodynamics in Geology, 365-379.

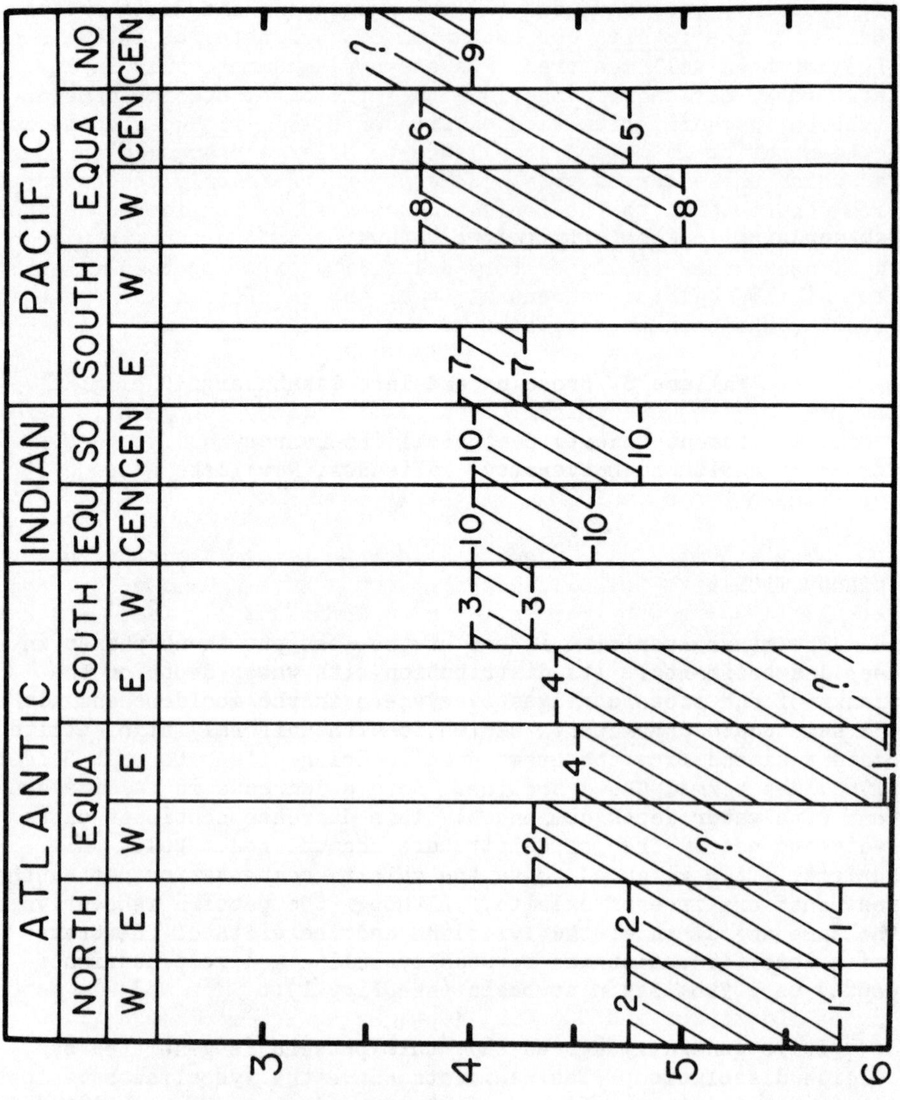

Fig. 1. Depth at top (lysocline) and base (compensation depth) of the transition zone between sediments subjected to little calcite dissolution and sediments which have lost virtually all their calcite to dissolution. The numbers on each line represent the source of the information: (1) Biscaye et al. (1976); (2) Kipp (1976); (3) Melguen and Thiede (1975); (4) Gardner (1975); (5) Bramlette (1961); (6) Parker and Berger (1971); (7) Broecker and Broecker (1974); (8) Valencia (1975); (9) Ben-Yaakov et al. (1974); (10) Kolla and Biscaye (1976).

reproducibly measured, and (2) the geographic and depth distribution of the in situ concentration of carbonate ion in the sea has not been well measured. As shown diagrammatically in Fig. 2, a 10 to 20 per cent change in either the carbonate ion concentration in equilibrium with calcite or the in situ carbonate ion concentration in ocean water leads to a large change in the depth at which these curves cross one another. Here only the carbonate ion contribution to the ionic product need be considered because the calcium ion content of deep sea water is constant to within ± 1% and either the laboratory solubility experiments were carried out at this calcium concentration or the saturation carbonate ion content has been recalculated to this concentration.

Recently one of these uncertainties was largely eliminated. As part of the Geochemical Ocean Section Study (GEOSECS), the carbonate ion distribution in the Atlantic and Pacific Oceans was determined with high accuracy. This survey will be extended to the Indian Ocean in 1978.

Two attempts have also been made to remove the other uncertainty. Calcite solubility measurements have been made by Ingle et al. (1973) and by Berner (1976). Unfortunately these estimates disagree by 35 per cent at 2°C and 16 per cent at 25°C. As we shall see the Ingle et al. value supports the contention that saturation horizon and lysocline are nearly coincident while the Berner value supports the contention that the saturation horizon lies a kilometer or two above the lysocline.

One other factor must be mentioned. The solubility of calcite increases with pressure. Based on the molar volume of calcite (36.9 cm^3 mol^{-1}) and the partial molar volumes of Ca^{2+} and CO_3^{2-} ions in sea water (-24.6 and 22.1 cm^3 mol^{-1} respectively at 25°C), Millero and Berner (1972) estimated the ΔV for this dissolution reaction to be -39.4 cm^3 mol^{-1}. Direct estimates based on solubility measurements made by Pytkowicz and his associates (Pytkowicz and Connors, 1964; Pytkowicz and Fowler, 1967; Pytkowicz et al., 1967; Hawley and Pytkowicz, 1969; Culberson, 1972), at various pressures up to 1000 atm yield a value of -31.0 cm^3 mol^{-1} at 25°C and -35.6 cm^3 mol^{-1} at 2°C. The more recent study of Ingle (1975) yields a value of -34.4 ± 0.1 cm^3 mol^{-1} at 25°C and -42.3 ± 1.4 cm^3 mol^{-1} at 2°C. Thus, the ΔV value estimated from the partial molar volumes of various ions in sea water at 25°C (-39.4 cm^3 mol^{-1}) is significantly larger than those obtained from the solubility experiments under high pressure (-31.0 and -34.4 cm^3 mol^{-1}). Were partial molar volumes of Ca^{2+} and CO_3^{2-} ion available at 2°C, a similar discrepancy would likely exist.

Thus the pressure correction needed to bring either the Ingle et al. or Berner values (measured at 1 atm pressure) to the

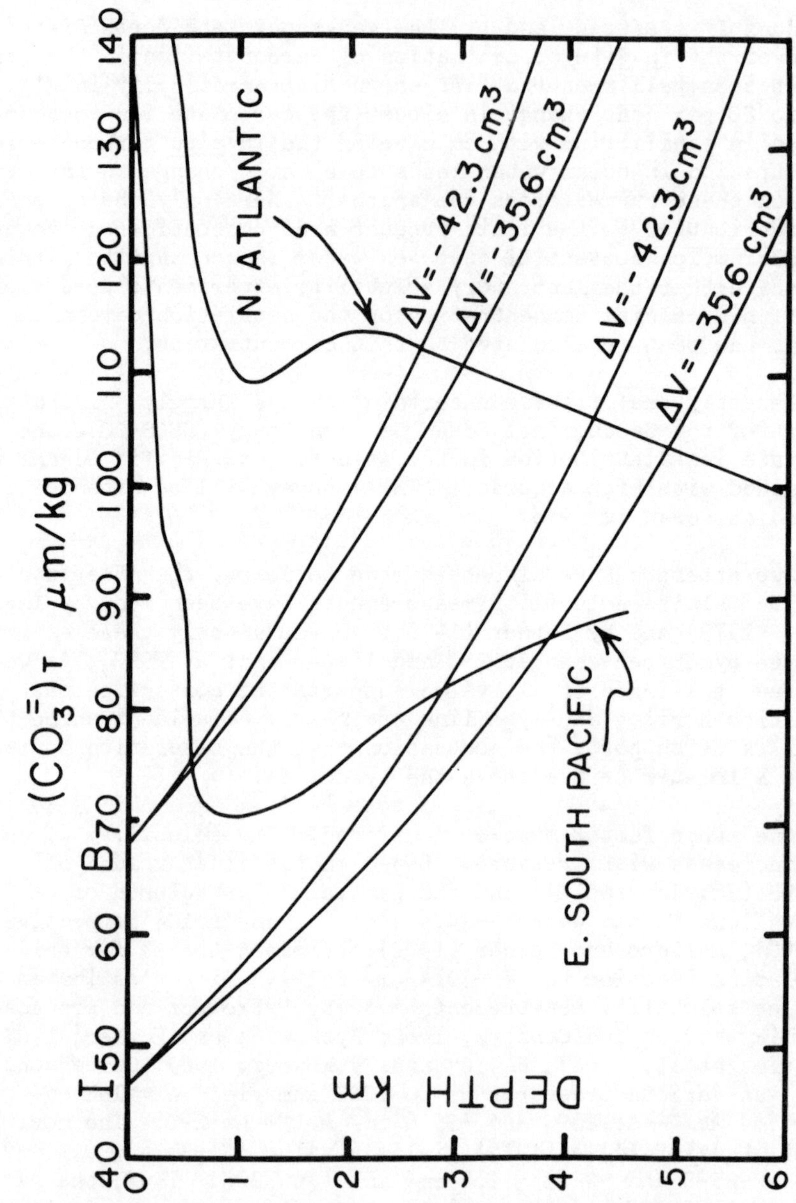

Fig. 2. $(CO_3^{2-})_T$ versus depth of ocean water. The solubility products of calcite obtained by Ingle et al. (1973) and Berner (1976) are used to calculate the saturation (CO_3^{2-}) value for a sea water of 35‰ salinity at 2°C. The value for Ingle et al. is marked with I and that for Berner with B. The effect of pressure was calculated using two ΔV values; i.e. -35.6 cm^3 of Culberson (1972) and -42.3 cm^3 of Ingle (1975) at 2°C. Typical (CO_3^{2-}) distributions in North Atlantic and eastern South Pacific oceans are shown (based on results of the Geochemical Ocean Sections Program).

lysocline pressure (300 - 500 atmospheres) adds yet another variable to the problem. Its importance is demonstrated in Fig. 2 where the intersection between the North Atlantic carbonate ion versus depth curve and saturation curves generated using the two solubilities and the limits on the pressure effect are shown.

In this paper we bring to bear other lines of evidence in an attempt to resolve this problem. These include the results of _in situ_ saturometry, the distribution of calcite and aragonite in the ocean sediments, and the results of kinetic experiments carried out in the laboratory.

IN SITU SATUROMETRY

Weyl (1961) developed a very simple technique for directly measuring the extent of supersaturation or undersaturation of a sea water sample. Ben-Yaakov (1970) and Ben-Yaakov and Kaplan (1971) adapted this technique for _in situ_ work in the deep ocean. Briefly, a pressure-compensated glass electrode is immersed in a cartridge containing calcite crystals (grain size $\sim 400\mu$). A pumping system is attached to the cartridge so that water can be drawn through it at so high a rate that interaction with the crystals is negligible. Under these circumstances the pH of the adjacent sea water is measured. When the pump is stopped the water trapped in the cartridge reacts with the calcite and the pH shifts toward the equilibrium value. As run by Ben-Yaakov, four-minute pumping periods are alternated with 12-minute periods during which the pore waters are isolated in the crystal pack. Typical traces using calcite (and using aragonite) are shown in Fig. 3.

Ben-Yaakov (1970) interprets the response to be exponential in character and assumes that the plateau reached in the case of aragonite and approached in the case of calcite represents equilibrium.

We (Broecker and Takahashi, in press) have combined Ben-Yaakov's field results with the GEOSECS data to obtain the _in situ_ carbonate ion content corresponding to this saturometer plateau. These are plotted as a function of water depth in Fig. 4. Included for comparison on this diagram are: (1) the solubility values (at 2^oC) of Ingle et al. (1973) and of Berner (1976), and (2) the carbonate ion content corresponding to the lysocline at five regions in the ocean. Clearly the saturometer, sediment, and the Ingle et al. (1973) solubility form a self-consistent data set. The slope of the "saturation" carbonate ion curve (equivalent to $\Delta V = -36\ cm^3\ mol^{-1}$) is consistent with the average of the two ΔV estimates for the dissolution of calcite at 2^oC given above ($-39\ cm^3\ mol^{-1}$). One of two conclusions might be drawn: (1) the results

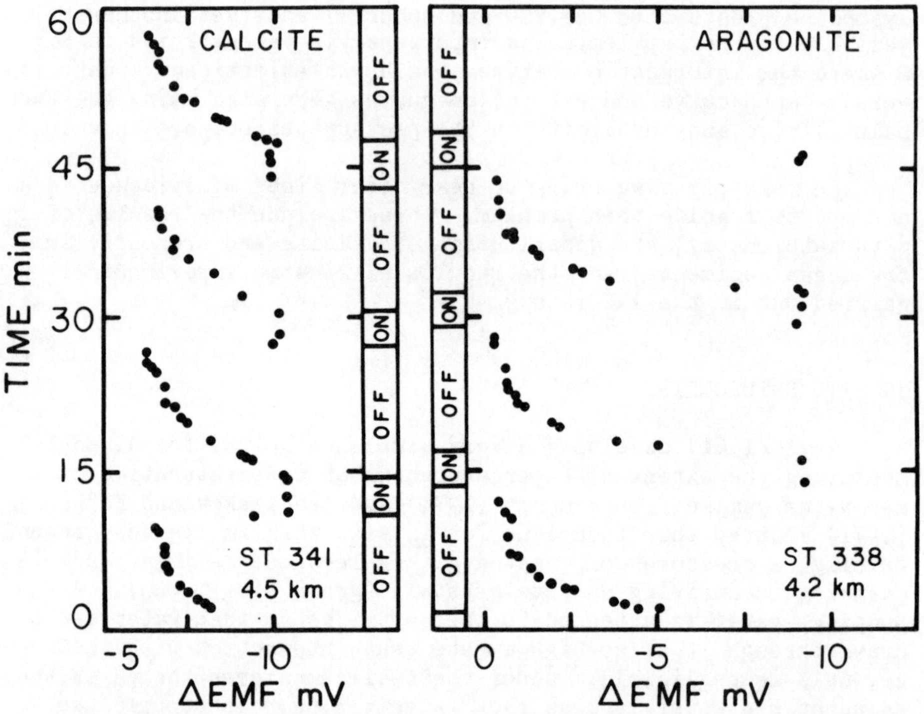

Fig. 3. Typical saturometer traces obtained using the Ben-Yaakov-Kaplan instrument (as modified for use by the Geochem. Secs. Prog.). Intervals during which sea water was flowing through the cartridge and during which the water in the cartridge was isolated are indicated.

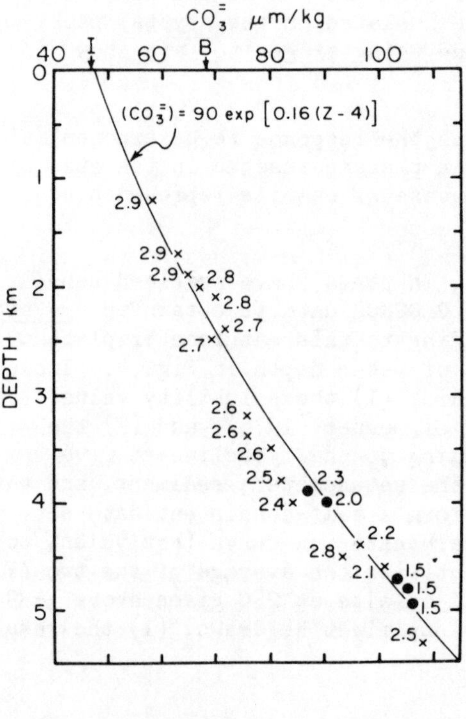

Fig. 4. The apparent solubility of calcite as indicated by <u>in situ</u> saturometry (x), by the oceanic lysocline ( ), and by laboratory studies at one atmosphere pressure (I for Ingle et al., 1973 and B for Berner, 1976). The numbers associated with the oceanic measurements are the phosphate content of the water (no phosphate was added to the artificial sea water used in the laboratory experiments). The equation fit through the points has the thermodynamic form; (CO_3^{2-}) has the units m/kg and Z is in km. The equivalent ΔV is -36 cm^3 mol^{-1}.

all correspond to the equilibrium solubility of calcite, or (2) some sort of kinetic impedance has shifted all the values by roughly the same amount yielding a pseudo-equilibrium curve. If the latter is correct it is necessary to call on a coincidence to generate the same kinetic shift in the sediments and in the Ingle et al. (1973) experiments as in the saturometer. It should be stated in this connection that the Ingle et al. experiments involved 24-hour equilibrations (as opposed to 0.2-hour equilibrations in the saturometer). Also if the second explanation is accepted, then the displacement is not strongly dependent on pressure (nor on phosphate content of the sea water as shown in Fig. 4).

Aragonite Results

A variety of laboratory studies have shown that aragonite should be about 1.47 ($\pm$ 0.02) times more soluble than calcite (see Berner, 1976). This difference should be the same in sea water as in fresh water (since neither the ionic strength nor the degree of major ion complexing will change when sea water is brought from equilibrium with calcite to equilibrium with aragonite). If there are major kinetic inhibitions to the equilibration of calcite with sea water then one would expect the magnitude of these inhibitions to be different for aragonite. In other words, the ratio of the apparent solubility of aragonite to the apparent solubility of calcite in sea water should not have the expected 1.47 value.

Two lines of evidence suggest that the apparent solubilities of the two phases are in the thermodynamic ratio. First if the lysocline for aragonite depth is predicted from the curve in Fig. 4, the result obtained is consistent with geologic observations along the mid-Atlantic Ridge (this cross-check cannot be made in the Pacific because the aragonite lysocline falls within the main thermocline where the in situ carbonate ion changes very rapidly with depth). Second, the saturometer results of Ben-Yaakov on aragonite also lie along the curve predicted using the thermodynamic ratio. Thus if these values do not represent the true solubility, then a kinetic shift occurs for aragonite which just matches that for calcite (i.e., such that the 1.47 solubility ratio is preserved).

Of course, the solubility experiments on which the 1.47 ratio is based could also be challenged. Similar kinetic effects could be called upon. Two arguments can be made in defence of values based on direct solubility measurements used by Berner (1976). The first is the ingenious experiment of Jamieston (1953) where he measured the conductivity of distilled water containing aragonite and containing calcite at various pressures. He found that at 3440 atmospheres (0°C) the two phases yielded the same conductivity and proposed this to be the equilibrium pressure. This corresponds to a free energy difference of 230 cal between the two phases at $0^{o}C$ and 1 atmosphere which in turn corresponds to a solubility ratio of 1.52.

The second confirmation comes from the results of direct conversion of calcite to aragonite. As summarized by Crawford et al. (1972) the $2^{o}C$ intercept of the phase boundary established in this way is 3110 atmospheres (corresponding to a ΔG of 210 cal and a solubility ratio of 1.47). As direct conversion has been accomplished at temperatures as low as $60^{o}C$ and as the slope of the phase boundary is well established, the extrapolation to $2^{o}C$ is firmly based.

LABORATORY KINETIC MEASUREMENTS

The kinetic studies of Morse and Berner (1974) suggest that the rate of calcite dissolution varies exponentially with the carbonate ion content undersaturation (see Fig. 5). They also suggest that the rate depends strongly on the phosphate content of the sea water. For the typical phosphate content of deep sea water (i.e. 2.0 $\pm$ 0.05 µm/kg) their rates follow the equation:

$$R \propto e^{-0.36\Delta CO_3^{2-}}$$

where ΔCO_3^{2-} is the difference between <u>apparent</u> calcite solubility value given above and the carbonate ion content present in their pH stat during a given experiment. The unit of ΔCO_3^{2-} is µm/kg. This strong exponential dependence has an important implication. The achievement of equilibrium will take very long times. Thus solubility estimates based on short term experiments could be misleading.

We have tried to assess the consequences of this to the saturometry experiments by contrasting equilibrium value obtained from the standard analysis of the saturometer curves with that obtained using the rate dependence given by the Morse-Berner experiments (see Fig. 6). In both cases best fits to experimental points were sought. As expected there is indeed a difference; the Berner-Morse kinetics yield an equilibrium CO_3^{2-} ion concentration about 4 µm/kg greater than that obtained from the linear kinetic treatment. If this treatment were extended to the Ingle et al. experiments it would suggest that even after 24 hours their solutions which approached equilibrium from undersaturation would fall 1 or 2 µm/kg short of the equilibrium carbonate ion content.

Although these calculations predict that short term experiments will lead to somewhat lower equilibrium carbonate ion

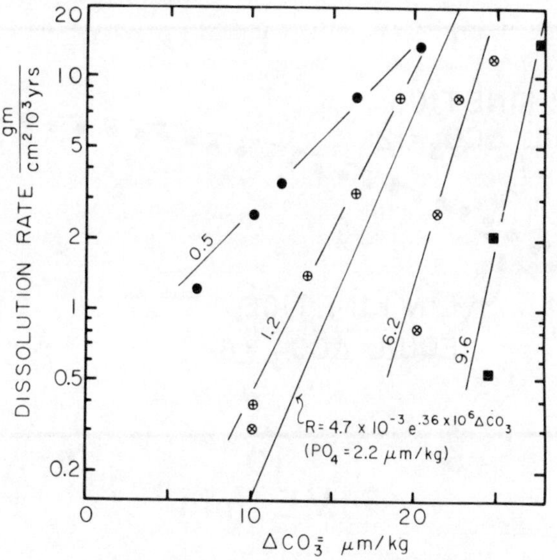

Fig. 5. Plot for the rate of dissolution (on a log scale) versus ΔCO_3^{2-} for the laboratory pH stat experiments of Morse (1974). The phosphate contents (in μm/kg) of the sea water used in each set of experiments are given on the best fit lines. ΔCO_3^{2-} is calculated using the "saturation" carbonate ion relationship presented in this paper. The interpolated curve for a phosphate content of 2.2 μm/kg adopted here is shown.

contents, this effect can in no way explain the 24 μm/kg difference between the 1 atm 2°C solubilities of Berner (1976) and of Ingle et al. (1973).

Rebuttal to Berner's Defence

In defence of his solubility results Berner makes two arguments with which we would like to take issue. First he states that since he calculated the solubility of calcite in sea water from measurements of the solubility of aragonite in sea water and the ratio of the solubility of calcite to aragonite as measured in fresh water that he avoids the serious kinetic effects associated with calcite in sea water. Whereas we do not challenge the legitimacy of Berner's approach, we do challenge his contention that direct measurements of the solubility of calcite in sea water cannot be made because of kinetic effects which are <u>not</u> shared by aragonite. In this connection we make four observations.

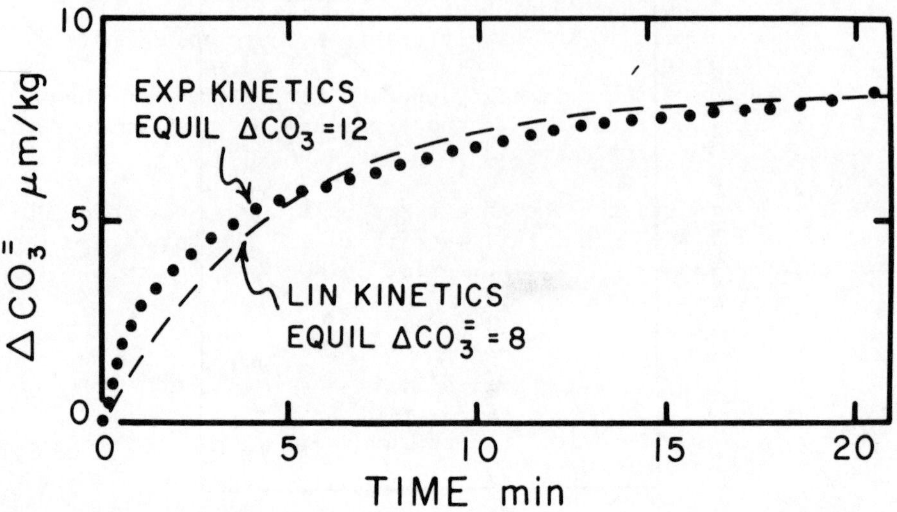

Fig. 6. Saturometer response models based on linear kinetics (dashed line) and on exponential kinetics (dotted line). The linear model gives a saturation ΔCO_3^{2-} of 8 μm/kg while the exponential model gives an equilibrium value of 12 μm/kg. Both curves are adequate fits to the actual saturometer results.

(1) The difference in depth of the calcite and aragonite lysoclines in the Atlantic is consistent with the 1.47 solubility ratio. (2) The apparent solubilities of calcite and aragonite as measured <u>in situ</u> saturometry are consistent with the thermodynamic ratio of 1.47 ± 0.02. (3) The response times for calcite and aragonite in the saturometer differ by no more than a factor of two (see Fig. 3). (4) Ingle et al. (1973) approached equilibrium from both under- and supersaturation for calcite in sea water and achieved the same result (i.e. they reversed the equilibrium).

Second, Berner uses the argument that his solubility for calcite is consistent with that calculated from the true thermodynamic constant whereas the value of Ingle et al. is not. The relationship between these constants is as follows:

$$K_C/K_{C'} = \gamma_{T\ Ca^{2+}} \cdot a_{H_2O} \cdot (\alpha_o/\alpha_s)(K_1 \cdot K_2/K_1' \cdot K_2') \quad (1)$$

and also

$$K_C/K_C' = \gamma_{T\ Ca^{2+}} \cdot \gamma_{T\ CO_3^{2-}} \quad (2)$$

K_C, K_1 and K_2 = the activity product for calcite, and the first and second dissociation constants for carbonic acid in infinitely dilute aqueous solutions.

K_C', K_1' and K_2' = the solubility product for calcite and the first and second apparent dissociation constants for carbonic acid sea water.

Table 1. The values for the parameters in Eq. (1) and the calculated value for the K_C/K_C' ratio at 25°C. All the values are for 25°C, and the sea water values are for 35.00°/oo salinity.

a_{H_2O}	0.982	
α_o	3.399×10^{-2} moles/ℓ.atm	Weiss (1974)
α_s	2.908×10^{-2} moles/ℓ.atm	Weiss (1974)
K_1	4.45×10^{-7}	Harned and Davis (1943)
K_2	4.65×10^{-11}	Harned and Scholes (1941)
K_1'	10.03×10^{-7} 10.09×10^{-7}	Mehrbach et al. (1974) Lyman (1956)
K_2'	7.70×10^{-10} 8.08×10^{-10}	Mehrbach et al. (1974) Lyman (1956)
$\gamma_{TCa^{2+}}$	0.225 ± 0.025	*/
K_C/K_C' (calculated)	$6.6\ (\pm 0.8) \times 10^{-3}$	K_1' and K_2' values of Mehrbach et al. (1974) were used in Eq. (1).
	$6.9\ (\pm 0.8) \times 10^{-3}$	K_1' and K_2' values of Lyman (1956) were used in Eq. (1).
K_C/K_C' (experimental)	$6.0\ (\pm 0.5) \times 10^{-3}$	Berner (1976) **/
	$7.1\ (\pm 0.6) \times 10^{-3}$	Ingle et al. (1973) **/

---

*/ The $\gamma_{TCa^{2+}}$ values in the literature range between 0.255 and 0.20 as reviewed by Whitfield (1973). A Brönsted-Guggenheim model calculation by Whitfield yields a value of 0.203.

**/ Berner (1976) stated the K_C/K_C' (or γ_{TCaCO_3}) value to be

0.060 for his K_C' value and 0.071 for the K_C' value of Ingle et al. (1973). Those values should be 0.0060 and 0.0071 respectively, and the corrected values are quoted here. The uncertainties quoted for the experimental K_C/K_C' values are derived from the errors in the K_C and K_C' values estimated by Berner (1976), i.e. $K_C = 3.56\ (\pm\ 0.10) \times 10^{-9}\ (\text{mol/kgH}_2\text{O})^2$ and $K_C' = 5.94\ (\pm\ 0.39) \times 10^{-7}\ (\text{mol}/\ell\text{SW})^2$ at 25°C.

α_o and α_s = The solubility of CO_2 gas in an infintely dilute aqueous solution and in sea water.

$\gamma_{TCa^{2+}}$ and $\gamma_{TCO_3^{2-}}$ = The total activity coefficient for Ca^{2+} and CO_3^{2-} ions. $\gamma_{TCa^{2+}}$ is a product of the fraction of uncomplexed Ca^{2+} ions and the single ion activity coefficient, and hence, yields the single ion activity for free Ca^{2+} ions when multiplied with the total Ca^{2+} ion concentration in sea water.

The dissolution constants are defined by:

$$K_1 = a_{H^+} \cdot a_{HCO_3^-}/\alpha_o \cdot pCO_2 \cdot a_{H_2O}$$

$$K_1' = a_{H^+} (HCO_3^-)/\alpha_s \cdot pCO_2$$

$$K_2 = a_H \cdot a_{CO_3^{2-}}/a_{HCO_3^-}$$

$$K_2' = a_H(CO_3^{2-})/(HCO_3^-)$$

where a_{H^+}. a_{H_2O}, $a_{HCO_3^-}$ and a $_{CO_3^{2-}}$ are respectively the activity of hydrogen ions, water, bicarbonate ions and carbonate ions; pCO_2 is the partial pressure of carbon dioxide gas exerted by the solution; and (HCO_3^-) and (CO_3^{2-}) are the total concentrations of HCO_3^- and CO_3^{2-} ion species including the free and complexed species.

The value for those parameters and the calculated K_C/K_C' values are listed in Table 1. The K_C/K_C' values which have been calculated using the K_1' and K_2' values of Lyman (1956) and Mehrbach et al. (1974) are listed. The K_C/K_C' value differs by 5% depending upon the choice of the apparent dissociation constant of carbonic acid in sea water. The literature values for $\gamma_{TCa^{2+}}$ vary from 0.255 to 0.20 (Whitfield, 1973). If this range is taken to represent the uncertainty in $\gamma_{TCa^{2+}}$, the K_C/K_C' value would be affected accordingly by as much as ± 12%. Berner (1976) chose the $\gamma_{TCO_3^{2-}}$ value of 0.030 by Pytkowicz (1975) and the $\gamma_{TCa^{2+}}$ value of 0.203 by Whitfield (1973), and obtained a K_C/K_C' value of 0.0061 which compares with 0.0060 for his K_C' value and 0.0071 for the K_C' value of Ingle et al. (1973). He considered that this close agreement of his K_C/K_C' value with this independent estimate supports his calcite solubility data. However, as shown in Table 1, the uncertainty for the calculated K_C/K_C', which mainly stems from the uncertainty in $\gamma_{TCa^{2+}}$ is so large that the calculated K_C/K_C' value cannot be effectively used for screening the experimental K_C' values. If a $\gamma_{TCa^{2+}}$ value of 0.237 obtained by Berner (1971) and by Simpson and Takahashi (1973) is used instead, Eq. (1) yields a K_C/K_C' value of 0.0073 and 0.0070 using respectively the K_1' and K_2' values of Lyman (1956) and of Mehrbach et al. (1974). These values tend to support the K_C' value of Ingle et al. (1973). At 2°C, the K_C' value of Berner differs from that of Ingle et al. by 35 per cent, and thus a comparison between the experimental and calculated K_C/K_C' ratios may be more effective for a selection of a reliable K_C' value. However, such a test is not presently possible because of a lack of the activity coefficient data for Ca^{2+} ion at 2°C.

POSSIBLE EXPLANATIONS FOR THE DIFFERENCES BETWEEN THE BERNER AND INGLE ET AL. SOLUBILITY RESULTS

Listed below are some of the possible explanations for the observed difference between the two modern solubility measurements.

1. Experimental errors in one or the other set of measurements. The observed difference of 35% in K_C' at 2°C corresponds to a difference of about 0.1 pH unit in the respective pH measurements. As both laboratories have long experience with the measurements involved, it is unlikely that the very large difference could stem from this source. At most a 5% difference could be explained in this way. As Berner recalculated all the values using the same set of constants, the difference cannot stem from errors in these constants.

2. Differences in the ingredients. Both laboratories used artificial sea water. Ingle et al. used the formula of Kester et al. (1967) (minus borate), while Berner used Turekian's compilation of sea water composition and included all species with concentrations greater than fluoride (hence borate was included). Neither group included phosphate. It is possible that the same calcite (or aragonite) exposed to different amounts of sea water, or the same amount of sea waters with different trace contaminant levels could yield different apparent solubilities due to different states induced into the outermost several molecular layers of the crystal surface. It must be noted that the crystal/sea water volume ratio employed in Berner's experiments is considerably smaller than that employed by Ingle et al. (1973) and for the saturometer experiments. This points out the need for experiments involving greatly different ratios of sea water volume to calcite surface area and experiments comparing real and artificial sea

water. In this connection it is unfortunate that both investigators did not measure the solubilities of <u>both</u> calcite <u>and</u> aragonite in sea water. Further calcite and aragonite "standard" crystals must be made available for interlaboratory comparisons.

3. Criteria for equilibrium: Although Ingle et al. (1973) obtained reproducible results by approaching saturation from undersaturated as well as from supersaturated solutions in the presence of calcite, and Berner (1976) did likewise for aragonite, it is not clear whether reversibility of the experiments can always be used as a criterion for equilibrium. It is possible that the results of Ingle et al. (1973) indicate an equilibrium between sea water and magnesian calcite coating on calcite crystals. In this connection it seems appropriate that solubility measurements be made in artificial sea water where the Mg^{2+} ion has been replaced by Ca^{2+} ion, as well as in sea water with the normal Mg^{2+}/Ca^{2+} ratio (i.e. 5 : 1).

CONCLUSIONS

Based on the facts assembled here we conclude that the laboratory solubility of Ingle et al. (1973) is consistent with observations made within the sea while that of Berner (1976) is not. However, it is possible that the Ingle et al. result does not represent thermodynamic equilibrium between calcite and sea water. Additional investigations must be made before this question can be answered.

ACKNOWLEDGEMENTS

We acknowledge the support of the Office of Naval Research for this research (through Grant N00014-75-C0210).

REFERENCES

Ben-Yaakov, S. Ph.D. Thesis, UCLA, 343 pp. (1970).
Ban-Yaakov, S. and Kaplan, I.R. J. Geophys. Res., <u>76</u>, (1971).
Berger, W.H. Deep-Sea Res., <u>15</u>, 31 (1968).
Berner, R.A. "Principles of Chemical Sedimentology", McGraw-Hill, N.Y., 240 pp., (1971).
Berner, R.A. and Morse, J.W. Am. J. Sci., <u>274</u>, 108 (1974).
Berner, R.A. Am. J. Sci., <u>276</u>, 713 (1976).
Biscaye, P.E., Kolla, V. and Turekian, K.K. J. Geophys. Res., <u>81</u>, 2595 (1976).
Kolla, V., Be', A.W.H. and Biscaye, P.E. J. Geophys. Res., <u>81</u>, 2605 (1976).

Bramlette, M.N. In Sears, M. ed., Oceanography: Am. Assoc. Adv. Sci. Pub., 67, 345 (1961).
Broecker, W.S. and Takahashi, T. Deep-Sea Res., (in press).
Broecker, W.S. and Broecker, S. In Studies in Paleo-Oceanography, SEPM Memoir 20, (W.W. Hay, editor, 44) (1974).
Crawford, W.A. and Hoersch, A.L. Am. Min., 57, 995 (1972).
Culberson, C.H. Ph.D. Thesis, Oregon State Univ., 178 pp. (1972).
Gardner, J.V. Cushman Foundation for Foraminiferal Research, Special Publication No. 13, 129-141 (1975).
Harned, H.S. and Scholes, S.R. J. Am. Chem. Soc., 63, 1706 (1941).
Harned, H.S. and Davies, R. J. Am. Chem. Soc., 65, 2030 (1943).
Hawley, J. and Pytkowicz, R.M. Geochim. Cosmochim. Acta, 33, 1557 (1969).
Ingle, S.E., Culberson, C.H., Hawley, J.E. and Pytkowicz, R.M. Marine Chem., 1, 295 (1973).
Ingle, S.E. Marine Chem., 3, 301 (1975).
Jamieson, J.C. J. Chem. Phys., 21, 1385 (1953).
Kester, D.R., Duedall, I.W., Connors, D.N and Pytkowicz, R.M. Limn. and Oceanogr., 12, 176 (1967).
Kipp, N.G. In Investigations of Late Quaternary Paleo-oceanography and Paleoclimatology, Geol. Soc. Amer. Memoir No. 145 (R. Cline and J. Hays, eds.), 3 (1976).
Lyman, J. Ph.D. Thesis, UCLA, 196 pp., (1956).
Mehrbach, C., Culberson, C.H., Hawley, J.E. and Pytkowicz, R.M. Limn. and Oceanogr., 18, 897 (1973).
Melguen, M. and Thiede, J. Marine Geology, (1974).
Millero, F.J. and Berner, R.A. Geochim. Cosmochim. Acta, 36, 92 (1972).
Parker, F.L. and Berger, W.H. Deep Sea Res., 18, 73 (1971).
Pytkowicz, R.M. and Connors, V.N. Science, 144, 840 (1964).
Pytkowicz, R.M., Disteche, A. and Disteche, S. Earth and Planet. Sci. Letters, 2, 430 (1967).
Pytkowicz, R.M. and Fowler, G.A. Geochem. J., 1, 169 (1967).
Pytkowicz, R.M. Limn. and Oceanogr., 10, 220 (1975).
Simpson, H.J. and Takahashi, T. In "Initial Reports of the Deep Sea Drilling Project", Vol. XX, National Science Foundation, Washington D.C., 877 (1973).
Valencia, M.J. Geochim. Cosmochim. Acta, 37, 35 (1973).
Weiss, R.F. Marine Chem., 2, 203 (1974).
Weyl, P.K. J. Geol., 69, 32 (1961).
Whitfield, M. Marine Chem., 1, 251 (1973).

NONEQUILIBRIUM THERMODYNAMICS IN METAMORPHISM[1]

George W. Fisher

Department of Earth and Planetary Sciences
Johns Hopkins University
Baltimore, Maryland 21218

INTRODUCTION

Recent studies of metamorphic structures provide a clear picture of the mechanisms involved in processes such as the replacement of one mineral assemblage by another at an isograd (Carmichael, 1969), the growth of reaction zones between incompatible assemblages (Vidale, 1969; A. B. Thompson, 1975), or the development of metamorphic segregations (Fisher, 1970). Taken together, these studies provide a basis for quantitative evaluation of the kinetics of metamorphic processes, with the twin goals of furthering our understanding of metamorphism and, eventually, using textures and structures in the rocks themselves as internal clocks to estimate the time involved in successive stages of a metamorphic cycle.

The first step is to formulate the necessary kinetic equations, and work out values for the appropriate kinetic coefficients from a combination of experiment, theory, and observations on natural rocks. There are several approaches we might use, but the formalism of nonequilibrium thermodynamics is particularly convenient for dealing with metamorphic processes, because it emphasizes the chemical potentials of components, which are often easier to evaluate than species concentrations. This paper reviews some ways in which the approach can be used to model the kinetics of metamorphic processes which can be studied in outcrop or thin-section; it will not consider regional processes such as heat flow or fluid convection (eg. Reed, 1970).

---

1 Financial support provided by Earth Sciences Section, National Science Foundation, NSF Grant DES 71-00406-A 02.

D. G. Fraser (ed.), Thermodynamics in Geology, 381-403.

FUNDAMENTAL RELATIONS

The majority of metamorphic processes can be visualized as the interplay of two simple mechanisms: diffusion of material from one assemblage to another, and chemical reactions between the diffusing species and adjacent mineral assemblages. Consequently, the kinetics of the overall process can be treated simply by evaluating the kinetics of these two steps, and their variations in space and time. In the formalism of nonequilibrium thermodynamics, diffusion rates are written in terms of the chemical potential gradients (Table 1, eq. (1.1)), and reaction rates are given in terms of reaction affinities, the difference between the chemical potentials of the participating species and their equilibrium values (eq. (1.2)). The link between diffusion and reaction is provided by the conservation equation (1.4), which expresses the fact that, at each point in a rock, any difference between the net loss of a component by diffusion and the net production by reaction produces a change in concentration.

The kinetic coefficients in eqs. (1.1) and (1.2) are matrices of phenomenological coefficients, defined so as to relate the fluxes to the corresponding potentials. Both matrices are symmetrical, so that $L_{ij}^D = L_{ji}^D$. The cross-terms, for which $i \neq j$, arise partly because of electrostatic interaction between ions and the presence of species containing more than one component, and partly because of motion of the reference frame used to measure diffusion. For systems involving diffusion of associated electrolytes and simple, one-component species, measured relative to a fixed reference frame, the cross-coefficients are small, and may usually be neglected. The straight coefficients are related to the conventional diffusion coefficients (D_{ii}) and dissolution rate constants (k_r) by expressions of the form

$$L_{ii}^D = \frac{c_i D_{ii}}{RT}, \quad \text{and} \quad L_{rr}^R = \frac{c_i^* k_r}{RT},$$

where c_i and c_i^* are the actual and equilibrium concentrations of the species involved (cf. Katchalsky and Curran, 1967). Both coefficients depend strongly on species concentrations, so relative L_{ii}^D and L_{rr}^R for species with comparable D_{ii} and k_r values can be estimated from solubility data alone. Order-of-magnitude estimates of typical coefficients in metamorphic rocks, based on the scanty experimental data available, are given in Table 2.

If the L_{rp}^R coefficients are much larger than the L_{ij}^D, local chemical reactions between diffusing species and adjacent minerals will occur almost instantaneously. Each mineral assem-

TABLE 1. PRINCIPAL KINETIC EQUATIONS IN METAMORPHIC PROCESSES

Rate of Diffusion of Component i

$$J_i^D = -\sum_{j=1}^{n} L_{ij}^D \nabla\mu j \qquad (1.1)$$

Rate of Reaction r

$$J_r^R = \sum_{p=1}^{m} L_{rp}^R A_p = \sum_{p=1}^{m} L_{rp}^R \sum_{i=1}^{n} \nu_i^r (\mu_i^* - \mu_i) \qquad (1.2)$$

Rate of Production of Component i

$$J_i^R = \sum_{r=1}^{m} \nu_i^r J_r^R \qquad (1.3)$$

Conservation Equation for Component i

$$\frac{dc_i}{dt} = -\frac{dJ_i^D}{dx} + \sum_{r=1}^{m} \nu_i^r J_r^R \qquad (1.4)$$

A_r - affinity of reaction r
c_i - concentration of component i
L_{ij}^D, L_{rp}^R - kinetic coefficients for diffusion and reaction
ν_i^r - stoichiometric coefficient of component i in reaction r
μ_i, μ_i^* - actual and equilibrium chemical potentials of i
$\nabla\mu_j$ - gradient in μ_j, equal to $d\mu j/dx$ in 1-dimensional diffusion, assumed throughout this paper.

blage will then be in local equilibrium (Fig. 1B), diffusion will be driven by the chemical potential gradients between assemblages (Fig. 1C), and the rate of the overall process will depend on the rate of diffusion between reaction sites. As discussed below, most metamorphic processes closely approximate a steady state, in which the potentials at each point in a growing structure remain constant. Eq. (1.4) therefore requires that the chemical reactions at each point must exactly balance the net influx by diffusion, dJ_i^D/dx. As a result, reactions occur mainly at the zone boundaries, where the fluxes change most rapidly (Fig. 1D), and diffusion-controlled structures characteristically have a strong spatial organization, with well-defined mineral zones showing sharp changes in composition at zone boundaries, all arranged in an orderly sequence of increasing or decreasing chemical potential (Fig. 1A).

TABLE 2. ESTIMATED VALUES FOR KINETIC COEFFICIENTS IN METAMORPHIS

Coefficient of Grain Boundary Diffusion

$$L_{ij}^{D} = \frac{2.538 \times 10^{-12}}{X^{grn}\ RT} \exp \frac{-8.37 \times 10^{4}}{RT} \quad mol^2/j \cdot m \cdot s$$

(estimated from published values for grain boundary diffusion in metals, and unpublished data of Fisher and Joesten on diffusion in calc-silicate nodules)

Coefficient of Lattice Diffusion in Silicates

$$L_{ij}^{D} = \frac{2.47}{RT} \exp \frac{-2.85 \times 10^{5}}{RT} \quad mol^2/j \cdot m \cdot s$$

(diffusion of K_2O in orthoclase, Foland, 1974)

Coefficient of Reaction Between Silicates and Aqueous Fluid

$$L_{rp}^{R} = \frac{4.59 \times 10^{3}}{X^{gb}\ RT} \exp \frac{-6.28 \times 10^{4}}{RT} \quad mol^2/j \cdot m^3 \cdot s$$

(estimated from values for silicate dissolution reactions at low temperature; Lerman et al., 1975; Hurd and Theyer, 1975)

X^{grn} - average grain diameter
X^{gb} - thickness of grain boundary film

On the other hand, if the L_{ij}^{D} are much larger than the L_{rp}^{R}, diffusion will be effectively instantaneous, virtually eliminating chemical potential gradients (Fig. 2B). The rate of the overall process will reflect the rates of the local chemical reactions, driven by the differences between the actual chemical potentials and their equilibrium values (Fig. 2C). Because the potential gradients are low, potential differences and reaction rates tend to be uniform over large volumes of rock (Fig. 2C), and reaction-controlled structures commonly lack the strong spatial organization of diffusion-controlled features. Newly precipitated minerals may be concentrated at favorable nucleation sites, but dissolution of preexisting minerals is much more widespread, and dissolution zones are poorly defined, without sharp boundaries (Fig. 2A). In multilayered reaction-controlled structures, the sequence of mineral zones will reflect variations in original bulk composition, distribution of favorable nucleation sites, or differences in the local reaction rate constants, rather than following a smooth progression of chemical potentials.

Many metamorphic structures have the strong spatial organization, sharp zone boundaries, and most importantly, the smooth

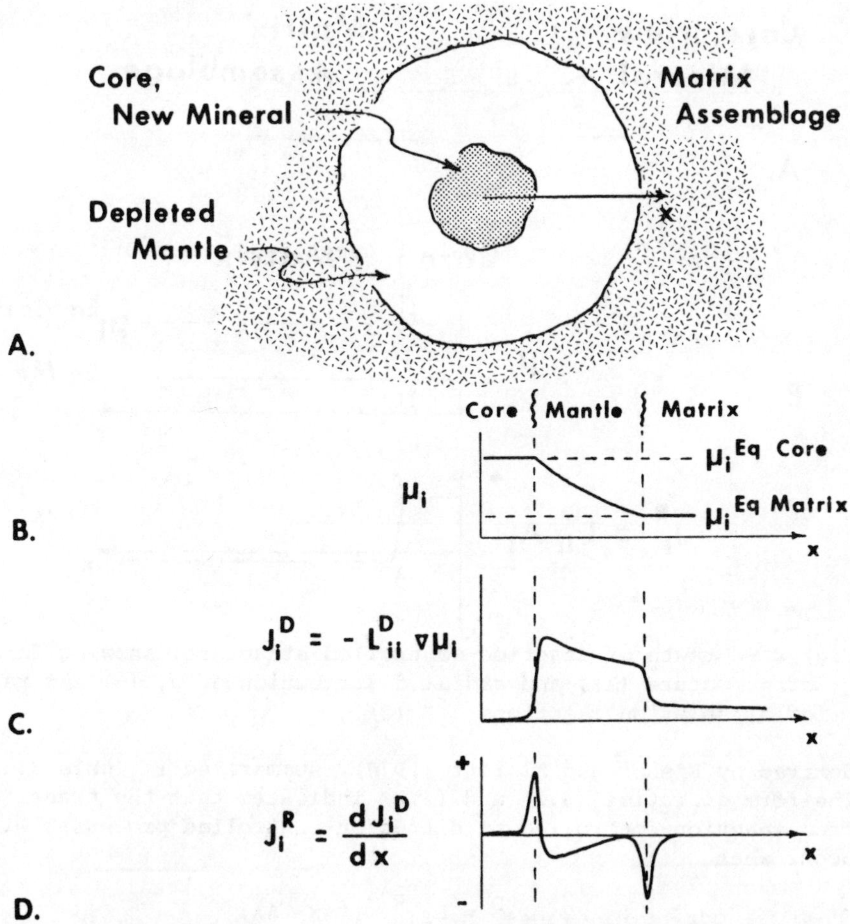

Fig. 1 - Growth of diffusion-controlled structure, showing form of structure (A), and radial distribution of μ_i (B), diffusive flux J_i^D (C), and rate of dissolution reactions J_i^R (D).

progression of assemblages characteristic of diffusion-controlled structures (eg. Loberg, 1963; Vidale, 1969; and A. B. Thompson, 1975), and it seems likely that most metamorphic processes are diffusion-controlled. In addition, many structures are preferentially developed parallel to a foliation or a lineation (eg. Loberg, 1963), suggesting that diffusion occurs mainly along grain boundaries, which tend to be more numerous and straighter parallel to the dominant fabric, rather than through crystal lattices.

The same conclusion emerges from a comparison of the estimated values for the kinetic coefficients in Table 2. Because L_{ij}^D and L_{rp}^R have different dimensions, direct comparison between the two can be misleading, and we must use the dimensionless ratios

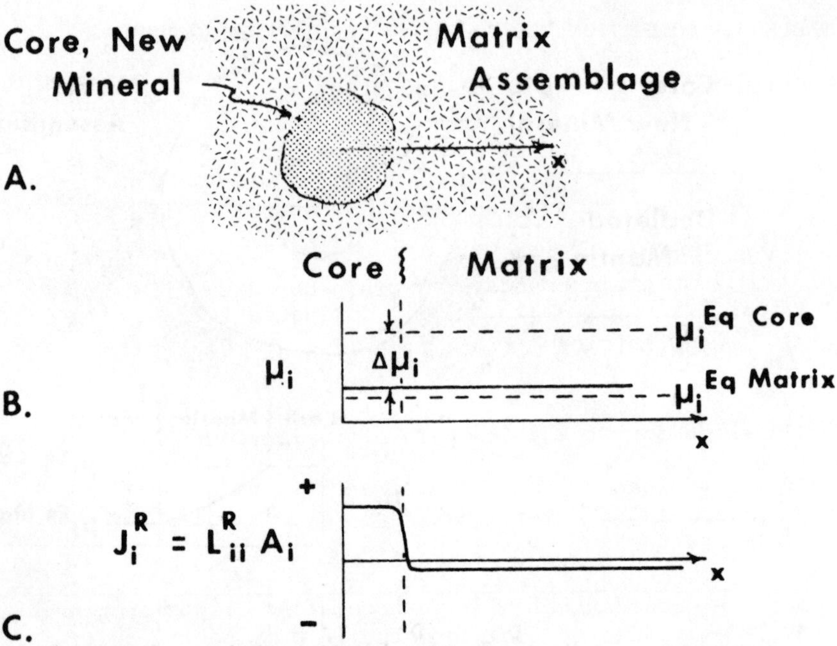

Fig. 2 - Growth of reaction-controlled structure, showing form of structure (A), and radial distribution of μ_i (B) and rate of dissolution reactions J_i^R (C).

derived by Fisher and Elliott (1974), summarized in Table 3. The form of ratios (3.1) and (3.2) indicates that the transition from reaction-controlled to diffusion-controlled processes will occur when

$$2\ L_{ij}^{D} = L_{rp}^{R}\ \nu_i^{r}\ X^{gb}\Delta X,$$

and the transition from grain-boundary to lattice diffusion will occur when

$$(L_{ij}^{D})^{\text{grain boundary}} = (L_{ij}^{D})^{\text{lattice}}$$

Using the estimated values for the coefficients (Table 2), we can use these expressions to delineate the range of temperature and structure size in which each mechanism will dominate (Fig. 3). It appears that reaction rates could control processes involving grains in the size range of crystal nuclei; but, Fig. 3 clearly suggests that grain boundary diffusion is the rate-determining step for any structure visible with a microscope. In order for reaction-controlled structures to form in rocks of normal grain size, it would be necessary to increase the grain-boundary diffusion coefficients drastically, reduce the reaction rate coefficients drastically, or both. The L_{ij}^{D} could be increased somewhat if penetrative deformation during recrystallization

TABLE 3. DIMENSIONLESS RATIOS GOVERNING METAMORPHIC PROCESSES

Criteria for Rate-Determining Step

$$\frac{2\,L_{ij}^{D}}{L_{rp}^{R}\,\nu_i^{r}\,X^{gb}\,\Delta X}\quad\begin{cases}\text{diffusion-controlled if} \ll 1\\ \text{reaction-controlled if} \gg 1\end{cases} \tag{3.1}$$

$$\frac{(L_{ij}^{D})_{\text{grain boundary}}}{(L_{ij}^{D})_{\text{lattice}}}\quad\begin{cases}\text{grain-boundary diffusion dominates if} \gg 1\\ \text{lattice diffusion dominates if} \ll 1\end{cases} \tag{3.2}$$

Criterion for Steady State in Diffusion-Controlled Processes

$$\frac{9\,X^{s}\,\Delta V_i\,\Delta c_i}{4\,\Delta X} \ll 1 \tag{3.3}$$

Δc_i - concentration difference along diffusion path

ΔV_i - volume of rock depleted in component i by transfer of one mole of i

ΔX - diffusion path-length

X^s - thickness of zone where reaction occurs between adjacent assemblages

decreased the effective grain size, or if the release of volatiles increased the grain-boundary diffusion coefficients. But it is difficult to see how these effects alone could increase the L_{ij}^D by the 8 or 9 orders of magnitude needed to induce reaction control. The only feasible way of inducing reaction control of metamorphic processes appears to be inhibition of the dissolution reactions, possibly by poisoning of the interfaces by impurities (Terjesen and others, 1961).

The relations and coefficients in Tables 1, 2 and 3 can be used to estimate the kinetics of many metamorphic processes, but the job will be much simpler if we make one additional assumption: that metamorphic processes closely approximate a steady state. Fisher (1973, p. 910-911; 1975, p. 115) showed that diffusion will automatically tend to shift potentials toward values such that the flux differences at every point in a structure just balance the local reactions, establishing a steady state. If this shift is rapid enough relative to growth of the overall structure, most of the diffusion will be driven by the steady-state potentials, and the small amount of mass transfer involved in attaining the steady state may be neglected.

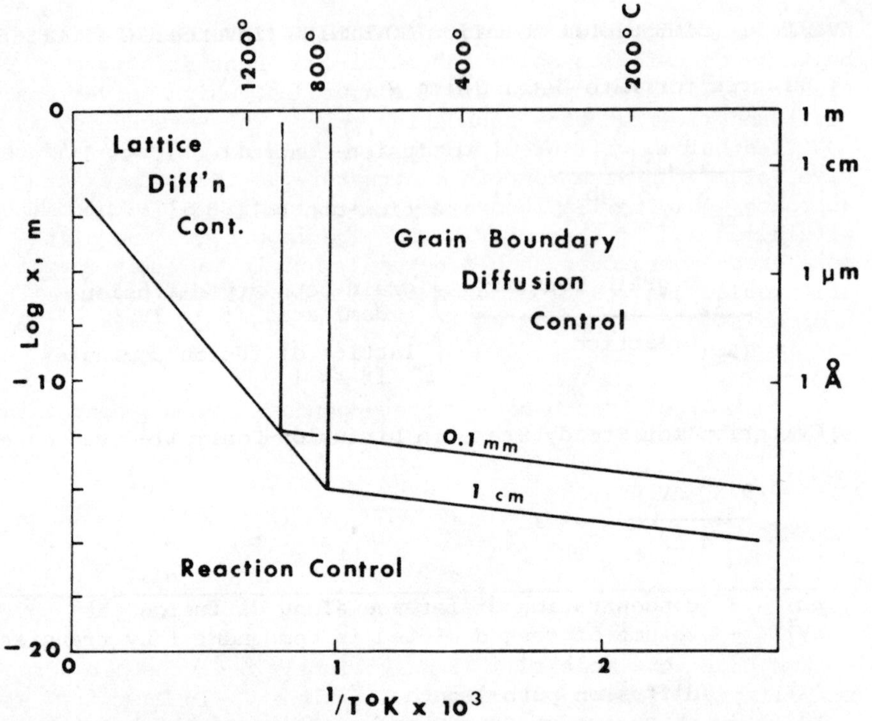

Fig. 3 - Plot showing variation in rate-controlling mechanism with Log of structure size (X) in m, and 1/T°K; boundaries of grain-boundary diffusion field contoured in grain size.

Physically, this assumption is reasonable, because the amount of mass transfer required to adjust potentials in the grain boundary film to steady state values is infinitesimal compared to the amount of diffusion required to grow a visible structure. Fisher and Elliott (1974) derived a dimensionless ratio which provides a more rigorous way of evaluating the steady-state assumption; for typical metamorphic segregations, the ratio has values on the order of 10^{-6}, well into the range of steady-state processes, confirming our intuition.

MODELS FOR DIFFUSION-CONTROLLED, STEADY-STATE PROCESSES

This reasoning suggests that we should be able to predict the kinetics of many metamorphic processes by simultaneous solution of eq. (1.1) for grain-boundary diffusion between assemblages in local equilibrium and eq. (1.4) for the steady state, where $dc_i/dt = 0$. Frantz and Mao (1974, 1975, 1976) and Weare and others (1976) have outlined analytical and numerical methods for solving these equations throughout a growing structure for a variety of boundary conditions. These approaches provide a nearly exact solution to many problems of interest, but can be time-consuming to solve for multi-component structures. Fisher (1975) developed a simpler approach, which solves eqs. (1.1) and (1.4) at zone boundaries only. This procedure, as modified below, will predict the same bulk composition and relative volume for each zone in a structure as the more exact approach, but it will ignore any variation in modal composition within zones. If these variations are important, one of the more exact approaches should be used, but in many metamorphic structures, the reaction zones are nearly uniform in composition, and the simpler approach will be sufficiently accurate. I will use it here.

Consider a hypothetical three-component system containing minerals A, B and C, with the molar compositions

Component	A	B	C
1	2	1	1
2	1	2	1
3	2	2	1

Assume that the assemblage AB has slightly over-stepped a reaction stabilizing C, due to sluggish nucleation of C. The first step in predicting the resulting structure is to find the sequence of mineral zones. At fixed T and P, each mineral will define a set of equilibrium potentials given by an equation of the form $\overline{G_A} = n_1^A\mu_1 + n_2^A\mu_2 + n_3^A\mu_3$, where $\overline{G_A}$ is the molar free energy of A, and n_i^A is the molar amount of component i in A. At constant T and P, the molar free energies are constant, so the three equations define a faceted equilibrium surface in chemical potential space (Fig. 4). The assemblage AB will initially lie at some point x on the metastable extension of the line AB. Once C nucleates, there may be a short stage of reaction-controlled growth, driven by the difference between the equilibrium potentials for ABC and the actual potentials, at point x. But as soon as C becomes large enough for growth to become diffusion-controlled, potentials in the grain-boundary network adjacent to ABC will shift to the invariant point, and radial potential gradients will develop between ABC and the more distant parts of the matrix, still at point x. Potentials along the diffusion path will be in equilibrium with A and B, so the potential gradients will be subject to the constraints

$$2\nabla\mu_1 + \nabla\mu_2 + 2\nabla\mu_3 = 0 \quad \text{and} \quad \nabla\mu_1 + 2\nabla\mu_2 + 2\nabla\mu_3 = 0,$$

obtained by differentiating the potential surfaces for A and B along the diffusion path. Assuming that the amount of overstepping is such that $\nabla\mu_1 = 1$ adjacent to C, letting $L_{11}^D = 1$, $L_{22}^D = 5$, $L_{33}^D = 10$, and assuming that all cross-coefficients

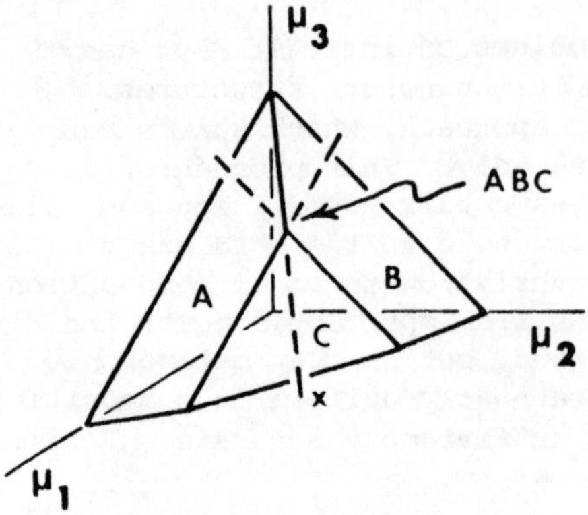

Fig. 4 - Equilibrium potential surfaces for hypothetical minerals A, B and C in system $\underline{1}$ - $\underline{2}$ - $\underline{3}$.

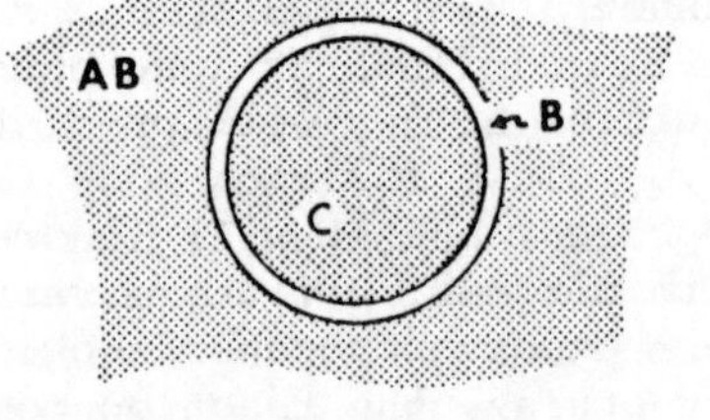

Fig. 5 - Form of structure resulting from growth of C in AB matrix, assuming molar volumes $\bar{v}_A = 50$, $\bar{v}_B = 50$, $\bar{v}_C = 30\ cm^3$.

are zero/2, the flux equations (1.1) become

$$J_1{}^D = -L_{11}{}^D \nabla\mu_1 \qquad = -1(1) \qquad = -1$$

$$J_2{}^D = -L_{22}{}^D \nabla\mu_2 = -L_{22}{}^D(\nabla\mu_1) \qquad = -5(1) \qquad = -5$$

$$J_3{}^D = -L_{33}{}^D \nabla\mu_3 = -L_{33}{}^D(-1.5\nabla\mu_1) = -10(-1.5) = +15.$$

In order for the assemblage ABC to remain at the invariant, these fluxes must be balanced by reactions among A, B and C. The net reaction must be a linear combination of the individual precipitation/solution reactions for A, B and C, each of the form $A = n_1{}^A\underline{1} + n_2{}^A\underline{2} + n_3{}^A\underline{3}$. The rates of these three reactions are controlled entirely by the diffusion rates, and may be found by solving eq. (1.4) for each of the three components, with $dc_i/dt = 0$,

$$-1 = 2J_A{}^R + 1J_B{}^R + 1J_C{}^R$$

$$-5 = 1J_A{}^R + 2J_B{}^R + 1J_C{}^R$$

$$+15 = 2J_A{}^R + 2J_B{}^R + 1J_C{}^R ,$$

which yields $J_A{}^R = 20$, $J_B{}^R = 16$, and $J_C{}^R = -57$, corresponding to the net reaction

$$20\ A + 16\ B + 1\ \underline{1} + 5\ \underline{2} \rightarrow 57\ C + 15\ \underline{3}\ .$$

/2 This assumption is made purely in order to simplify the arithmetic for presentation here, and is in no way necessary to carry out the calculations; if non-zero cross-terms are present, they may simply be incorporated in eq. (1.1) in the normal way.

This reaction consumes slightly more A than B, so if the rock initially contained equal amounts of A and B, a thin mantle of pure B, from which all of the A has been consumed, will begin to form around the growing grain of C as soon as the reaction becomes diffusion-controlled (Fig. 5).

The second step is to work out the fluxes within each zone, and the reactions at the zone boundaries. Assuming that $\nabla\mu_1$ still has the value 1 adjacent to C, and introducing the constraints imposed on the $\nabla\mu_i$ in the mantle by equilibrium with B, the flux equations in the inner part of the mantle become

$$J_1^D = -1(1) = -1$$

$$J_2^D = -5(\nabla\mu_2) = -5\nabla\mu_2$$

$$J_3^D = -10(-\nabla\mu_2 - 0.5) = +10\,\nabla\mu_2 + 5 \quad .$$

Consequently, we have three unknowns in the system, $\nabla\mu_2$, J_B^R, and J_C^R, which may be found from the three conservation equations. In the steady state, eq. (1.4) equates the local reaction rates to the changes in the fluxes at successive points along the diffusion path. In the problem considered here, all fluxes must vanish at the center of the C core, so the net change in the fluxes between the B/C boundary and the center of the core must equal the fluxes in the inner part of the mantle. The calculation can be simplified by assuming that the fluxes are zero throughout the core, so that the changes in the fluxes, and hence the reactions, occur only at the B/C boundary. This procedure is equivalent to integrating eq. (1.4) over the radius of the core, and it will yield an exact solution to the bulk composition of the core. But it will not reveal any local variations in core composition due to internal precipitation. The near uniformity of core compositions in many natural segregations suggests that a detailed analysis of the distribution of internal precipitation will rarely be needed; but if it is, the procedures of Frantz and Mao (1976) and Weare and others (1976) may be used.

Substituting the appropriate values into eq. (1.4),

$$1\,J_B^R + 1\,J_C^R = -1$$

$$2\,J_B^R + 1\,J_C^R = -5\nabla\mu_2$$

$$2\,J_B^R + 1\,J_C^R = 10\nabla\mu_2 + 5,$$

which gives $J_B^R = 2.667$, $J_C^R = -3.667$, and $\nabla\mu_2 = -0.333$, corresponding to the reaction

$$2.667\ B + 1\ \underline{1} \rightarrow 3.667\ C + 1.667\ \underline{2} + 1.667\ \underline{3}\ ;$$

and the fluxes in the inner part of the mantle are $J_1^D = -1.000$, $J_2^D = 1.667$, and $J_3^D = 1.667$, using the convention that outward fluxes are positive.

Using eq. (1.4), we can next find the reactions at the AB/B boundary in terms of the flux differences across the boundary,

$$(J_i^D)^{matrix} - (J_i^D)^{mantle} = n_i^A J_A^R + n_i^B J_B^R \ .$$

Both A and B are present in the matrix, so the fluxes there are subject to the corresponding constraints, and become $J_1^D = -\nabla\mu_1$, $J_2^D = -5\nabla\mu_1$, and $J_3^D = +15\nabla\mu_1$, where $\nabla\mu_1$ is the value in the matrix. J_A^R and J_B^R are also unknown, so in order to solve the conservation equations, we need to determine the fluxes in the outer part of the mantle. Again we can simplify the calculation by assuming that the fluxes are uniform throughout the mantle, so that we may use the values already found for the fluxes in the inner mantle/3. This procedure is equivalent to integrating eq. (1.4) across the entire mantle, including its outer (but not its inner) boundary, and treats the precipitation reactions as if they occur only at the mantle-matrix boundary. As in the case of the core/mantle reaction, this approach will yield an exact solution to the bulk composition of the mantle, but it will not provide information on any local variations in mantle composition due to internal precipitation. Making the appropriate substitutions in the conservation equation at the boundary, we find $J_A^R = 1.286$, $J_B^R = -1.638$, and $\nabla\mu_1 = +0.064$, corresponding to the reaction and matrix fluxes shown in Fig. 6. This cyclical reaction is a simple way of representing the entire process, and represents the amount of reaction that occurs during the time required to transfer one mole of component $\underline{1}$ inward across the mantle.

In these calculations, I have tacitly assumed that the L_{ij}^D are constant throughout the growing structure. This assumption is not strictly correct, because the L_{ij}^D depend upon c_i. It should hold to a good approximation for many metamorphic structures,

/3 In a planar segregation, we may directly equate the fluxes at the inner and outer mantle boundaries, each of which is measured per unit area; but in spherical segregations, we must equate the total flux passing through a concentric shell in the inner mantle to the total flux passing through a concentric shell in the outer mantle. That is, we must assume $4\pi r^2 \cdot J_i^D =$ constant. We can do so by including an implicit $4\pi r^2$ term in J_i^D and $\nabla\mu_i$ for spherical structures; with this understanding, the equations presented here may be applied to both spherical and planar segregations.

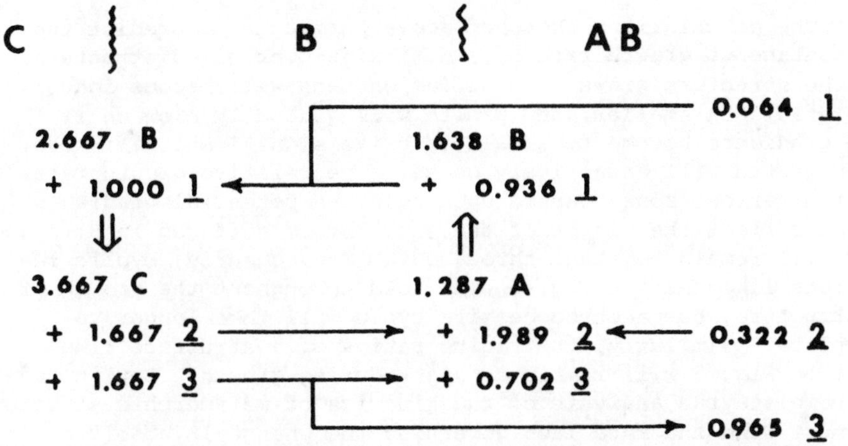

Fig. 6 - Cyclic reaction for growth of structure in Fig. 5; fluxes are measured with reference to a set of inert markers.

where the potential differences between interacting assemblages are small (eg. Fisher, 1970), but in structures involving large potential differences, this assumption may lead to serious errors (cf. Frantz and Mao, 1976) and different $L_{ij}{}^D$ values should be used at each zone boundary. In the steady state, the potentials and therefore the $L_{ij}{}^D$ will remain constant at each boundary. However, the use of different $L_{ij}{}^D$ values at successive boundaries will impose flux differences across each zone, which must be balanced by internal precipitation. The net amount of this internal precipitation is easily found by a simple modification of the procedure outlined above: simply solve eq. (1.4) for the total flux difference across the mantle, in effect integrating the internal precipitation throughout the mantle, using the restrictions imposed on the $\nabla\mu_i$ by equilibrium with B, and remembering that any reaction within the mantle must involve mineral B only. This procedure will also give a local value for the fluxes at the AB/B boundary, which may be used in eq. (1.4) to obtain the reaction at the AB/B boundary as before.

These methods are easily generalized to systems containing any number of components and minerals. In a system of c components, a total of c $\nabla\mu_i$ must be found in each zone; and at each zone boundary involving p minerals, a total of p reaction rates must be found. But provided that the amount of overstepping is known, there are p restrictions on the $\nabla\mu_i$ (although in some cases it is necessary to solve the equations for two or more zones simultaneously in order to find them all), so only (c + p - p) unknowns are independent. Consequently, the c conservation equations will always be sufficient to determine all the variables. If the over-stepping is not known, an arbitrary value may be assumed, and the ratios of the diffusion and precipitation rates determined.

The calculations sketched above permit us to predict the instantaneous growth rate of a diffusion-controlled structure. As the structure grows, the diffusion path will become longer, the gradients smaller, and growth will gradually slow until the gradients become too small to drive significant diffusion, when growth will effectively cease. The relative growth rates of the various zones depend upon relative potential gradients, which reflect the slopes of the equilibrium surfaces in μ_i space, and will remain constant throughout. Consequently, cyclic reactions like that in Fig. 6 will hold throughout the growth of a structure, but each successive cycle will take longer to complete. Similarly, the radius ratios of a structure like that in Fig. 5 will remain constant during growth. But in order to complete the analysis of the kinetics of metamorphic structures, we must find the rate laws governing the change in absolute size of the structure. This is easily done by using the reaction at the core/mantle boundary, together with the molar volumes of the minerals, to write an expression for the growth rate of the core in terms of the flux of one component through the mantle, then integrating over time. Table 4 presents the rate laws for the processes and structures considered here; all assume that growth is isothermal and isobaric, so that the potential differences driving the process are constant. Comparable equations for closely related processes have been developed by Helgeson (1971), Ildefonse and Gabis (1976) and others.

THE TRANSFORMATION TO PRACTICAL REFERENCE FRAMES

So far, all of our calculations have been made relative to a reference frame fixed on a set of inert markers (Darken, 1948; Brady, 1975). This frame corresponds most closely to our intuitive view of diffusive behavior, and is useful in theoretical work. But few metamorphic rocks contain suitable inert markers, and we are frequently forced to choose a more practical reference frame, such as one moving with the mean velocity of all components, or with the velocity of a single component. The conversion from one frame to another is simplest if we rewrite the fluxes in the form $J_i^F = v_i^F C_i$, where F represents any reference frame, and v_i is the velocity of i. Then, to convert from frame 1 to frame 2,

$$J_i^{F2} = J_i^{F1} - C_i v_{F2}^{F1} = C_i (v_i^{F1} - v_{F2}^{F1}),$$

where v_{F1}^{F2} is the velocity of frame 2 with respect to frame 1, and C_i is the local concentration of i in the rock. For example, the velocities in the mantle of the structure in Fig. 6 are $v_1 = -1.0/0.2 = -5.0$, and $v_2 = v_3 = 1.667/0.4 = 4.168$. Consequently, the flux of component 2 relative to a frame fixed on component 1 is $J_2^1 = 0.4(4.168 + 5.0) = 3.667$, and similarly for J_3^1. Fig. 7 shows the cyclic reaction of Fig. 6, converted

TABLE 4. RATE LAWS FOR METAMORPHIC PROCESSES

Diffusion-Controlled Processes

Planar Structures

$$t = \frac{X_c^{\,2}(\alpha-1)}{2\beta L_{ii}{}^{D}\Delta\mu_i} \qquad (4.1)$$

Spherical Structures

$$t = \frac{R_c^{\,2}(\gamma-1)}{2\beta L_{ii}{}^{D}\gamma\Delta\mu_i} \qquad (4.2)$$

Reaction-Controlled Processes

$$t = \frac{R_c}{2\beta L_{rr}{}^{R}\Delta\mu_i X^{gb}} \qquad (4.3)$$

α = X_{mantle}/X_{core}

β = (d core volume/dt)$/J_i{}^D$

γ = R_{mantle}/R_{core}

$\Delta\mu_i$ = potential difference across mantle

R_c, X_c = radius or thickness of core

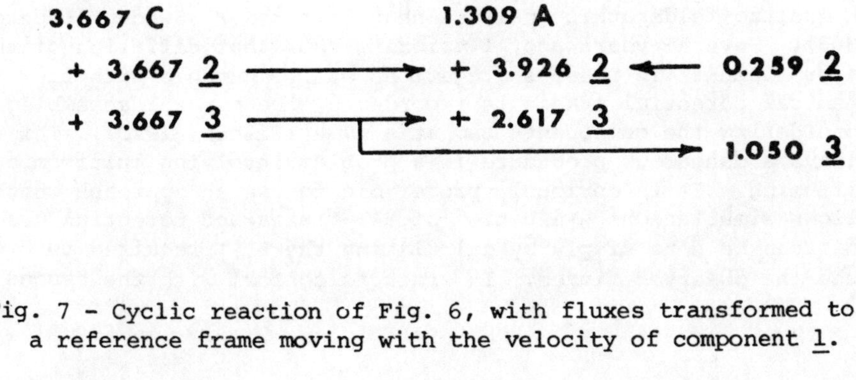

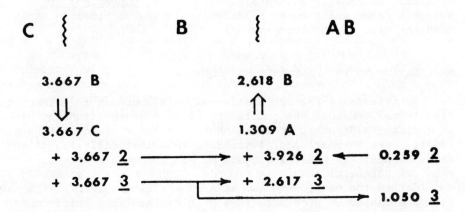

Fig. 7 - Cyclic reaction of Fig. 6, with fluxes transformed to a reference frame moving with the velocity of component 1.

to a frame fixed on component 1. Note that in this frame, the flux of 1 is zero by definition, so that the reactions at each zone boundary must be written to conserve 1. But the overall reaction produces a mantle with 0.260 moles of B, and a core with 3.667 moles of C, just as in the cyclic reaction of Fig. 6.

The coefficients relating the J_i^1 to the $\nabla\mu_i$, which we will write L_{ij}^1, are related to the L_{ij}^D in the inert frame by the expression

$$L_{ij}^1 = L_{ij}^D - \frac{c_i}{c_1} L_{1j}^D - \frac{c_j}{c_1} L_{i1}^D + \frac{c_i c_j}{c_1^2} L_{11}^D$$

(Kirkwood and others, 1960). Note that, unlike the L_{ij}^D, the L_{ij}^1 depend on the concentrations of the components in the rock, and will therefore vary markedly from one zone to another. For example, the L_{ij}^1 in the matrix of the structure in Fig. 5 are $L_{22}^1 = 6.00$, $L_{23}^1 = L_{32}^1 = 1.33$, and $L_{33}^1 = 11.78$, while those in the mantle are $L_{22}^1 = 9.0$, $L_{23}^1 = L_{32}^1 = 4.0$, and $L_{33}^1 = 14$. All of the previous calculations could equally well have been carried out with these coefficients, and would have given the cycle of Fig. 7 directly.

There are many possible choices of reference frame; most of the common ones are reviewed by Kirkwood and others (1960) and by Brady (1975). It is obviously essential to use a reference frame which can be measured accurately, so it is often preferable to use a single easily-measured component, rather than using a frame which depends on all of the fluxes, some of which may be difficult to measure accurately. It is also convenient to select a frame which moves relatively slowly, in order to minimize the size of the cross-terms.

APPLICATION TO THE VASTERVIK SEGREGATIONS

To illustrate the application of this approach to metamorphic structures, consider the andalusite-biotite segregations developed in quartzo-feldspathic gneisses near Vastervik, Sweden (Loberg, 1963). Several years ago, I tried to show that diffusion of the major elements in these segregations was compatible with the chemical potential gradients recorded by the mineral assemblages, considering the components two at a time (Fisher, 1970). This is always a dangerous procedure in a problem involving multicomponent diffusion. It is obviously preferable to use an approach which allows simultaneous evaluation of all fluxes and potential gradients. This can be done simply by calculating the L_{ij}^D required to produce the observed fluxes. In order to conform with the second law

TABLE 5. ANALYSIS OF VASTERVIK SEGREGATIONS

	Chemical Compositions[1] cation percent			Observed Fluxes J_i^{Al}=(3)-(1)	L_{ii}^{Al}
	matrix	mantle			
	(1)	(2)	(3)[2]	(4)	(5)[3]
SiO_2	72.75	71.73	64.12	-8.63	----
$AlO_{3/2}$	13.32	14.90	13.32	----	----
FeO	1.53	0.28	0.25	-1.28	50
MgO	1.70	0.04	0.04	-1.66	32
CaO	0.30	0.26	0.23	-0.07	.006
$NaO_{1/2}$	3.95	5.15	4.60	+0.65	5
$KO_{1/2}$	6.47	7.64	6.83	+0.36	1
	100.00	100.00	89.40		

1 - Recalculated from Fisher, 1973, Table 3.
2 - Normalized to Al content of matrix.
3 - Based on assumed Ca flux of +0.01.

of thermodynamics, the local entropy production in any irreversible process must be positive, which in turn requires that the diagonal L_{ii}^D coefficients be positive, and that the off-diagonal L_{ij}^D terms be smaller than $\sqrt{L_{ii}^D L_{jj}^D}$ (Katchalsky and Curran, 1967, p. 91). If a set of coefficients conforming to these conditions can be found, the diffusion is compatible with the observed potential gradients; if not, it is incompatible.

The first step is to measure the fluxes. Using Al as a reference frame, we can measure the change in flux at the mantle/matrix boundary simply by comparing the composition of the matrix (Table 5, col. 1) with that of the mantle, normalized to the Al content of the matrix (col. 3). There was little exchange of material with the matrix during growth (Loberg, 1963), so these flux differences approximate the fluxes in the mantle (col. 4). The fluxes were constrained by equilibrium with microcline, plagioclase, biotite, and quartz; writing these equilibria using the mineral compositions given by Fisher (1973, p. 920), and conserving Al to conform with our choice of reference frame, we obtain three constraints on the $\nabla\mu_i$,

$$\nabla\mu_{Ca} = 6.15\ \nabla\mu_K - 4.00\ \nabla\mu_{Na}$$

$$\nabla\mu_{Mg} = -0.27\ \nabla\mu_K + 0.23\ \nabla\mu_{Na} - 0.62\ \nabla\mu_{Fe}$$

$$\nabla\mu_{Si} = 0,$$

where the subscripts represent abbreviations for oxide components.

Substituting these into eq. (1.1), we can attempt to find a set of L_{ii}^{Al} terms compatible with the observed fluxes. We cannot obtain a unique solution, because we have nine unknowns (six L_{ii}^{Al} and three $\nabla\mu_i$) and only six equations. But, by assuming successive values for three of the L_{ii}^{Al}, and solving for the other three, we can find whether or not a set of positive L_{ii}^{Al} exists.

Using the fluxes as given in Table 5, it is impossible to find a set of positive L_{ii}^{Al}, even allowing the assumed values to range over 20 orders of magnitude. Difficulties arise with Si and Ca. The problem with Si is that $\nabla\mu_{Si} = 0$ because quartz is ubiquitous, so no L_{Si}^{Al} coefficient can account for the observed flux. Assuming that the flux is real, several possibilities exist: J_{Si}^{Al} could reflect a cross-coefficient coupling it to some other $\nabla\mu_i$ term (for example, $\nabla\mu(H_2O)$, due to the presence of $Si(OH)_4$); it could have been driven by a pressure gradient reflecting local volume changes; or it could reflect an infiltration mechanism. The problem with Ca is that the coefficient turns out to be negative for a very broad range of assumed coefficients. This could be avoided by invoking cross-coupling with $\nabla\mu_{Mg}$, but the problem is more likely uncertainty in the CaO data. The CaO difference between mantle and matrix is probably within analytical uncertainty, so the sign of the flux is uncertain. A reasonable set of positive L_{ii}^{D} coefficients can be found by setting $J_{Ca}^{Al} = +0.01$ instead of -0.07 (Table 5, col. 5). Accordingly, the behavior of SiO_2 remains to be clarified, but the fluxes of the other major components are compatible with the observed gradients within analytical uncertainty, and a diffusion-controlled mechanism is a realistic possibility for these segregations.

If we had independent estimates of the L_{ij}^{Al}, we could use them to predict a cyclic reaction, and compare the predicted segregation with the observed one. In this case, of course, the effort is futile, because agreement would merely confirm the correctness of our arithmetic and the consistency of our data.

We can, however, use the estimated L_{ij}^{D} values of Table 2, together with the observed geometry and composition of the segregations and Fisher's (1970) calculation of the K potential difference across the mantle, to calculate the growth rates of these segregations from the rate laws in Table 4. Fig. 8 is a log-log plot of the resulting growth curves, and provides a useful way of visualizing a variety of metamorphic processes. The form of this plot depends on the relative slopes of the two curves; the reaction-controlled curve has a slope of one, because size varies linearly with t, while the diffusion-controlled curve has a slope of 1/2, because size varies with $\sqrt{t}$. The effective curve at any stage of a process will be the lower one,

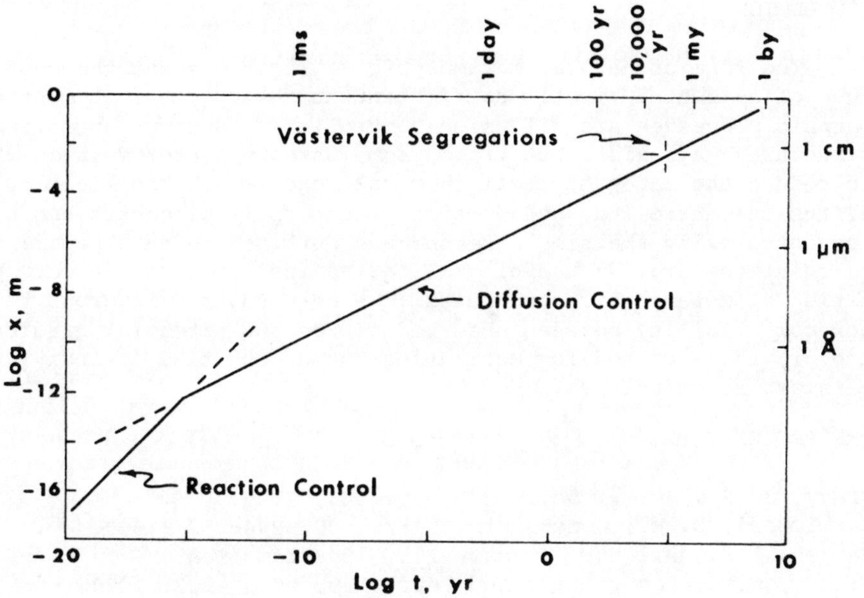

Fig. 8 - Plot of Log of core radius (X) vs. Log time (t) for Vastervik segregations.

so if a process involves a reaction-controlled stage, it must always precede the diffusion-controlled stage. This plot strongly supports the interpretation that the segregations were diffusion-controlled. The curves intersect at a size slightly less than 1 Å. In this size range, our model obviously loses its relevance, but if the coefficients in Table 2 are at all correct, any reaction-controlled stage must have ended long before the segregations grew to recognizable size, and the relevant curve must have been the diffusion-controlled one. This curve reaches the observed core radius of 0.5 cm at a time of about 66,000 years, which seems to be a reasonable time estimate for a single stage in a metamorphic episode.

To the extent that these segregations are typical metamorphic structures, this plot also reinforces the general conclusion that most metamorphic structures are diffusion-controlled, and that the only feasible way to produce reaction-controlled structures is to reduce L_{rp}^R. If L_{ij}^D were raised enough to allow processes to follow the reaction-controlled curve, structures a meter thick could form in a few hundred years. The position of the diffusion-controlled curve agrees much better with geologic experience; it indicates that structures should not grow much larger than 10 to 50 cm, even in the time span of an orogenic episode, say 100 my.

CONCLUSION

The form of natural metamorphic segregations and the magnitude of estimated kinetic coefficients both suggest that most metamorphic processes are diffusion-controlled; reaction-controlled structures may exist, but if so, some unknown factor must operate to reduce the rates of local chemical reactions. The kinetics of diffusion-controlled, steady-state metamorphic processes can be predicted using the simple techniques outlined in this paper; the calculations provide useful qualitative insight into the interpretation of metamorphic structures; can be used as a rigorous test for compatibility between observed fluxes and potential gradients; and provide a method for estimating the growth time of individual metamorphic structures.

REFERENCES CITED

Brady, J. B., Am. J. Sci., 275, 954-983, 1975.

Carmichael, D. M., Contr. Min. Petr., 20, 244-267, 1969.

Darken, L.S., A.I.M.E. Trans., 175, 184-201, 1948.

Fisher, G.W., Contr. Min. and Petr., 29, p. 91-103, 1970.

_____, Am. J. Sci., 273, 897-924, 1973.

_____, in Cooper, A. R. and Heuer, A. H., Mass Transport Phenomena in Ceramics, Plenum Press, 111-122, 1975.

_____, and Elliott, D., Carnegie Inst. Washington Publ. 634, p. 231-241, 1974.

Foland, K.A., Carnegie Inst. Washington Publ. 634, p. 77-98, 1974.

Frantz, J.D. and Mao, H.K., Carnegie Inst. Washington Yearbook 73, 384-392, 1974.

_____, Carnegie Inst. Washington Yearbook 74, 417-424, 1975.

_____, Am. J. Sci., in press, 1976.

Helgeson, H.C., Geochim. Cosmochim. Acta. 35, 421-469, 1971.

Hurd, D.C. and Theyer, F., in Gibb, T.R.P., Analytical Methods in Oceanography, Am. Chem. Soc., 211-230, 1975.

Ildefonse, J.P. and Gabis, V., Cosmochim. Geochim. Acta, 40, 297-303, 1976.

Katchalsky, A and Curran, P.F., Nonequilibrium Thermodynamics in Biophysics, Harvard, 1967.

Kirkwood, J.G., Baldwin, R.L., Dunlop, P.J., Gosting, L.J. and Kegeles, G., J. Chem. Physics 33, 1505-1513, 1960.

Lerman, A., Mackenzie, F.T., and Bricker, O,P., Earth Planet. Sci. Lett. 25, 82-88 , 1975.

Loberg, B., Geol. Foren. Stockholm Forh., 85, 3-109, 1963.

Reed, W.E., J. Geophys. Res., 75, 415-430, 1970.

Terjesen, S.G., Erga, O., Thorsen, G. and Ve, P., Chem. Eng. Sci., 14, 277-288, 1961.

Thompson, A.B., J. Petrol., 16, 314-346, 1975.

Vidale, R., Am. J. Sci., 267, 857-874, 1969.

Weare, J.H., Stephens, J.R., and Eugster, H.P., Am. J. Sci., in press, 1976.

STUDY PROBLEMS

1) Given the following mineral compositions and L_{ij} values,

	1	2	3	Molar Volume (cm^3)
A	2	1	2	50
B	1	2	2	50
C	1	1	1	30
D	0	0	1	10
L_{ij} (x 10^{12} mol^2/m.j.3)	1	5	10	

a) Calculate the form of a diffusion-controlled reaction zone formed between two planar rock layers with the assemblages ABC and ABD, under conditions such that CD is less stable than AB, neglect all cross-coefficients, and assume that both rocks initially control equal modal proportions of their constituent minerals.

b) Repeat 1a, assuming that the cross-coefficients have the maximum values possible, $L_{ij} = \sqrt{L_{ii}L_{jj}}$.

c) Predict the structure formed when B grows in a rock containing the metastable assemblage ACD, assuming that B nucleates on an AD grain boundary. Repeat, assuming nucleation on AC and CD boundaries.

2) P.C. Carman (Journal of Physical Chemistry, v.42, p.1707-1721, 1968) extended Darken's (1948) model for diffusion to aqueous solutions containing a variety of dissolved species. The essence of the approach is that the L_{kk} for each species k is first obtained from the equation

$$L_{kk} = c_k q_k,$$

where c_k is the concentration of species k, and q_k is the intrinsic mobility of k; cross-terms coupling the diffusion of one species to the potential gradient of another are assumed negligible. Next, the L_{kk} for species are converted to the L_{ij} for components, using mass-balance and equilibrium relations between components and species; cross-terms between components arise whenever species with more than one component are present. Use this approach to predict the L_{ij} for diffusion of all components in a hydrothermal fluid belonging to the system $KO_{1/2}$ - $AlO_{3/2}$ - SiO_2 - H_2O - HCl, when the species present and their concentrations are KCl (1.5 mol/l), $Si(OH)_4$ (2 mol/l), H_2O (52 mol/l), HCl (1 mol/l), assuming that $q = 10^{-12}$ $mol^2/m.j.s.$ for all species.

SOLUTIONS TO PROBLEMS

1a. Using constraints imposed on potentials by assemblage AB, $\nabla\mu_1 = 0.667\ \nabla\mu_3$, and $J_1^D = +0.667$, $J_2^D = +3.333$, $J_3^D = 10.000$. The reactions needed to produce these fluxes are

At ABC: 38.004C + 10.000(3) → 13.335A + 10.667B + 0.667(1) + 3.333(2),

At ABD: 0.667A + 12.668D + 0.667(1) + 3.333(2) → 2.001B + 10.000(3).

Assuming that both rocks initially consist of equal proportions (by volume) of the minerals present, the reaction zones will form in the following proportions, normalized to 1 cm AB adjacent to ABD.

Zone	ABD	AB	AB	ABC
Thickness (cm)	Infinite	1	10.87	Infinite
Vol. % A	33.3	29.2	51.9	33.3
B	33.3	70.8	48.1	33.3
C				33.3
D	33.3			
	100.0	100.0	100.0	100.0

1b. In doing this problem, it is essential to retain at least seven significant figures. The zones have the form (normalized to 1 cm of AD adjacent to ABD)

Zone	ABD	AD	A	AB	ABC
Thickness (cm)	Infinite	1.000	0.00000338	3.05871314	Infinite
Vol. % A	33.3	56.4	100.0	59.2	33.3
B	33.3			40.8	33.3
C					33.3
D	33.3	43.6			
	100.0	100.0	100.0	100.0	100.0

Note: In order to solve for the reactions and fluxes in the zones of AD and A, the two zones should be solved simultaneously.

1c. For all three nucleation sites, the structure is (normalized to 1 cm^3 of B)

Zone	B	D	AD	ADC
Vol. (cm^3)	1.000	7.848	33.510	Infinite
Vol. % A			65.6	33.3
B	100.0			
C				33.3
D		100.0	34.4	33.3
	100.0	100.0	100.0	100.0

2. $L_{ij} \times 10^{12}$ $mol^2/m.j.s$

i \ j	1	2	3	4	5
1	1.5	0	0	-.75	1.5
2	0	0	0	0	0
3	0	0	2	4	0
4	-.75	0	4	60.37	-.75
5	1.5	0	0	-.75	2.5

Note: Components are 1, $KO_{1/2}$; 2, $AlO_{3/2}$; 3, SiO_2; 4, H_2O; 5, HCl.

SUBJECT INDEX